AF683138

WESTHOLLOW RESEARCH CENTER LIBRARY
SHELL DEVELOPMENT COMPANY
P. O. BOX 1378
HOUSTON, TX. 77001, U.S.A.

JAN 29 1981

ATLAS OF INTERFERENCE LAYER METALLOGRAPHY

by Hans-Eugen Bühler
and Hans Paul Hougardy

in collaboration with
Günter Bartz (†), Gerhard Elßner, Gabriele Engler, Isolde Gräf,
Emil Hillnhagen, Wolf-Ulrich Kopp, Günter Robusch and Hans Zogg

WESTHOLLOW RESEARCH CENTER LIBRARY
SHELL DEVELOPMENT COMPANY
P. O. BOX 1378
HOUSTON, TX. 77001, U.S.A.

Published by the Metallography Subcommittee
Deutsche Gesellschaft für Metallkunde

ISBN 3-88355-016-7
© 1980 by Deutsche Gesellschaft für Metallkunde, Adenauerallee 21, D-6370 Oberursel 1, Phone Nr. (06171) 4081
All rights of reproduction, either of the text in any language whatsoever, or of the photographs, for any purposes including non-scientific, are reserved by the publishers.

Printed in W-Germany by Pharma-Druck Inge Hartmann und Sohn Oberursel (Taunus)

Introduction

In order that the phases and microstructural features of metallic and non-metallic materials can be observed under the reflection microscope the specimen must, in most cases, undergo a special preparation. This is because the reflectivities of the individual phases and microstructural constituents rarely differ sufficiently to produce a well-contrasted image. Most metallographic investigations involve the use of etching in which, as a rule, the different rate of chemical reaction of the different microstructural constituents is exploited to reveal their structure. The result of etching is either a shadowing effect due to the production of a surface relief, or selective tinting, or different interference colours on different crystallites due to the formation of surface layers of different thicknesses. Although the formation of a relief usually allows the different microstructural constituents to be distinguished clearly, the areal proportion of the phases in the plane of section and their shape are generally not faithfully reproduced under these conditions. Only in very few cases can contrast be produced by optical methods. The use of polarised light is one example.

W. Pepperhoff (1960) was the first to suggest a technique which allowed the observation of the "true" microstructure by a purely optical method, without the necessity for chemical attack. The deposition of interference layers on polished specimen surfaces not only decisively improves the light-dark contrast between different phases but also effectively enhances the colour differences between the different microstructural constituents. In many cases several phases in a specimen can be observed and identified simultaneously where previously several successive etches were necessary. Nor does the deposition of interference layers alter the areal proportions, so that this physical method of accentuating the contrast permits a high experimental accuracy when using techniques for quantitative microstructural analysis. This advantage can be used in other fields. In work with the microprobe, for example, the specimens must be flat if systematic experimental errors are to be avoided. The use of chemical etching to reveal the phases to be measured is therefore unsuitable. Again, in this case, the phases can be contrasted by means of interference layers since no relief is produced. The technique can also be used for the quantitative determination of the optical constants – refractive index and absorption coefficient – of metallic and non-metallic phases.

In recent years there have been numerous publications describing the application of this method of revealing microstructures not only in the case of metallic phases but also for poorly reflecting and weakly absorbing materials, e.g. oxides or ceramics. A review of interference microscopy and its contribution to the metallography of solids appeared in 1970 under the title "Interferenzschichten-Mikroskopie" by W. Pepperhoff and H.-H. Ettwig. This standard work aimed at microscopists active in a wide range of scientific disciplines presents the theoretical principles of the technique in an excellent fashion and illustrates its practical application by means of numerous examples.

Soon after the introduction of the interference layer technique, however, it became clear that in many cases the practical manipulation was proving difficult for the metallographer not only because particular care is necessary in the preparation of specimens but also because the new work published in many individual papers was presented too theoretically. This "Atlas of Interference Layer Metallographie" attempts to close this gap. A working party from the Metallography Subcommittee of the Deutsche Gesellschaft für Metallkunde has taken over the task of presenting and evaluating the data and knowledge hitherto accumulated in the field of interference layer microscopy so that the ordinary metallographer working in the field can apply the different techniques reliably and with ease. Only sufficient of the theory of evaporated interference layers is given in this atlas for an understanding of the possible applications. The most important equations are given without derivation for readers who wish to use calculations to estimate the possibilities of contrast enhancement by interference layers. These passages are printed in italics and can be omitted by the reader without any loss of continuity.

The book is in two parts: an introduction giving the essentials of the theoretical basis and information necessary for the practical application of the different techniques, and an illustrated section giving characteristic examples from the different groups of materials showing the possibilities of interference layer metallography. The coating conditions are given for all the examples so that the metallographer has the possibility of reproducing the contrast without great theoretical knowledge. With this book, the Metallography Subcommittee of the Deutsche Gesellschaft für Metallkunde hopes to have provided information concerning the preparation of metallic and non-metallic materials not previously available in many cases.

Oberursel, May 1980

Metallography Subcommittee of the Deutsche Gesellschaft für Metallkunde

Hans-Eugen Bühler and Hans Paul Hougardy

Contributors

Contributions to the text and pictures included in this book have been made by:

Günther Bartz (†), Dipl.-Phys.
Ernst Leitz Wetzlar GmbH, Postfach 2020, 6330 Wetzlar

Hans-Eugen Bühler, Prof. Dr.-Ing.
Dechema-Institut, Theodor-Heuss-Allee 25, 6000 Frankfurt/Main 97

Gabriele Engler
Thyssen AG, Forschung, Postfach 110067, 4100 Duisburg

Gerhard Elßner, Dr. rer. nat.
Max-Planck-Institut für Metallforschung, Seestraße 92, 7000 Stuttgart

Isolde Gräf, Institut für angewandte Materialforschung (IFAM)
Lesumer Heerstraße 36, 2820 Bremen 77

Emil Hillnhagen, Dr. rer. nat., Krupp-Forschungsinstitut GmbH,
Münchener Straße 100, 4300 Essen

Hans Paul Hougardy, Dr.-Ing., Max-Planck-Institut für
Eisenforschung GmbH, Max-Planck-Straße 1, 4000 Düsseldorf

Wolf Ulrich Kopp, Dr. rer. nat.
Peckhauser Straße 57, 4020 Mettmann

Günter Robusch
Thyssen AG, Forschung, Postfach 110067, 4100 Duisburg

Hans Zogg, Dr. sc. nat.
Institut f. technische Physik, ETH Hönggerberg CH-8093 Zürich, Schweiz

The authors have received photographic material, ideas and suggestions as to the form of this book from numerous colleagues. Their thanks are due to:

Dr. P. Brezina, Gebr. Sulzer AG, Winterthur, Schweiz

R. Burghardt, Daimler Benz AG, Mannheim

K. Dieser, Farbwerke Hoechst AG, Frankfurt/Main

Dr. C. Hergert, Th. Goldschmidt AG, Essen

Dr. E. Horn, Thyssen AG, Forschung, Duisburg

Dr. H. Kiesheyer
Thyssen-Edelstahlwerke AG, Forschungsinstitut, Krefeld

Dr. H. Knosp, Degussa, Pforzheim

Dr. D. Kossel, Ernst Leitz GmbH, Wetzlar

B. Kranefeld, Carl Zeiss, Zweigniederlassung Köln

D. Menzel, W.C. Heraeus GmbH, Hanau

Prof. Dr. G. Ondracek, Kernforschungszentrum Karlsruhe

Prof. Dr. W. Pepperhoff
Mannesmann Forschungsinstitut GmbH, Duisburg

Dr. H.-J. Schüller, Allianz Zentrum für Technik GmbH, Ismaning

S. Weber, Vacuumschmelze GmbH, Hanau

I. Werner, IBM Deutschland GmbH, Sindelfingen

U. Wilke-Dörfurt, Vacuumschmelze GmbH, Hanau

Dipl.-Ing. F. Berentzen
Thyssen Edelstahlwerke AG, Forschungsinstitut, Krefeld

The authors are particularly indebted to the Secretary of the Deutsche Gesellschaft für Metallkunde, Dr. Volker Schumacher, whose assistance has greatly facilitated the rapid and careful production of this volume.

The English translation is by Dipl.-chem. Janet Mordike to whom thanks are due.

The authors are also indebted to Professor Peter Ryder for his critical reading of the text.

They also wish to express their thanks to the Stifterverband Metalle for financial support of their work.

Instructions for the use of this book

This Atlas of Interference Layer Metallography is in two main parts:

1. Principles and experimental procedure for the different interference layer techniques in metallography (Chapter 1) and

2. practical examples taken from a series of different materials (Chapter 2).

Chapters 1 and 2, with their subsections, are so constructed that a list of literature references of importance to the reader is given at the end of each section. It may be, therefore, that the same literature reference appears in several places in the book. This has been done deliberately in order to give the reader all the necessary information appertaining to the section that interests him. The literature references are restricted to the review articles or papers from which pictures were taken, since these themselves have the full list of pertinent literature references. As well as descriptions of the micrographs, the second half of the book contains brief reviews for each of the individual alloy systems describing the practical questions associated with metallographic examination and identification. The commentaries to the micrographs do not only set forth the advantages of interference layer metallography. They also indicate, where appropriate, when and where other metallographic investigatory techniques can be used with less effort or greater prospects of success. To this extent, the following discussion does not only describe the real possibilities of the interference layer technique, it also attempts a demarcation from the usual, well-established preparative methods.

At the end of the book the reader will find author and subject indexes, and Chapter 4 gives a source list. In addition to the introductory notes, therefore, the reader who has a query regarding special preparative techniques has a chance of making contact with workers who already have experience in the corresponding field.

Contents

	Introduction		3
	Contributors		5
	Instructions for using this book		7
1.	Principles and experimental procedure in the interference layer technique		9
1.1	Principles of the interference layer technique		10
	1.1.1	Effect of an interference layer	10
	1.1.2	Determination of the optical constants	20
	1.1.3	Light-dark and colour contrasts	24
	1.1.3.1	Colour contrasts	24
	1.1.3.2	Light-dark contrasts	39
1.2	Experimental details of the interference layer technique		41
	1.2.1	Preparation of polished sections	41
	1.2.2	Production of interference layers	52
	1.2.2.1	Deposition of layers by evaporation	52
	1.2.2.2	Production of layers by sputtering	53
	1.2.2.3	Devices for the measurement of layer thickness	56
	1.2.2.3.1	Measurement by means of vibrating quartz	56
	1.2.2.3.2	Optical layer thickness measurement	56
	1.2.3	Properties of layer materials and of layers	59
	1.2.3.1	Materials for evaporation and evaporated layers	59
	1.2.3.2	Materials for sputtering and sputtered layers	61
2.	Examples of the application of the interference layer technique		67
2.1	Differentiation of phases		68
	2.1.1	Differentiation of the three modifications of iron (α, γ, ε)	68
	2.1.2	Iron alloys and steels	84
	2.1.3	High temperature nickel base alloys	111
	2.1.4	Hard metals	114
	2.1.5	Copper alloys	122
	2.1.6	Aluminium alloys	128
	2.1.7	Other non-ferrous alloy systems	141
	2.1.8	Composites (metallic and non-metallic coatings, plasma layers, metal-ceramic composites)	146
	2.1.9	Brazed joints	158
	2.1.10	Ceramic and metal-ceramic materials, plastics	173
	2.1.11	Magnetic materials	177
2.2	Applications independent of the material		183
	2.2.1	Concentration differences in solid solutions (diffusion, segregation)	183
	2.2.2	Reinforcement of anisotropy effects	192
	2.2.3	Pore volume	197
	2.2.4	Corrosion and oxidation	200
	2.2.5	Observation of states of magnetic order	210
3.	Author and subject index		221
4.	Source list		227
5.	Optical constants of phases and layer materials		229

1. PRINCIPLES AND EXPERIMENTAL PROCEDURE IN THE INTERFERENCE LAYER TECHNIQUE

1.1 PRINCIPLES OF THE INTERFERENCE LAYER TECHNIQUE

1.1.1 Effect of an interference layer

The interference layer technique in reflection microscopy is based on the production of reflection interference filters by coating the polished specimen surface with interference layers using suitable techniques. The effect of such a (in some cases non-absorbing) layer is that light waves incident on the specimen surface are attenuated by multiple reflection at the metal/layer and layer/air interfaces. The consequence of this is an accentuation of the contrast between the microstructural constituents, dependent on the optical constants of two neighbouring phases. This contrast enhancement consists not only of an increase in the difference in the intensity of light reflected from the two phases but also of an increase in the colour contrast.

In this context light can be described as electromagnetic waves. Fig. 1 shows the change in amplitude, i.e. inlight intensity, with time. It repeats itself after a period designated "wavelength" in Fig. 1, which represents one vibration of the light. The number of vibrations per second, in other words the frequency, is perceived by the human eye as colour. At the high frequency of $7.5 \cdot 10^{14}$ vibrations per second, a human observer sees "blue", at $4.3 \cdot 10^{14}$ vibrations per second, "red". The only prerequisite is that the amplitude exceeds a minimum value. Unless this value is reached the eye can differentiate only between light and dark.

Light propagates with a velocity of ca. 300000 km/s. The end of a vibration is displaced with respect to the beginning in the direction of propagation of the light. A wave can be described by three quantities, two being sufficient to characterise it. These quantities are: 1) the velocity of propagation, 2) the period of vibration, or its reciprocal, the frequency, i.e. the number of vibrations per second, 3) the distance propagated per vibration, i.e. the wavelength. Since the velocity of ligth in vacuum is constant, the wavelength is sufficient to describe the light. In this generally acceptable representation "red" is light with a wavelength of ca. 700 nm (10^9 nm (nanometres) = 1 m), "blue" is light with a wavelength of ca. 400 nm. These are the limits of the spectrum visible to the human eye.

The corresponding frequency f in Hz (1 Hz (Hertz) is 1 vibration per second) can be calculated for any wavelength λ *from the equation* $c = \lambda \cdot f$ *where* c = *velocity of light, ca. 300000 km/s. From this the frequencies are found to be* $7.5 \cdot 10^{14}$ *Hz corresponding to 400 nm and* $4.3 \cdot 10^{14}$ *Hz corresponding to 700 nm.*

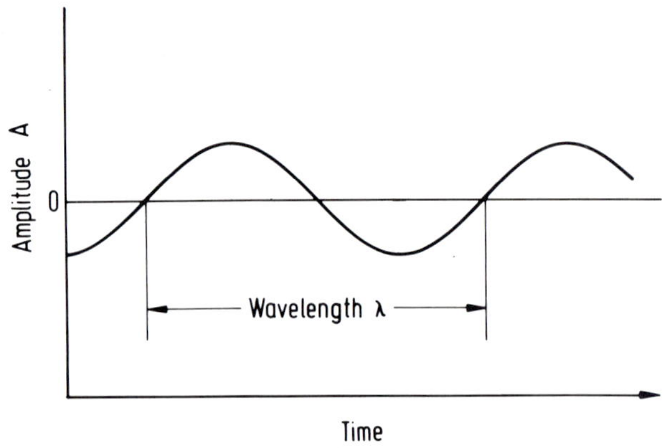

Fig. 1: Description of a wave.

The effect of an interference layer on a metallic or non-metallic specimen surface, and the conditions it must satisfy to achieve the desired accentuation of the contrast is illustrated by Fig. 2. 1)2). If a beam of light E of wavelength λ is incident on a coated surface, the reflected beam P consists of superposed beams of light reflected from the different interfaces.

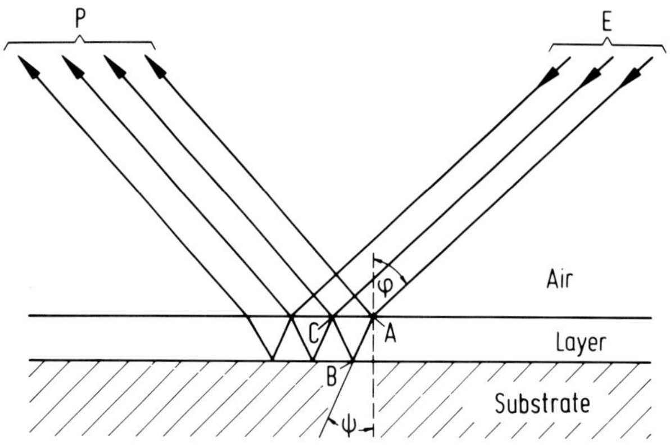

Fig. 2: Reflection of a beam of light E at a coated metal surface. See text for an explanation of the symbols.

Some of the light incident at an angle φ is reflected at the air/layer interface (A) (wave M in Fig. 3). The remainder is refracted at an angle ψ and then reflected from the layer/metal interface (B). Some of this light is reflected back at the layer/air interface (C) but the rest emerges (wave N in Fig. 3) and becomes superposed on the light reflected at this interface (A) to give the resultant wave P. In addition to the change in amplitude accompanying reflection at the layer/metal interface the light also suffers a change in phase or "phase difference", designated δ_r. The path difference between the light traversing the layer and that reflected at the air/layer interface together with the phase difference δ_r produces a phase shift β between the light waves M and N (Fig. 3). Superposition of the two wave trains yields a new wave P.

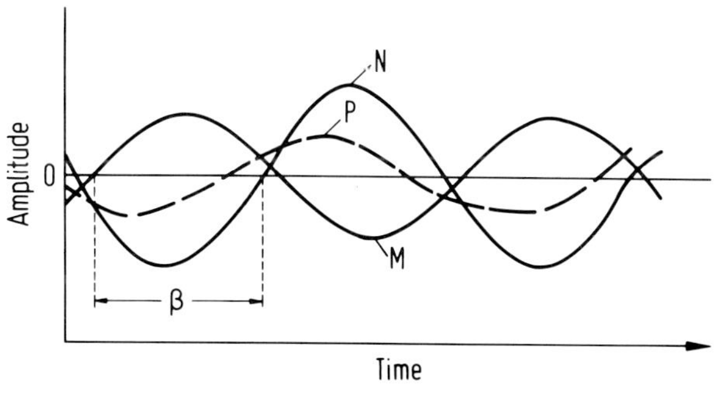

Fig. 3: Superposition of two light waves with different amplitudes and a phase difference β.
M: wave reflected at the air/layer interface
N: wave emerging from the layer after reflection at the layer/metal interface
P: resultant of M and N

If the phase difference β is exactly half a wavelength and the amplitudes A_M of M and A_N of N equal, mutual extinction of the two waves occurs. The phase condition ($\beta = \lambda/2$) and the amplitude condition

($A_M = A_N$) are satisfied. The reflectivity of the phase is zero at this wavelength so that it shows up black and well-contrasted against the other phases for which these two conditions are not fulfilled. The contrast is defined by the value

$$K = \frac{R_1 - R_2}{R_1} \quad (1)$$

$$\text{with} \quad R_1 > R_2$$

where
K = contrast
R_1 = reflectivity of phase 1
R_2 = reflectivity of phase 2

The values of K lie between 0 and 1. If $R_2 = 0$, $K = 1$, which is the maximum achievable contrast. If $R_2 = R_1$, $K = 0$, and the two phases cannot be differentiated. It is assumed here that the contrast is determined for a given wavelength, c.f. Fig. 4. For the same value $\Delta R = R_1 - R_2$, the contrast K increases with decreasing R_2 as seen from Table 1.

R_1	R_2	R_1-R_2	K
0.6	0.5	0.1	0.17
0.2	0.1	0.1	0.5

Table 1. The influence of the reflectivity R_1 and R_2 of two phases on the contrast K for a constant difference $\Delta R = R_1 - R_2$ between the reflectivities.

If the reflectivities of two phases are reduced by differing amounts, the contrast between them is increased. This can be effected by the production of an interference layer as shown in Figs. 2 and 3. If **both** the amplitude and phase conditions can be satisfied for one phase, $R_2 = 0$ and the maximum contrast $K = 1$ is achieved.

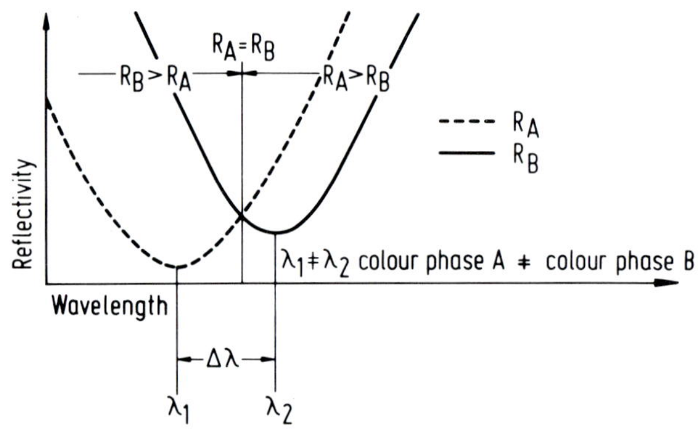

Fig. 4: Dependence of the reflectivity of two coated phases A and B on the wavelength of the illumination 6).

The illumination system of a metallurgical microscope emits light consisting of a mixture of light waves with different wavelengths. By placing a commercial interference filter in the ray path of the microscope a monochromatic beam can be produced of any wavelength from 400 nm (blue) to 700 nm (red) (c.f. explanation of Fig. 18). By means of a microscope-photometer the reflectivity of a single or multiphase specimen coated with an interference layer can then be measured as a function of the wavelength of the illumination. Fig. 4 shows the plot of reflectivity against wavelength for two phases A and B. At λ_1 phase A has a minimum reflectivity $R_{min\,1}$. Due to the different optical properties of phases A and B, the reflection minimum of phase B occurs at the wavelength λ_2.

If $R_1 = R_B$ and $R_2 = R_A$ are substituted in Equation (1), K is a maximum at λ_1. If $R_A = R_B$, K = 0. In the case of λ_2, $R_1 = R_A$ and $R_2 = R_B$ are substituted in Equation (1). K then has a second maximum.

Fig. 5 shows the curve for contrast as a function of wavelength. The contrast is a maximum at λ_1.

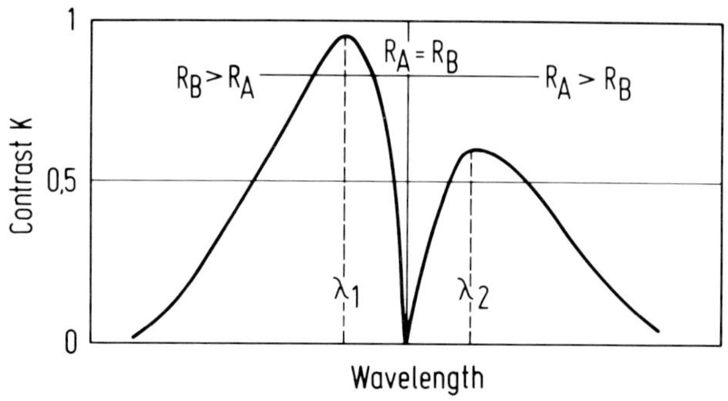

Fig. 5: Curve of the contrast K (c.f. Equation (1) page 12) as a function of wavelength for the phases and coating conditions given in Fig. 4.

The curve for the contrast in Fig. 5 is valid only for observation in monochromatic light. In white light differences in colour as well as in intensity can be observed between the individual phases. These colours are complementary to the light waves extinguished at the reflection minimum. This is considered in more detail in section 1.1.3.

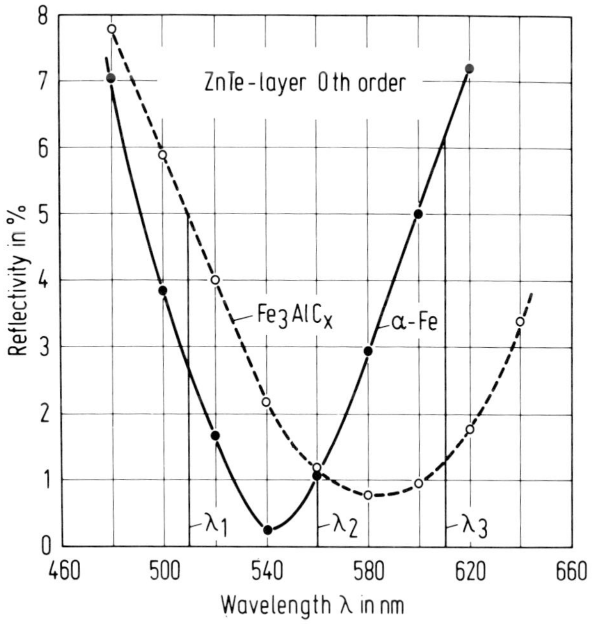

Fig. 6: Reflectivity of α-iron and of iron aluminium carbide Fe_3AlC_x as a function of the illumination wavelength. Vacuum deposited ZnTe layer.

The dependence of the reflectivity of the phases on the wavelength of the illumination illustrated in Fig. 4 and the possibility it offers for the production of different degrees of contrast, is considerably enhanced by interference layers. In the uncoated condition the reflectivity of many materials is often only very slightly dependent on the wavelength of the illumination (c.f. Fig. 67 on page 71, Figs. 118 and 119 on page 91).

As a practical example, Fig. 6 shows the curve of the reflectivity of α-iron and a carbide after deposition of a ZnTe layer (c.f. section 2.2.1). In this case the reflection minimum of the α-iron lies in the green-yellow wavelength range at 540 nm and so the phase appears purple on the section. The curve for the carbide Fe_3AlC_x has a flat minimum at about 590 nm, and the phase appears bluish under the microscope. Details are given in the explanation of Fig. 19, page 26. Fig. 7 shows the contrast calculated from Equation 1 plotted as a function of wavelength. At 540 nm, the reflection minimum of α-iron, the contrast is a maximum and at 560 nm it is zero. If light of this wavelength is used as illumination the two phases are indistinguishable. At 590 nm, the second maximum, the α-iron is brighter than the carbide but at 540 nm it is darker. By suitable choice of illumination, the contrast between the phases can be varied over a wide range even to contrast reversal (c.f. Figs. 347 to 349).

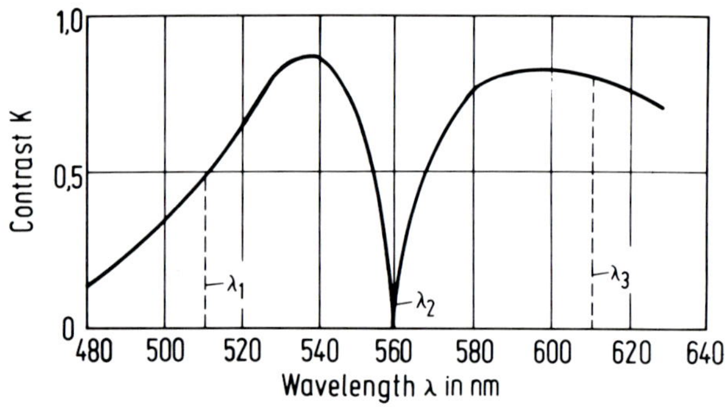

Fig. 7: Contrast between α-iron and Fe_3AlC_x as a function of wavelength. Calculated from the measurements in Fig. 6.

λ_{min} is the wavelength at which the phase condition is satisfied for one phase. It can be fixed by choice of the thickness d_s of the vacuum deposited interference layer.

The condition is:

$$d_s = m \cdot \frac{\lambda_{min}}{4 \cdot n_s} + \frac{\lambda_{min}}{4 \cdot n_s} \cdot \left(\frac{\delta_r}{\pi} - 1 \right) \qquad (2a)$$

$m = 1$: zero-order
$m = 3$: 1st order
$m = 5$: 2nd order
etc.

where
n_s = refractice index of the non-absorbing vacuum deposited interference layer
λ_{min} = wavelength of light for which the phase condition holds
δ_r = phase difference of light beam after reflection at the metal surface
m = 1,3,5 gives order (c.f. Fig. 8).

Using Equation (2a) it is possible to calculate the thickness d_s of the layer for a given value of n_s, i.e. for a particular layer material, giving rise to the desired location of λ_{min}. If, for example, the layer thickness is preset for the coating process, the wavelength of the minimum λ_{min} can be calculated by rearranging (2a) to:

$$\lambda_{min} = \frac{4 \cdot n_s \cdot d_s}{m + \frac{\delta_r}{\pi} - 1} \qquad (2b)$$

As a rule first order layers are used, i.e. the value of 3 should be inserted for m. Values for n_s and δ_r are given in Figs. 62 to 65, page 60 to 63, and in the tables in Chapter 5. If n_p is the refractive index of the coated phase and k_p the absorption coefficient of the coated phase, the phase change is given by the equation:

$$\tan \delta_r = \frac{2 \cdot n_s \cdot k_p}{n_s^2 - n_p^2 - k_p^2} \qquad (2c)$$

The reflectivity is given by:

$$R = \frac{q_1^2 + q_2^2 - 2 \cdot q_1 \cdot q_2 \cdot \cos \alpha}{1 + q_1^2 \cdot q_2^2 - 2 \cdot q_1 \cdot q_2 \cdot \cos \alpha} \qquad (3a)$$

$$\text{with } q_1 = \frac{n_s - 1}{n_s + 1} \qquad (3b)$$

$$q_2^2 = \frac{(n_p - n_s)^2 + k_p^2}{(n_p + n_s)^2 + k_p^2} \qquad (3c)$$

$$\alpha = d_s \cdot \frac{4\pi \cdot n_s}{\lambda} - \delta_r \qquad (3d)$$

If the wavelength of the illumination is constant, the reflectivity R varies periodically with the layer thickness d_s. This is exemplified by the variation in the reflectivity of α-iron with the layer thickness in Fig. 8. For this periodic variation and for non-absorbing layers, the magnitude of the reflectivity at the minimum and maximum is independent of the order. Different layer thicknesses do however result in different interference colours, c.f. Fig. 24 to 35, page 35 and 37 and Fig. 72 page 71. A more detailed discussion of colour perception and the problems associated with the description of colour intensity and shade is given in section 1.1.3.

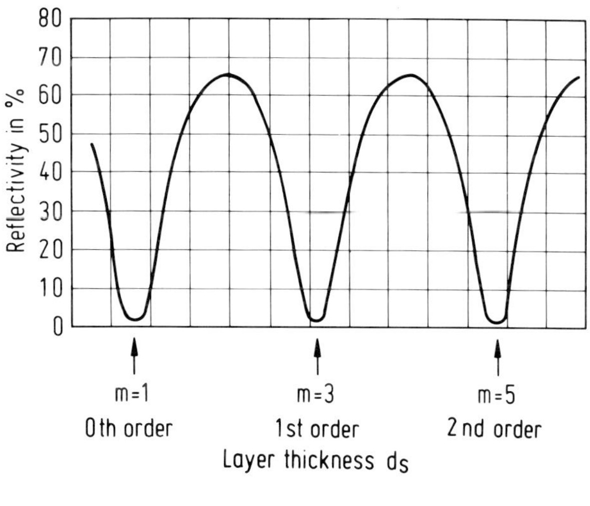

Fig. 8: Influence of the thickness of a ZnS layer on the reflectivity of α-iron. Wavelength of illumination λ = 540 nm. On the x-axis, one unit corresponds to 20 nm.

At the maxima of the curves the reflectivity is higher than in the uncoated condition. This effect can be used to increase the reflectivity of weakly reflecting materials. The difference in reflectivity between the two components of a metal-ceramic composite is sometimes so great that the fine structure of both cannot be shown up simultaneously in an optical micrograph. With a suitable coating it is possible to increase the reflectivity of the ceramic component and diminish that of the metal, so reducing the differences in brightness to a level which can be reproduced by a photographic emulsion (c.f. section 2.1.8, Figs. 277 to 279).

Fig. 9 shows that if weakly reflecting materials are coated with ZnSe, the minimum reflectivity decreases and the maximum increases as the refractive index decreases (c.f. also Fig. 379, section 2.2.4).

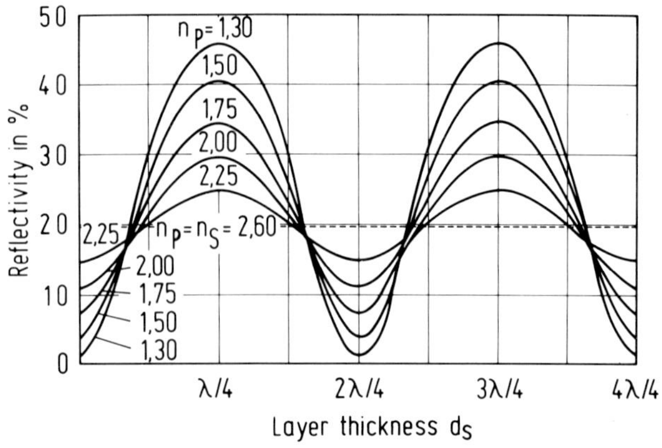

Fig. 9: The change in reflectivity of non-absorbing, weakly reflecting materials (n_p = 1.3 to 2.6) after coating with ZnSe (n_s = 2.6). The layer thickness is plotted as a fraction of the wavelength of the illumination 3). For λ = 520 nm, $\lambda/4$ corresponds to a layer thickness of 130 nm.

If d_s from Equation (2a) is substituted in Equation (3d) a value of π (m-1) is obtained for α. Equation (3a) then gives the reflectivity R_{min} which satisfies the phase condition.

$$R_{min} = \left[\frac{q_1 - q_2}{1 - q_1 \cdot q_2}\right]^2 \qquad (4)$$

If the amplitude condition is also satisfied, $R_{min} = 0$, i.e. $q_1 = q_2$ (Equation 4). This is possible only with a non-absorbing material with a particular refractive index n_s, given by

$$n_s^2 = n_p + \frac{k_p^2}{n_p - 1} \qquad (5)$$

For each phase, the amplitude **and** the phase conditions can be satisfied simultaneously only by means of a layer with the value of n_s given by Equation (5) and at certain thickness. If the desired value of λ_{min} is fixed, this layer thickness is given by Equation (2a). In the case of α-iron if average values of n_p = 2.0 and k_p = 3.2 taken from Table 9, page 233 are substituted in Equation (5) a value of 3.49 is obtained for n_s, i.e. approximately the value fpr GaP, for which n_s = 3.45. In the case of γ-iron Table 9, page 233 gives values of n_p = 2.0 and k_p = 3.5. The value of n_s for which the phase condition and amplitude condition can simultaneously be satisfied is thus 3.77.

A non-absorbing layer material with a refractive index as high as this is not at present available. In this case the amplitude and phase conditions can be satisfied only by absorbing layers (c.f. section 1.2.3.2). The influence of the refractive index of the deposited layer on the contrast is illustrated in Figs. 10 and 11. Curves of reflectivity of austenite and ferrite as a function of the wavelength of the illumination corresponding to Fig. 4 were measured or calculated for every material tested, i.e. for every value of n_s, and the contrast at the reflectivity minimum was determined.

If the contrast is defined according to Equation (1), Fig. 10 shows that as the refractive index increases, the contrast between the two modifications of iron, ferrite and austenite, increase steadily up to a refractive index of 3.4 at which value it reaches a magnitude of 1. This occurs when the amplitude condition and the phase condition are satisfied for ferrite. As n_s is increased to even higher values, a reversal in the contrast is observed.

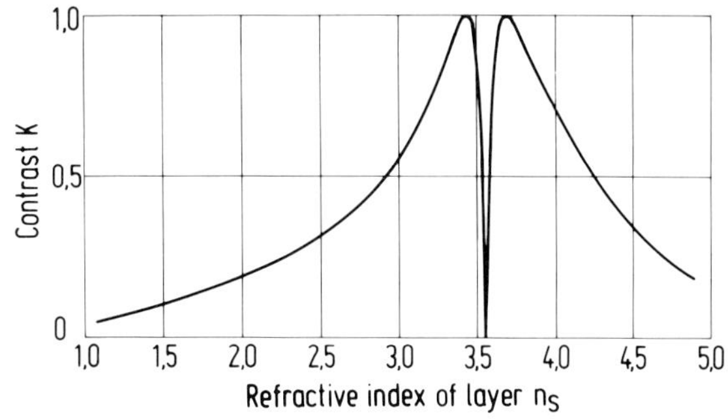

Fig. 10: Change in the contrast between ferrite and austenite at the reflection minimum R_{min} as function of the refractive index of the coating layer n_s. The illumination has the wavelength of the reflection minimum λ_{min} in each case (c.f. Fig. 4, page 12).

At a refractive index of 3.7 the phase and amplitude conditions are satisfied for austenite and the contrast is again 1, c.f. Fig. 87 page 77, Figs. 88 to 91, page 77. The deviations from the values calculated from numerical values of n_s and k_p arise from the uncertainty in the derivation of the optical constants, c.f. section 1.1.2. A practical application is described in section 2.1.1.

In Fig. 11, the quotient $R_{min\ \alpha\text{-Fe}}/R_{min\ \gamma\text{-Fe}}$, plotted as a function of the refractive index of the layer, gives a measure of the contrast. This representation is frequently found in the literature. It is used in this atlas in the case of pictures taken from the literature. The quotient increases with increasing refractive index until, at a refractive index of about 3.4, the amplitude condition for α-iron is satisfied, i.e. the quotient has a value of infinity which represents maximum contrast between austenite and ferrite. The curve then falls of steeply to a value of n_s of approximately 3.7, the refractive index at which the amplitude condition for γ-iron is satisfied. Here again the contrast K is unity and the quotient zero, but now the austenite appears dark against the ferrite. The representation in Fig. 11 has the disadvantage that at maximum contrast the quotient can have the value zero or infinity. It has the advantage that as well as information about contrast it also shows which phase is the brighter.

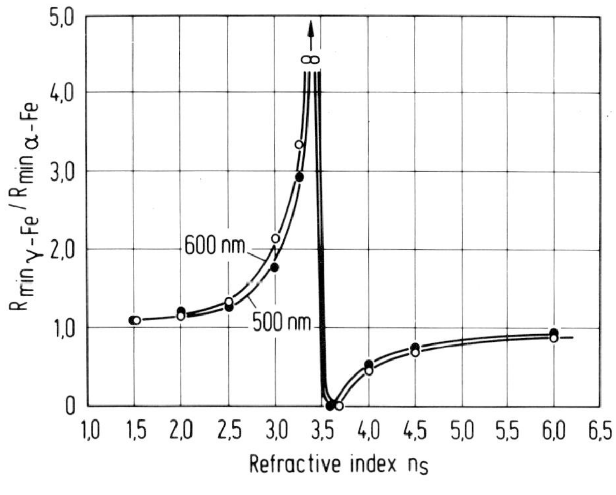

Fig. 11: Ratio of the reflection minima of ferrite and austenite as a function of the refractive index n_s of the layer.

According to the foregoing discussion, maximum contrast is achieved between two microstructural constituents when the reflected wave from one microstructural constituent suffers extinction by interference and the reflectivity is equal to zero. For this to occur it is necessary that both the phase condition **and** the amplitude condition be satisfied. If non-absorbing layers are used, especially if the metallic substrate is highly reflecting, refractive indices of 4 and more would be needed for the latter condition to be satisfied. Since such high refractive indices are not at present technically feasible (c.f. data in sections 1.2.3.1 and 5), other measures must be employed to accentuate the contrast. One possibility of enhancing the contrast on highly reflecting metals is the deposition of absorbing layers 1)4). The amplitude of light reflected at the specimen surface (c.f. Fig. 2 page 11) is altered even more as it passes through an absorbing layer. Thus the same effect can be achieved by absorption due to the layer as by an increase in the refractive index of the layer. In the case of vacuum deposited absorbing interference layers, reflection minima of the same form as those shown in Fig. 4, page 12, still occur but an analytical solution of the reflectivity at the reflection minimum R_{min} is no longer possible and a numerical treatment becomes necessary. Reference should be made to the cited literature for details of the computations, 1).

In the case of non-absorbing layers, the reflectivity at the reflection minimum is independent of the order of interference, the only difference is that the curves become steeper at higher orders (c.f. Figs. 8, 9 and 14). In the case of absorbing layers on the other hand the reflectivity changes with the order of interference, c.f. Fig. 67, page 71. This means that even with highly reflecting metals, the amplitude condition can be satisfied by increasing the layer thickness. Thus the contrast of highly reflecting

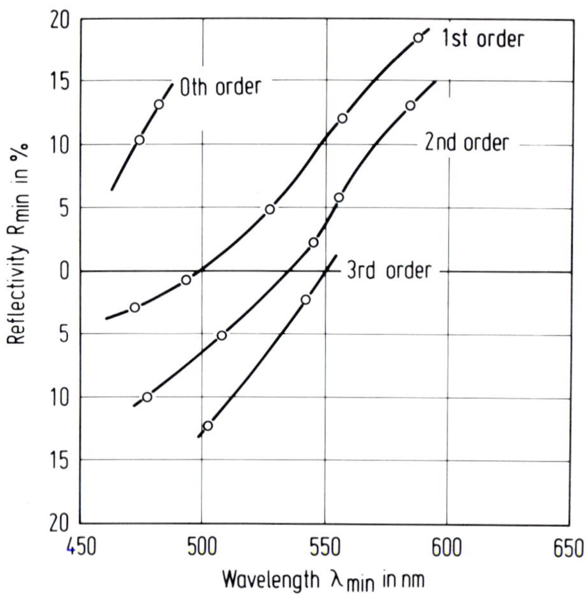

Fig. 12: R_{min} for Al as a function of wavelength of illumination for different orders of interference. Vacuum deposit of Sb_2S_3. For the sake of clarity, the reflectivity, which is always positive, is plotted symmetrically about zero.

microstructures can be enhanced by suitable choice of layer thickness, in other words of the order of interference, and of wavelength. Fig. 12 shows that by increasing the layer thickness (zero to third order) the reflectivity at the interference minimum can be decisively reduced, the wavelength remaining constant. If the curve lies in the upper half of the diagram the amplitude condition is under-fulfilled. If it lies in the lower half of the diagram it is over-fulfilled. The point at which the curve intersects the zero line, i.e. the fulfilment of the amplitude condition, occurs at a different wavelength for each order.

Up to now it has been assumed that the optical constants of the layer materials are independent of the wavelength of the illumination. This is a simplification. For exact calculations, the variation in the optical constants with wavelength must be taken into account, c.f. Figs 62 to 65, pages 60 to 63 and Chapter 5.

The size of the phases sets a limit on the use of the interference layer method. Under certain circumstances, particles smaller than the limit of resolution of the optical microscope can be shown up by chemical etching techniques. Since the interference layer technique always reveals particles in their true dimensions, they must have a size of at least 0.2 µm in order to be seen. Fig. 174, page 109, shows the example of a material in which the very fine precipitate of σ-phase could not be seen after coating but became visible on etching.

Literature to 1.1.1

1) Pepperhoff, W. and Ettwig, H.-H.: Interferenzschichten-Mikroskopie. Wissenschaftliche Forschungsberichte, Vol. 73, Dr. Steinkopff-Verlag Darmstadt 1970

2) Pepperhoff, W.: Gefügeentwicklung durch Interferenz-Aufdampfschichten. Arch. Eisenhüttenwes. **32** (1961) 269

3) Pepperhoff, W. and Pusch, R.: Kontraststeigerung in der Auflichtmikroskopie durch reflexionserhöhende Aufdampfschichten. Mannesmann-Forschungsberichte **140** (1961) 88

4) Ettwig, H.-H., Bühler, H.-E. and Jackel, G.: Absorbierende Schichtwerkstoffe in der Interferenzschichten-Mikroskopie. Arch. Eisenhüttenwes. **41** (1970) 957

5) Semlitsch, M.: Interferenz-Aufdampfschichten bei Mikrogefügeanalysen von Werkstoffen mit der Elektronenstrahl-Mikrosonde. Prakt. Metallographie **8** (1971) 259

6) Jackel, G., Bühler, H.-E. and Robusch, G.: Interferenzschichten-Mikroskopie an opaken und transparenten Werkstoffen. Thyssenforschung **4** (1972) 67

7) Bühler, H.-E.: Ein Beitrag zur Gefügeentwicklung durch Interferenz-Aufdampfschichten. Radex-Rundschau **3/4** (1967) 67

8) Bühler, H.-E., Jackel, G. and Thiemann, E.: Möglichkeiten der quantitativen Beschreibung von oxydischen Phasen mit Hilfe der ätzfreien Gefügeentwicklung nach Pepperhoff. Arch. Eisenhüttenwes. **41** (1970) 409

9) Pepperhoff, W. and Kohlhaas, E.: Das optische Verhalten von Metallen. In: Metallphysik. published by Verein Deutscher Eisenhüttenleute, Düsseldorf (1967) 444

10) Pepperhoff, W. and Ettwig, H.-H.: Polarisations-Mikroskopie durch aufgedampfte Interferenzschichten. Radex-Rundschau **3/4** (1967) 667

1.1.2 Determination of the optical constants

In order to describe a microstructure quantitatively it is necessary to state the nature of the phases present. As a rule this is possible only by chemical and crystallographic analysis. A metallographic technique is based on the determination of the optical constants. The refractive index n_p and the absorption coefficient k_p of a phase can be determined by measurement of the reflectivity in the polished and in the coated condition. In many cases, a phase can be identified by these two quantities 1)2).

Equations (3), page 15, describe the relationship between the reflectivity and the optical constants. If the refractive index of the layer is known, two equations containing the two unknown optical constants can be constructed from two measurements of the reflectivity (with and without surface layer) and solved. The equations can be further rearranged so that if the absorption coefficient k_p is plotted against refractive index n_s, the curves of equal reflectivity are circles, 1).

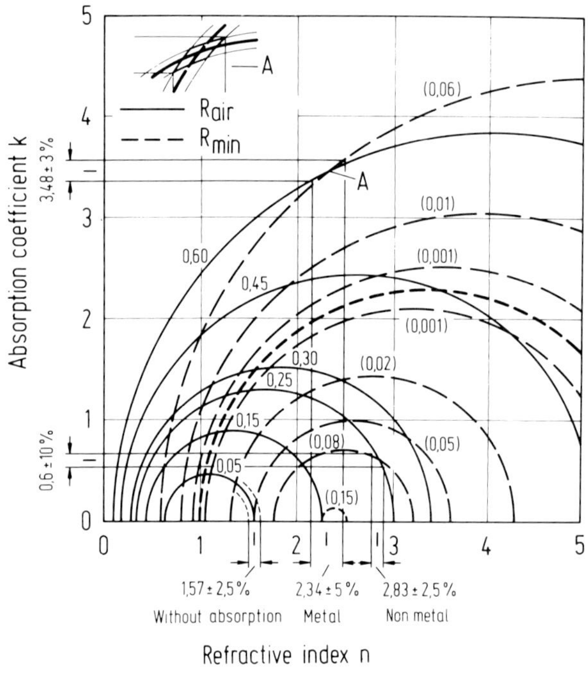

Fig. 13: Relationship between the reflectivity R and the optical constants k_p and n_p after coating with zinc sulphide. R_{air} is the reflectivity in the uncoated state, R_{min} the reflectivity after coating. The numbers on the circles are the reflectivity values. An enlargement of point "A" is shown at the top left. The scatter bands for the reflectivity in the uncoated and coated state are drawn in to demonstrate the uncertainty in the determination of the point of intersection.

Fig. 13 shows the dependence of the reflectivity of a phase on its absorption coefficient k_p and refractive index n_p. The curves of equal reflectivity are circles. A graphical determination of the optical constants is possible from these curves. The errors associated with this are smaller the steeper the angle of intersection between the reflectivity curves for the coated and uncoated state. The limits of error for the measurement of the reflectivity are given for several groups of materials. The errors for the determination of the values of k_p and n_p can be obtained from these.

For metals with a high reflectivity they are about ±3% for k_p and ±5% for n_p. For ceramic materials and non-metallic inclusions in steels with a relatively low reflectivity, errors of ±10% for k_p and ±2.5% for n_p can be expected. For transparent phases with $k_p = 0$, n_p can be determined to within ±2.5%. In comparison, the use of immersion oils for the determination of the optical constants of steels involves errors up to ten times greater than those associated with the interference layer technique.

In the determination of the optical constants, care must be taken to measure the true reflectivity at the interference minimum. Fig. 14 shows that the interference bands become narrower at higher orders.

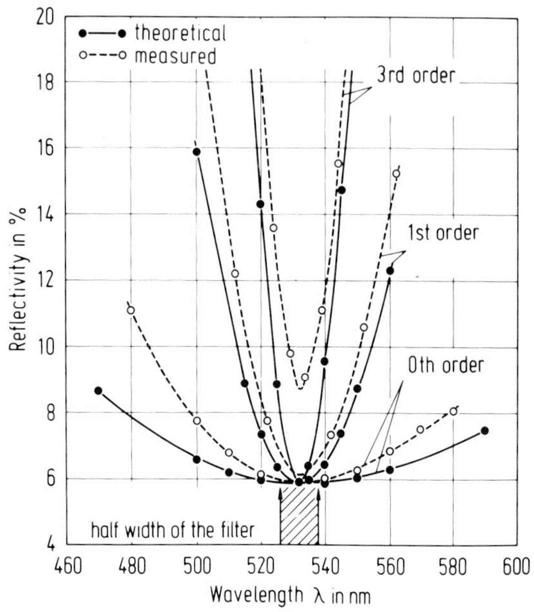

Fig. 14: Comparison of the reflectivities of Armco iron at different orders of interference obtained by measurement with an interference filter (half value width 12 nm) and by calculation. ZnS evaporated layer. The theoretical value of R_{min} lies at 532 nm, 3).

In principle this should permit a more accurate measurement of R_{min} at high orders. Since, however, the interference filters used to produce a monochromatic beam have a finite band width (c.f. explanation of Fig. 18, page 25) i.e. transmit light over a range of wavelengths, the error in the determination of R_{min} is increased. It is not always practicable to use filters with a very narrow band width or grating monochromators because they let through very little light with the result that the measurement of the light intensity and hence of the reflectivity becomes very inaccurate. Fig. 14 further shows that, contrary to theoretical prediction, the value of R_{min} increases at higher orders. Let the value at which the width of the reflectivity curve is equal to the half value width of the filter (c.f. Fig. 18) be designated $R_{50\%}$ and the difference between this and the reflectivity at the minimum R_{min} be

$$\Delta R = R_{50\%} - R_{min}. \qquad (6)$$

This value is indicated for the curves in Fig. 15. The examples of Armco iron, glass and wustite demonstrate that as the reflectivity curves become steeper, the quantity defined in Equation (6) increases. It is a qualitative measure of the error in R_{min} at higher values and characterizes the spectral range either side of R_{min} included in the measurement of the reflection minimum.

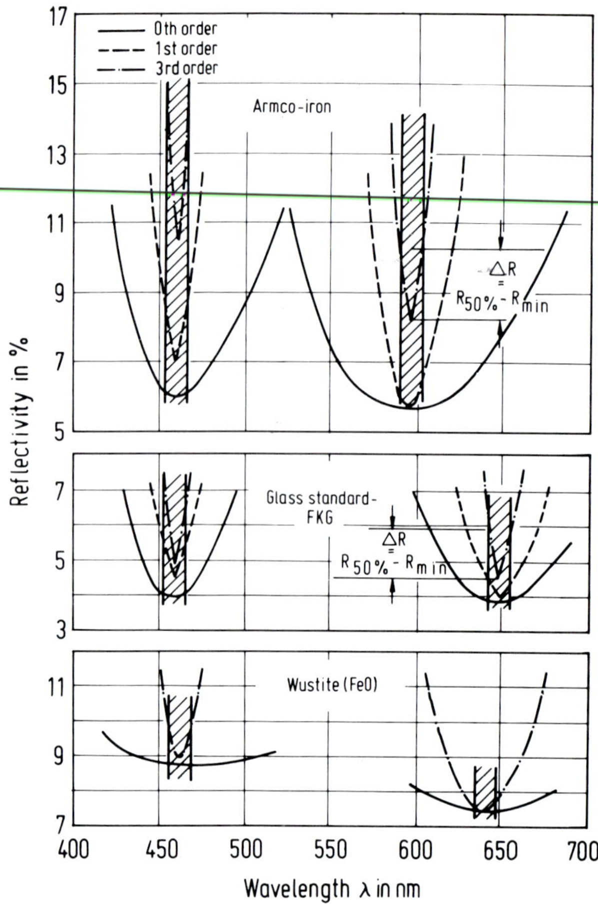

Fig. 15: Relationship between the half value width of the filter and the steepness of the reflectivity curves at different orders of interference 3).

If the increase in R_{min} for the steeper reflectivity curves produced by thicker vacuum deposited layers is characterized by the quotient

$$R_{min}(\text{zero order}) \, / \, R_{min}(\text{nth order}) \qquad (7)$$

which is then plotted against the value ΔR defined in Equation (6) a straight line as in Fig. 16 is obtained.

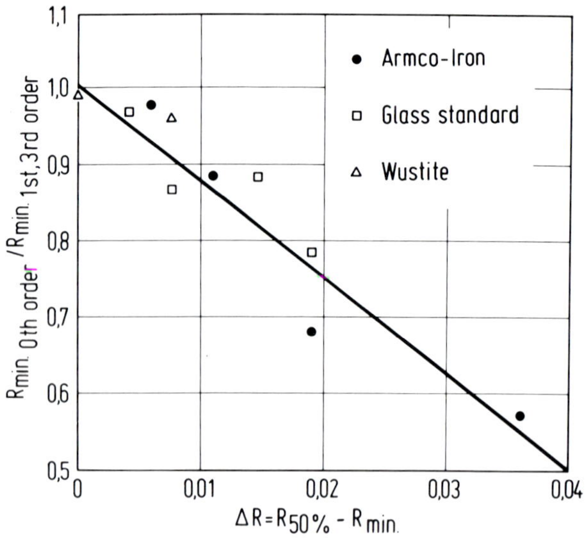

Fig. 16: Influence of the order of interference and the bandwidth of the interference filter on the error in the determination of the optical constants. (According to Equation (6) high values of ΔR signify a larger error in the determination of the optical constants).

This shows that when $\Delta R = 0$ there is no shift of R_{min} to higher values: the quotient has a value of 1. As ΔR increases, i.e. as the layer thickness is increased or the reflectivity curves become steeper the value of the reflection ratio decreases. This means that the error in the determination of R_{min} due to the half value width of the filter increases. Since steep reflectivity curves are desirable for the exact location of the wavelength of R_{min} a very exact determination of R_{min} is possible only by making the light beam more nearly monochromatic i.e. by using a monochromator with a very narrow half value width as far as the light intensity permits.

Literature to 1.1.2

1) Pepperhoff, W. and Ettwig, H.-H.: Interferenzschichten-Mikroskopie. Wissenschaftliche Forschungsberichte, Vol. 73. Dr. Steinkopff-Verlag, Darmstadt 1970

2) Pepperhoff, W.: Quantitative Auflichtmikroskopie mit Hilfe aufgedampfter Interferenzschichten. Arch. Eisenhüttenwes. **36** (1965) 941

3) Jackel, G., Bühler, H.-E. and Robusch, G.: Interferenzschichten-Mikroskopie an opaken und transparenten Werkstoffen. Thyssenforschung **4** (1972) 67

4) Bühler, H.-E.: Die Bestimmung optischer Konstanten durch Mikroreflexionsmessung mit Hilfe des Interferenzschichten-Verfahrens. Archiv für Lagerstättenforschung in den Ostalpen **10** (1970) 70

5) Bühler, H.-E., Jackel, G., Thiemann, E. and Baumgartl, S.: Über optische Eigenschaften des hexagonalen Epsilon-Eisens in Eisen-Mangan-Legierungen. Prakt. Metallographie **6** (1970) 279

6) Pepperhoff, W. and Kohlhaas, E.: Das optische Verhalten von Metallen. In: Metallphysik. Published by Verein Deutscher Eisenhüttenleute Düsseldorf (1967) p. 444

7) Mitsche, R.: Die Anwendung der Mikroreflexionsmessung in der Metallographie. Berg- und Hüttenmännische Monatshefte **107** (1962) 27

8) Demirsoy, S.: Die Entwicklung des Mikroskop-Photometers mit besonderer Berücksichtigung der Reflexionsmessungen. Zeiss-Mitteilungen **4** (1967) 254

9) Caye, R. and Medenbach, K.: Die Bestimmung des Brechungsindex transparenter Minerale durch Messung ihres Reflexionsvermögens. Leitz-Mitteilungen für Forschung und Technik, Vol. 5 (1971) 155

10) Scheidl, H. and Mitsche, R.: Neue Beispiele aus der Mikroreflexionsmessung. Radex-Rundschau **3/4** (1967) 596

1.1.3 Light-dark and colour contrasts

1.1.3.1 Colour contrasts

After coating, the phases have a pronounced reflection minimum dependent upon the wavelength of the illumination (Fig. 6). The separation $\Delta\lambda$ (c.f. Fig. 4, page 12) of the reflection minima of two phases viewed in white light results in a difference in colour between the phases which appear in the colours complementary to λ_{min}.

In many cases a visual assessment and general description of the colours is the only means of locating the reflection minima and gauging the distance between them. The concept of colour will be discussed briefly in the following in order to understand the procedure for uniquely defining the colours on coated layers and colour photographs with regard to the colour contrasts 1).

Three concepts are necessary for the unique description of colour as perceived by the eye (colour sensitivity) namely hue, brightness (value) and saturation (chroma), Fig. 17.

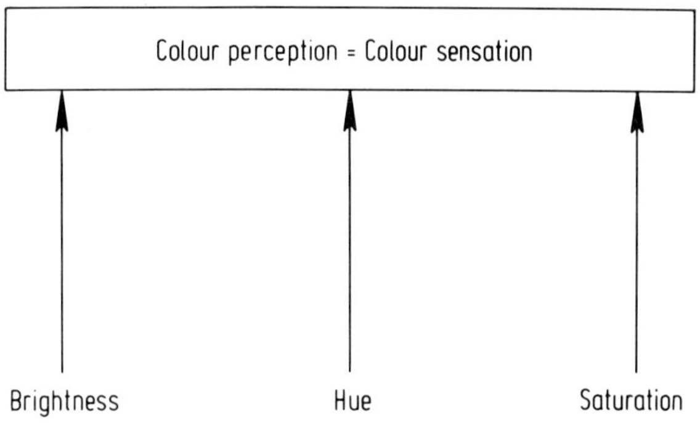

Fig. 17: Components of colour perception.

In common parlance, hues are described as red, green, blue etc. and the "colours" black, white, grey are considered "without colour". By strict definition, the hue is the property that distinguishes a chromatic colour from an achromatic colour 2). In physical terms the hue of light is given by its wavelength λ (c.f. Fig. 1, page 10). Light of a single wavelength is described as monochromatic. Perfectly monochromatic light is however impossible to achieve. Other frequencies are always present in the band width or half value width (Hwb) (Fig. 18). Monochromatic light can also be produced by diffraction at gratings in a so-called grating monochromator with which half value widths down to 1 nm can be obtained. The narrower the half value width, the lower is the intensity of the monochromatic light. Thus half value widths of less than 5 nm are frequently impracticable for optical investigations.

The **brightness** is determined by the amplitude of the light wave (c.f. Fig. 1, page 10). If the illumination is white light, the brightness of a transparent object increases with its transmissivity and that of an opaque object with is reflectivity. In white light the hue is independent of the intensity of illumination, i.e. of the brightness. The darkness level is frequently used as a measure of the brightness 3). Darkness level 0 is the maximum brightness perceivable by the eye for a given hue. Darkness level 10 corresponds to perfect blackness. The intermediate levels are fixed so that a human observer has the impression of equally spaced changes in brigthness.

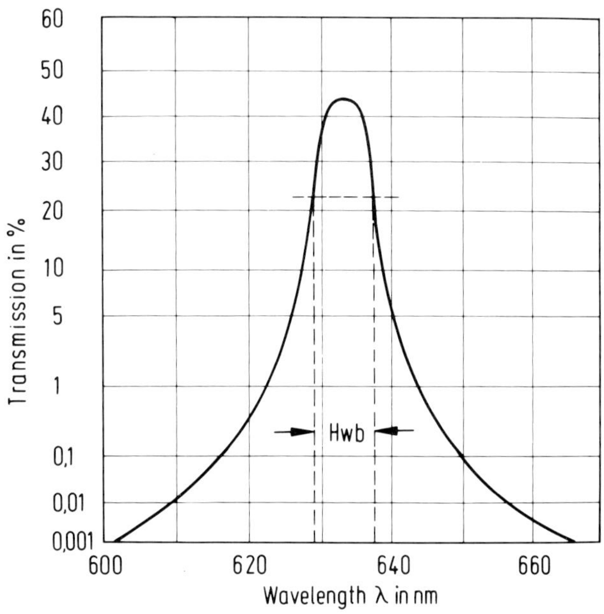

Fig. 18: Transmission of a colour filter as a function of the wavelength. The half value width at 22% transmission is approximately 10 nm.

Let A_0 be the brightness of the optimum colour, i.e. the maximum brightness perceivable by the eye for a given colour. $h = A/A_0$ is then the relative brightness of a colour 5) 6). The darkness level D is then given by

$$D = 10 - 6.1723 \log (40.7\, h + 1)$$

The **saturation** is a measure of the so-called white component. If all the wavelengths of the visible spectrum i.e. 400 to 700 nm appear to the eye to be equally bright, the effect is white i.e. without colour. Light of wavelength 550 nm appears green. It has the maximum possible saturation. As white light is added to this it becomes increasingly unsaturated, in other words is becomes increasingly paler until it is without colour. Its saturation is zero. These relationships are illustrated in Fig. 22. Following the rows from left to right hue and brightness remain unchanged but saturation increases. Following the columns from top to bottom hue and saturation remain unchanged but brightness, as defined by the darkness level, decreases. The colour photographs in Figs. 114 and 116 on page 89 taken with colour filters show that different brightnesses can create very different impressions of colour.

Body-colours are the colours of objects which do not themselves emit light. They reflect only a fraction of the incident light; the remaining wavelengths are absorbed. A red body reflects only the red component of the light, the other components are absorbed. This explains why the colour of a pigment or dye is dependent on the spectral composition of the source of illumination. A red body looks black in blue light because the illumination contains no red light which could be reflected. It is particularly important to bear this in mind when working with interference layers. The hue depends on the nature of the illumination. The colours in Figs. 22 and 23 are body-colours.

The colour sensitivity of the eye to the components red (ca. 700 nm), blue (ca. 435 nm) and green (ca. 545 nm) must be taken into account if an exact description of a colour is to be given in terms of hue, brightness and saturation. The co-ordinates x and y in Figs. 19 and 20 follow on from this but will not be explained in detail here except to say that every colour is unambiguously defined by the values of x, y and z where $x + y + z = 1$ 2).

In the USA and most English-speaking countries, the **colours are classified according to the Munsell system.** In the DIN colour charts used for this atlas, the standard colour values, the new Munsell values and the Ostwald units are given for each hue in addition to the colour designation according to DIN 6164. Transformation of the colour coordinates (x, y, z) of the DIN 6164 system into Munsell units and vice versa is therefore possible without difficulty.

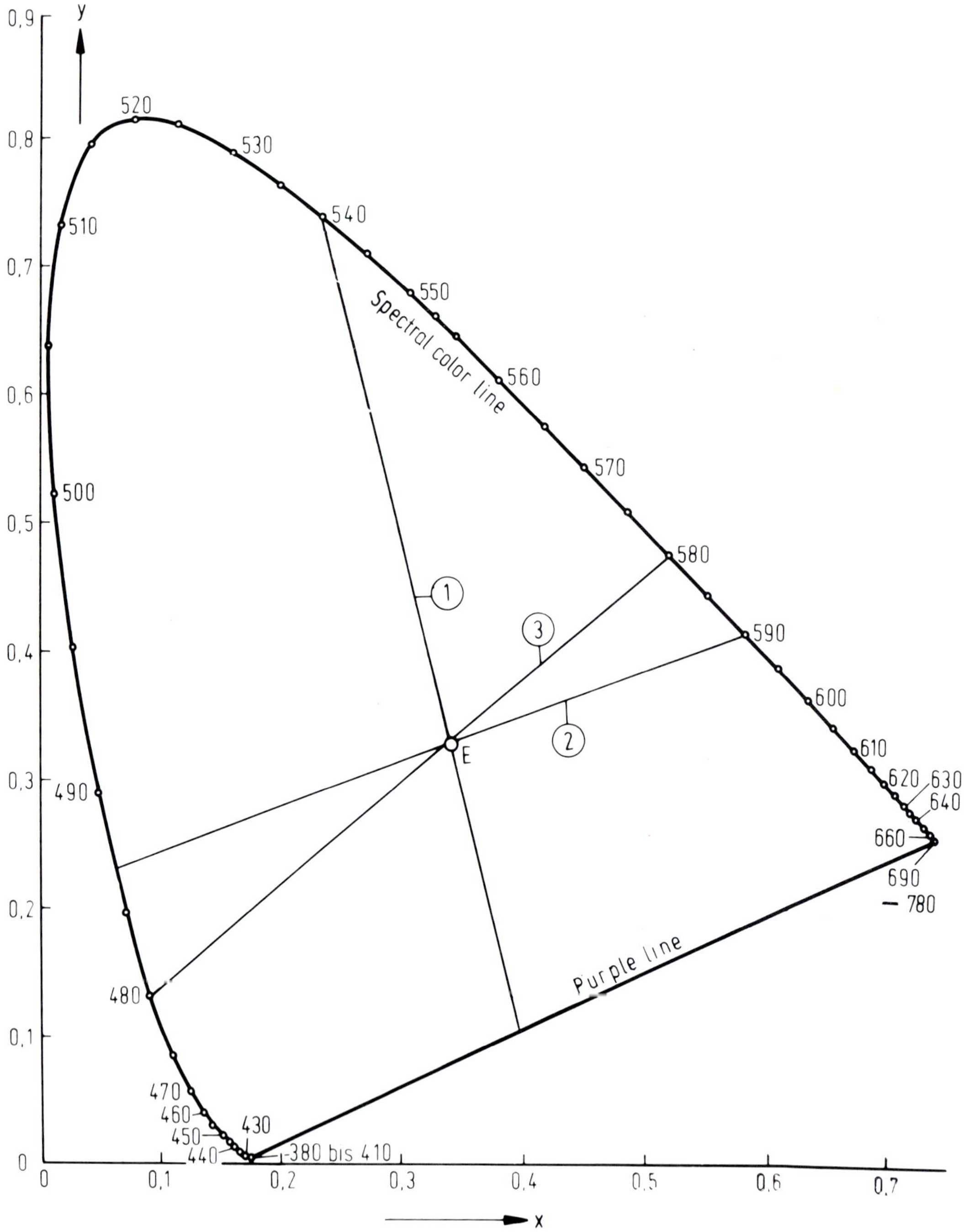

Fig. 19: Colour chart with lines of constant colour, also illustrating the meaning of complimentary colours (see text). E is the achromatic point.

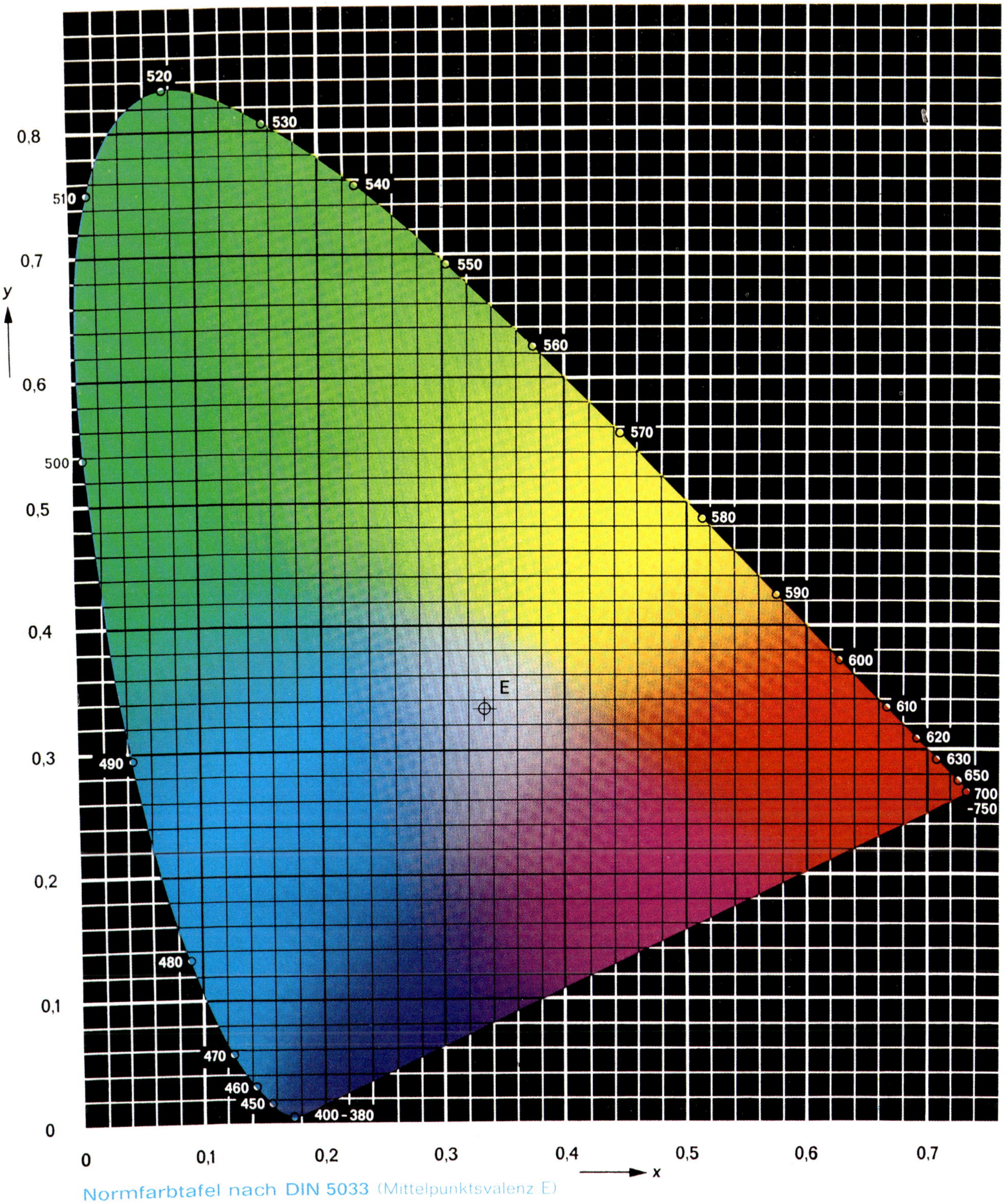

Fig. 20: Colour chart. The colours serve only as a rough guide to the position of the different colour types. E is the achromatic point.

The boundary curves in Figs. 19 and 20 represent the fully saturated colours denoted by their wavelengths. They enclose the region of all possible colours. Point E is the so-called point of no colour (achromatic point). This point represents the impression of "white" for a specified illumination. Colours of the same hue lie on straight lines from the boundary curve (spectral colour sequence) to the point of no colour. The closer to the point of no colour, the lower is the saturation. These colours constitute one colour type. The hue of a colour type is characterized by a number from the DIN comparative chart (c.f. Figs. 22 and 23) or by the equivalent wavelength λ_f. This is the wavelength given by the intersection of the straight line representing the colour type with the boundary curve (the spectral colour sequence). In the case of purplish colours, the wavelength of the complementary colours is used preceded by a minus sign. The saturation as a measure of the white component of a colour decreases in the direction of the point of no colour. Figs. 19 and 20 are valid only for one darkness level. They are a section through a three dimensional colour solid taken at constant darkness level. Other darkness levels lie along the z-axis which lies perpendicular to the plane of the page 6). The different darkness levels for the point of no colour give the grey levels between black and white. Fig. 22 is a section through the colour solid at constant hue.

Colours are complementary if they combine to give no colour, i.e. white. They are given by the intersection of the straight line through E with the opposite section of the boundary curve. If the wavelengths around 480 nm (blue) are filtered out of spectrally pure white light, the eye perceives the complementary colour "yellow" with equivalent wavelength ca. 580 nm. This colour can be obtained from Fig. 19 by joining the point corresponding to 480 nm on the spectral colour sequence curve with the point E and extrapolating this line to cut the opposite section of the spectral sequence curve (line 3 in Fig. 19).

If a coloured object with a reflection minimum at a particular wavelength is illuminated with white light, the eye perceives it in the complementary colour. This can be obtained by extrapolating the line representing the colour type to which the reflection minimum belongs beyond the point of no colour.

According to Fig. 6 (page 13), α-iron coated with a certain thickness of ZnSe has a reflection minimum at 540 nm. This phase would thus appear in the complementary colour purple if viewed in white light. The line from 540 nm through the point of no colour E is denoted 1 in Fig. 19. The line from 590 nm to 485 nm through E (line 2 in Fig. 19) shows that Fe_3AlC_x, which has a reflection minimum at 590 nm (c.f. Fig. 6, page 13), would appear blue in white light, c.f. section 2.2.1. These data refer only to the colour type and not to the saturation or brightness.

In order that body colours can be assessed, the colour types 1 to 24 marked in Fig. 21 are presented as colour plates for several degrees of saturation (1 to 7) and several darkness levels 4)7). Fig. 22 is a reproduction of the plate for colour type 3. By this means each colour can be defined uniquely by three numbers. The first is the hue T, the second the saturation S, and the third is the darkness level D. The third colour in the second row in Fig. 22 has the code number 3:3:2, i.e. hue 3, saturation 3 and darkness level 2, pronounced three to three to two. For the achromatic colours i.e. the grey tones, the first number denoting hue is replaced by –. The assessment is valid only for the original object under specified conditions with illumination type C i.e. approximately white light 7). The gradings in hue, saturation and brightness are chosen so that an observer with normal vision has the impression of equal spacing between the individual hues, degrees of saturation and darkness levels over the whole colour system 3)4). The hues are designated by the appropriate number or the equivalent wavelength.

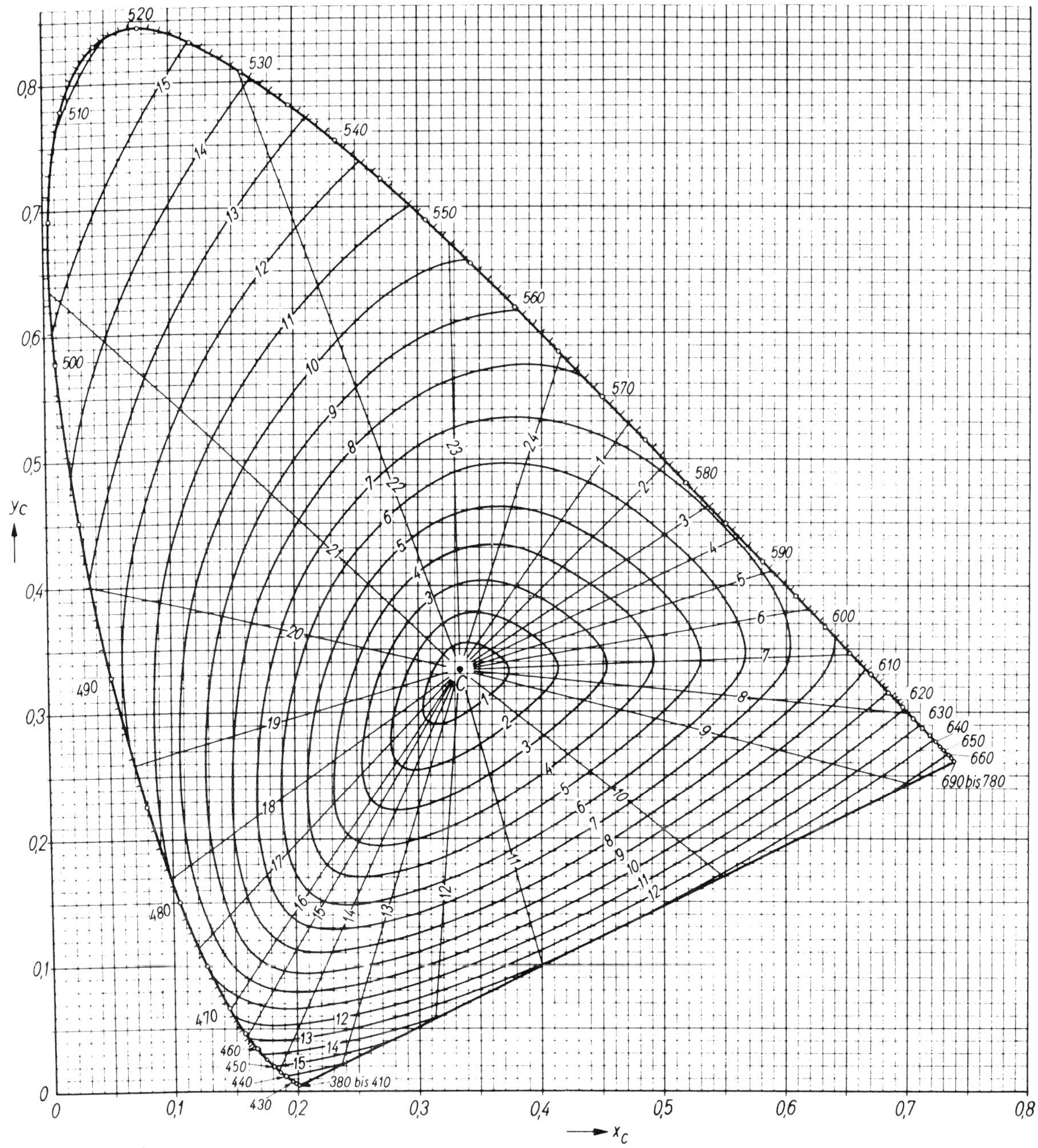

Fig. 21: Colour chart with the colours (straight lines) shown in the colour plate and curves of constant saturation. Illumination type C 7).

Fig. 22: Designation of body-colours by hue, saturation and brightness. Hue 3. See text for explanation of the numbers.

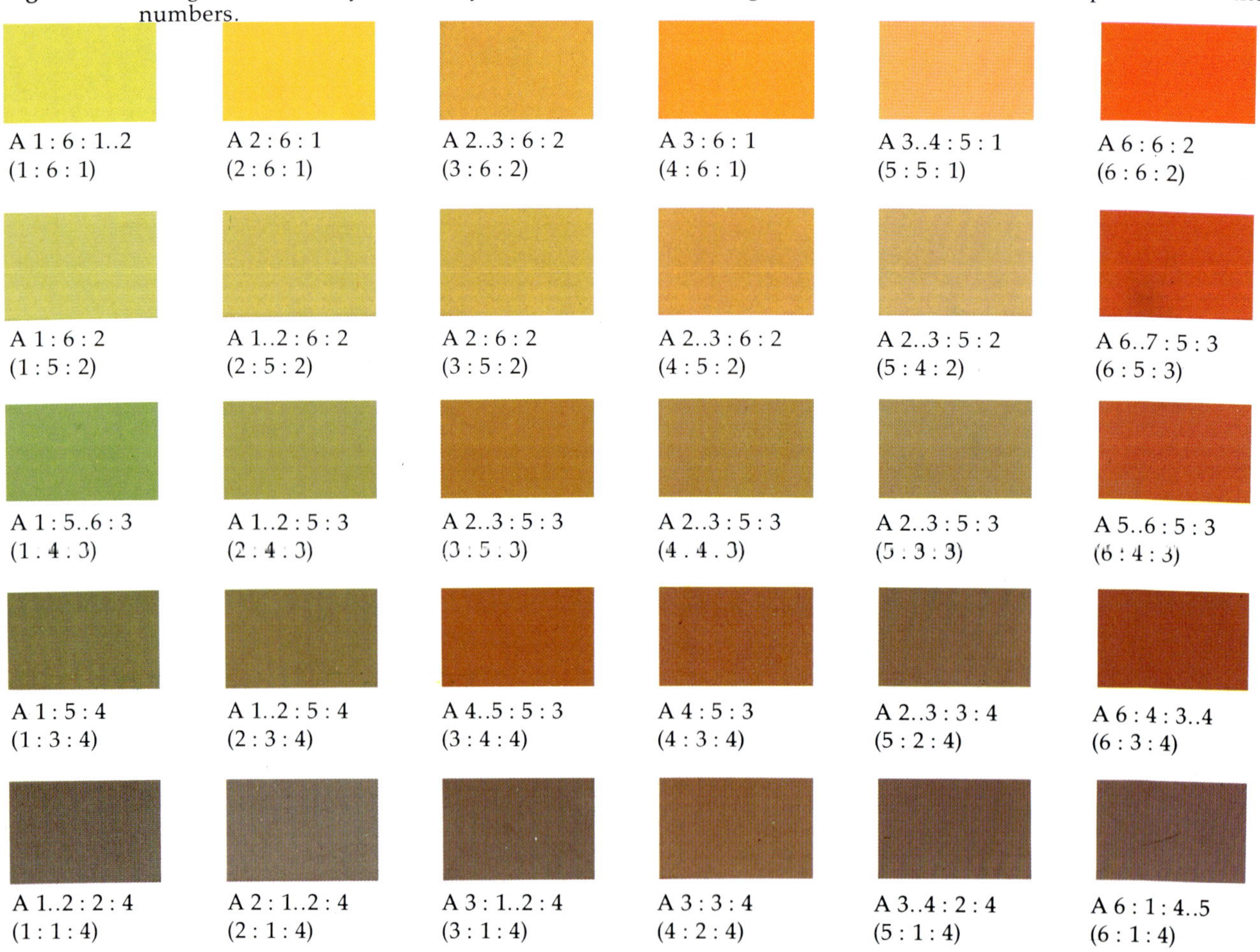

Fig. 23a: Selected colour samples corresponding to the colours most frequently observed on coated specimens. Each hue is shown with decreasing saturation and increasing darkness level. The 5th sample of each hue shows the least saturation, mostly at darkness level 4. The numbers in brackets indicate the hue, saturation and darkness level of the originals.

Fig. 23b: Selected colour samples corresponding to the colours most frequently observed on coated specimens. Each hue is shown with decreasing saturation and increasing darkness level. The 5th sample of each hue shows the least saturation, mostly at darkness level 3 or 4. The numbers in brackets indicate the hue, saturation and darkness level of the originals.

An extract from the colour plates is shown in Fig. 23. The hues, degrees of saturation and darkness levels selected are among those most frequently observed on coated specimens. Colour data from these plates should always be amended to "similar to colour code n:n:n according to DIN 6164" because the colours have been slightly altered in the printing process. An exact assessment requires the use of the original colour plates 7). In order to make this clear, all the colour codes in this atlas are given the prefix "A" (Atlas). If this prefix is omitted it means that after printing the picture under discussion was compared with the original colour plates. The colour changes which may occur during printing are indicated in Figs. 22 and 23. Besides the description A n:n:n the coordinates according to DIN 6164 which describe the colour of the original **before** printing are given in brackets. It is seen that a shift of one or two numbers in the hue, saturation or darkness level with respect to the original may occur during printing. The fact that the selected samples shown in Fig. 23 may be used to describe most interference colours quite well is illustrated for example in Figs. 215–221, 223, 224, 230, 234, 235, 243, 248, 249, 251 and 253.

In the descriptions of the pictures, the colours are referred to by their common names to allow speedy identification. More accurate classification is possible by comparison with the colour plates.

This system of describing body-colours permits an improved assessment and an objective numerical description of the colour contrasts on coated specimens. Figs. 25 to 29 show part of the microstructure of a steel St 37, coated with ZnS, i.e. a non-absorbing layer. The value n_s is practically independent of the wavelength, c.f. Fig. 62, page 60. Under these conditions the difference between the wavelengths of the reflection minima for two phases is independent of the layer thickness (c.f. Fig. 9, page 16). In Table 2 the colours of the matrix and the two phases are given according to DIN 6164. If the complementary colours to these hues and the wavelength for the hue with the highest degree of saturation are determined from Fig. 21, the values given in Table 3 are obtained.

Figure	Matrix	Phase 1	Phase 2	Order
25	2:5:2	22:2:6	2:4:4	1
26	3:5:2	23:2:6	5:3:4	1
27	9:2:3	22:3:5	4:1:7	1
28	22:5:4	2:2:7	22:2..3:6	1
29	2:5:2	4:2:6	1:2..3:5	2

Table 2: Colours of the matrix and the phases in Figs. 25 to 29. In this case the colours were compared with the original colour plates 7) after printing. For this reason the prefix "A" has been omitted.

Figure	Matrix hue	Matrix λ_{min1} nm	Phase 1 hue	Phase 1 λ_{min2} nm	Phase 2 hue	Phase 2 λ_{min3} nm	$\lambda_{min1}-\lambda_{min3}$ nm
25	2	477	22	–529	2	477	0
26	3	482	23	–533	5	486	4
27	9	493	22	–529	4	485	–
28	22	–529	2	477	22	–529	0
29	2	477	4	485	1	468	9

Table 3: Wavelengths of the complimentary colours of the matrix and the phases in Figs. 25 to 29 corresponding to Table 2.

The colours of Fig. 29 are similar to those of Fig. 25, but they differ by one order. The values of $\lambda_{min1}-\lambda_{min3}$ are not constant. The variations are due to the fact that the determination of the hue by simple colour comparison is not accurate enough, and the colour prints sometimes show colour displacements relative to the original. In addition, the brightnesses of the micrographs are reduced to approximately the same level by the photographic process, independently of the brightness of the microstructure. For dark phases, e.g. Phase 1 in Fig. 25, the hue can be determined only approximately. In some cases there is no corresponding sample in DIN 6164. The saturation of the colours is, however, according to Table 2, practically independent of the layer thickness, as one would expect. An exact determination of λ_{min} is possible only with photometric methods.

The optical constants of layers deposited by sputtering (see Section 1.2.3.2) vary not only with the wavelength but also with the layer thickness. Thus the value of $\lambda_{min1}-\lambda_{min2}$ changes with the layer thickness. Figs. 30 to 35 show the microstructure of the brazing alloy BNi5 coated by sputtering, i.e. with an **absorbing** layer, see Section 2.1.9. From these micrographs the values given in Tables 4 and 5 are obtained.

Figure	Matrix	Phase
30	2:4:1	2:4:2
31	2..3:5..6:1..2	3..4:5:2..3
33	3:3..4:1..2	3:3:2
32	4..5:2:2	6:2..3:3
34	9..10:0..1:2	10..11:2:4
35	–:0:2	9:1:4

Table 4: Colours of the matrix and the cubic phase in Figs. 30 to 35. In this case the colours were compared with the original colour plates 7). For this reason the prefix "A" has been omitted.

Figure	Matrix		Phase		
	hue	λ_{min1} nm	hue	λ_{min2} nm	$\lambda_{min2}-\lambda_{min1}$ nm
30	2	477	2	477	0
31	2–3	477–482	3–4	482–485	~4
33	3	482	3	482	0
32	4–5	485–486	6	488	~2
34	9–10	493–500	10	500	~4
35	–	–	9	493	–

Table 5: Wavelengths of the complimentary colours of the matrix and the cubic phase in Figs. 30 to 35.

The colour data in Table 4 show that the colour saturation in Fig. 35 is less than that in Fig. 30. This demonstrates the greater absorption by the thicker layer. The brightness is comparable in all the pictures because any differences are usually cancelled out by suitable exposure when the pictures are taken. The variation of $\Delta\lambda$ with layer thickness is according to Table 5 within the experimental error.

Fig. 21 contains an important indication for the assessment of colour differences. The spacing between hues 1 to 24 appear equal to a normally sighted observer. This means that a difference of 10 nm between the reflection minima of two phases does not give the visual impression of the same difference in hue over the whole colour range. It can therefore be expedient in visual observations of colour shade differences to coat the specimens until they appear red (colour shades 2 to 5) 11), because in this range even small differences in the equivalent wavelength give the impression of a different hue (c.f. Figs. 24 to 35 and 133 to 138, page 97).

Fig. 24: Specimen coated with wedge-shaped evaporated layer. The interference colours are shown from the 1st to the 3rd order. By means of the wedge-shaped layer a continuous increase in layer thickness is attained. Photographed in white light. The same colours are repeated in each order (c.f. Fig. 8).

Figs. 25–29: Steel St 37, oxidized. Coated with ZnS, 1st and 2nd order. The layer thickness increases from Fig. 25 to Fig. 29. Taken from the area shown at low magnification in Fig. 24.

24 5 : 1

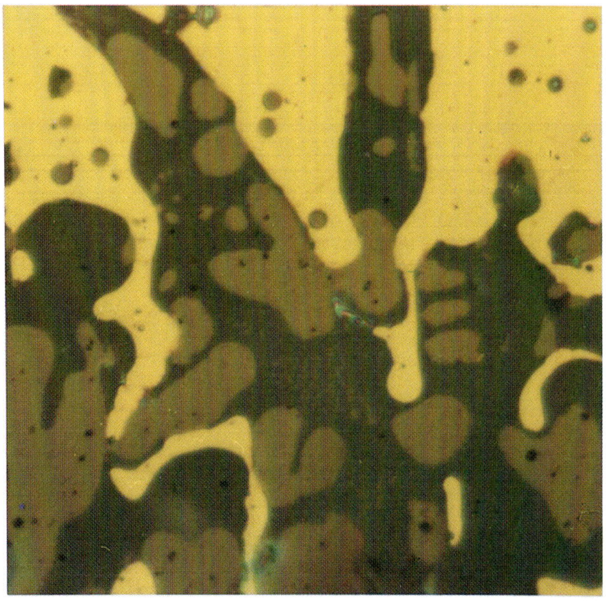

25 200 : 1

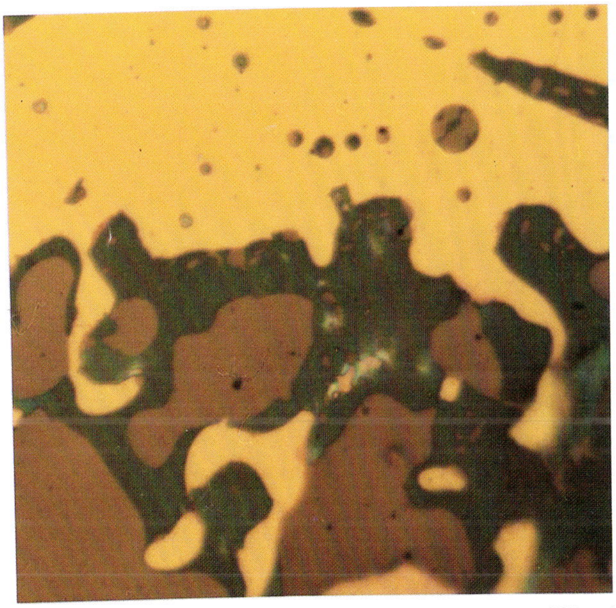

26 200 : 1

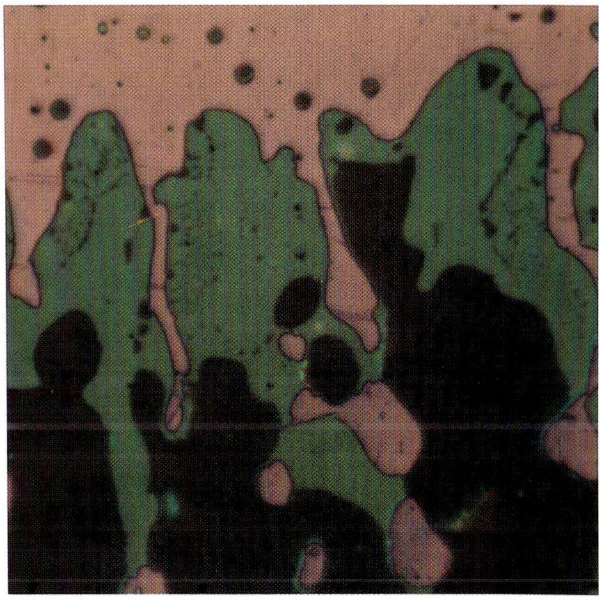

27 200 : 1

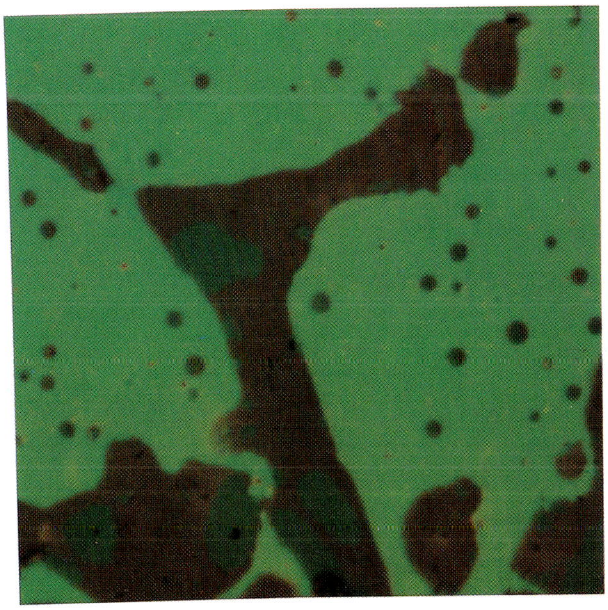

28 200 : 1

29 200 : 1

Figs. 30–35:	Brazing alloy BNi5 (c.f. Section 2.1.9), coated by sputtering with an iron cathode and oxygen, zero-order. The layer thickness increases in the order 30, 31, 33, 32, 34, 35. Differential interference contrast.
30:	coating time approx. 300 s
31:	coating time approx. 305 s
33:	coating time approx. 310 s
32:	coating time approx. 315 s
34:	coating time approx. 320 s
35:	coating time approx. 325 s

30 500 : 1

31 500 : 1

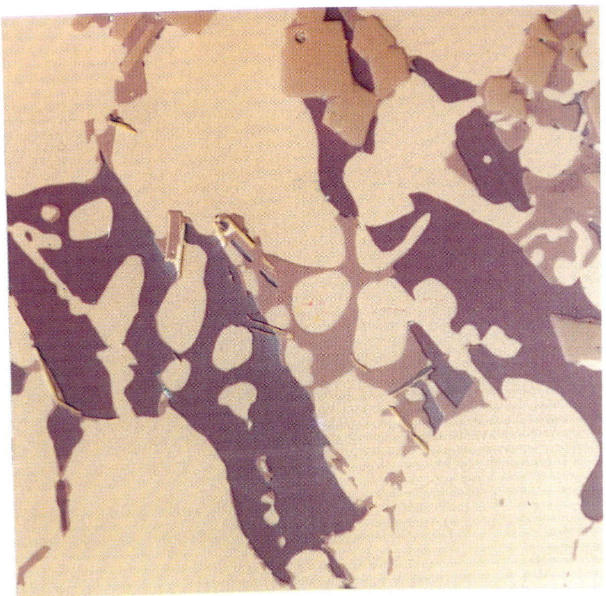

32 500 : 1

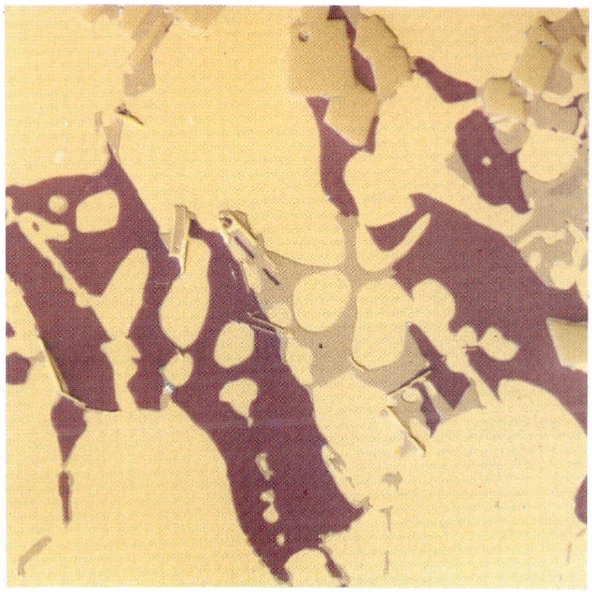

33 500 : 1

34 500 : 1

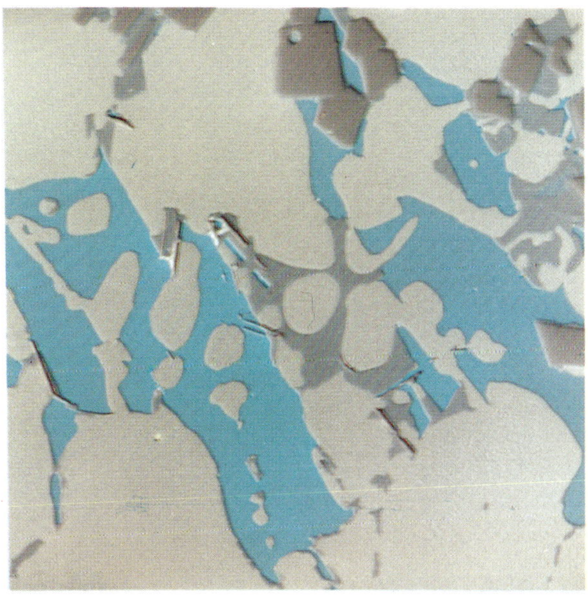

35 500 : 1

It has been mentioned on several occasions that body colours depend on the nature of the illumination. This must be taken into account both in the observation of coated specimens under the microscope and in the classification of colours. For example low voltage lamps emit light with a different spectral composition from that of xenon lamps. Particularly in the case of colour photographs care must be taken when using low voltage lamps always to use the same voltage (or better the same current) namely that at which the best colour reproduction is achieved for a given film type, since the spectral composition of the light emitted by these lamps varies with voltage. Preferred sources of illumination are those matched to the emulsion of the selected colour film since only in this case are the colours faithfully reproduced 10).

For this reason, the colour temperature of the illumination source should be checked repeatedly. The exposure must be adhered to exactly because colour sensitive emulsions have a much smaller exposure tolerance than black and white. If these rules are not observed colour displacement will result. In taking prints or making enlargements, the filter must be carefully matched to the paper. Experience shows that this precaution is not always observed in amateur projects. By giving the work to specialist laboratories, especially if model colours are provided, satisfactory colour reproductions can generally be achieved. In this case the colour deviations are generally less than one hue.

The photographs reproduced in the present volume were produced using various equipment and film emulsions. Colour displacement also arose during copying and printing. In many cases therefore the colours will not correspond with those observed under the microscope on a similar specimen coated with an identical thickness of the same coating material. This error is however not fundamental because to a first approximation all colours suffer a similar shift. This means that their relative position is unchanged. In the descriptions of the micrographs in this book the colours are given as they appear **after** printing on comparison with the original DIN colour plates. The colour characterization is thus uniform, since in some cases colours occur which are not included in Figs. 22 and 23. On page 32 some micrographs are listed, for which the colours were determined in addition from Figs. 22 and 23.

As a consequence of the printing process, displacements of one hue, one saturation step or one darkness level may occur in the judgement of the colour micrographs used. This is also true for a comparison of different copies of this book.

1.1.3.2 Light-dark contrasts

The purpose of a coating layer is to accentuate the contrast between two or more phases in a specimen. With the aid of Fig. 21 it is possible to determine the values of λ_{min} required to obtain a particular colour difference with the type of illumination upon which Fig. 21 is based. A light-dark or black-white contrast is however sufficient for a quantitative evaluation of the microstructure. For this, the maximum contrast is given, according to Equation (1), when the reflection minimum of one phase is zero and that of the second phase is as high as possible. This has already been explained with the help of Figs. 4 to 7. For R_{min} to be zero, both the phase and amplitude conditions must be satisfied, i.e. the reflection is completely extinguished by interference. This optimum condition cannot always be fulfilled because the choice of layer materials is not unlimited (c.f. chapter 5) so that a particular refractive index or desirable ratio between refractive index and absorption cannot be guaranteed in every case. Working with monochromatic light, however, sufficient contrast can be achieved even in this case 8)9). According to Fig. 6, a contrast close to 1 is achieved at λ_1. According to Fig. 7 maximum contrast is achieved at 540 and 590 nm. In white light, these contrasts are poorer, (c.f. Figs. 96 to 100, page 79, 103 to 111, page 83, and 163 to 165, page 105). In some cases the black-white contrast achieved in monochromatic light is better than the colour contrast (c.f. Figs. 151 to 156, page 103).

The use of colour sensitive systems offers no advantage over monochromatic light. Such systems (e.g. colour television receivers) work with three signals in the colours red, green and blue. For observation purposes these signals are mixed and the original colour impression is recreated. The example in Fig. 7 shows that at 480 nm (blue), 560 nm (green) and 640 nm (red) the contrast is poorer than at 540 nm and 590 nm. For any transmission system, however, the loss of information, i.e. experimental accuracy, is always greater if the contrast is poor than if it is good. The use of three signals thus offers no advantage over monochromatic light. In the example given in Fig. 7 the use of monochromatic light with a wavelength of 540 or 590 nm ensures the best possible contrast and hence the most reliable evaluation. Only when several phases must be represented simultaneously for a visual assessment can the use of a colour sensitive system be advantageous. For quantitative evaluations, however, the use of monochromatic light is always preferable. In general the aim is to coat the specimens so that λ_{min} lies in the region of 540 nm (green) because the sensitivity of the eye is a maximum for green light and optical instruments are optimally corrected at this wavelength. The appropriate phase then appears in the complementary colour purple when viewed in white light (c.f. Fig. 21, page 29). Only in special cases may it be expedient to work at other wavelengths.

Literature to 1.1.3

1) Schultze, W.: Farbenlehre und Farbenmessung. Springer-Verlag Berlin, Heidelberg, New York 1977

2) DIN 5033: Farbmessung

3) Richter, M.: Das System der DIN-Farbenkarte. Die Farbe **1** (1952/53), 85

4) Richter, M.: Über Entstehung, Aufbau und Anwendung der DIN-Farbenkarte DIN 6164. DIN-Mitt. **42** (1963) 269

5) Rösch, S.: Die Kennzeichnung der Farben. Physikal. Z. **29** (1928) 83

6) Rösch, S.: Ein Farbkörpermodell als Hilfsmittel bei Farbmessungen nach DIN 6164. Die Farbe **10** (1961) 315

7) DIN 6164: DIN-Farbenkarte

8) Kohlhaas, E. and Fischer, A.: Kontraststeigerung beim Interferenzschichten-Verfahren durch Verwendung eines Verlauf-Linienfilters. Prakt. Metallographie, **6** (1969) 339

9) Bühler, H.-E. and Meyer, L.: Kontraststeigerung in der Auflichtmikroskopie mit Hilfe aufgedampfter Interferenzschichten unter Verwendung monochromatischen Lichtes. ZEISS-Information, **15** (1967) 118

10) Frühling, H. G., Münch, W. and Richter, M.: Die Eignung der Xenon-Lampe als Standard-Lichtquelle für Strahlungs- und Farbmessungen. Die Farbe **5** (1956) 41–68

11) Gräf, I. and Draugelates, U.: Phasenidentifizierung in Nickelloten mit modernen metallographischen Techniken. Prakt. Metallographie, special Volume **7** (1977) 311

12) Heide, E.: Eine Methode zur Bestimmung der farbartgleichen Optimalfarbe und ihres Hellbezugwertes. Die Farbe **17** (1968) 101

13) Hougardy, H. P. and Bühler, H.-E.: Die Beschreibung von Farben im Bereich der Metallographie. Prakt. Metallographie, special Volume **10** (1979) 583

1.2 EXPERIMENTAL DETAILS OF THE INTERFERENCE LAYER TECHNIQUE

1.2.1 Preparation of Polished Sections

1.2.1.1 General procedure and guide lines

In all previously published work on the subject of interference layer microscopy particular stress was laid on the decisive influence of specimen preparation on the success or failure of the experiments. This is equally true for both evaporation and sputtering.

A very clean and representative specimen surface is essential because, due to the strong absorption of metallic and to some extent also of non-metallic materials, light cannot penetrate very far into the specimen. Thus the interaction of the light with the specimen and the deposited layer occurs only in very thin surface zones. These surface zones, which are only a fraction of a wavelength of light thick, are freely exposed to external influences which do indeed affect them strongly.

The properties of the surface are altered by impurity layers formed for example by the reaction or absorption of oxygen or water vapour from the atmosphere or also during polishing, e.g. water (drying) stains. During grinding and polishing fine swarf is removed mechanically from the surface and the underlying material suffers deformation. This deformed layer masks the true microstructure of the specimen and so falsifies the assessment. In the case of metallic materials the thickness of these deformation layers, which consist of very fine crystallites and can have completely different optical properties from the unstressed crystal, is many times greater than the penetration depth of the light (Fig. 36). The deformation layers formed during mechanical grinding and polishing must be removed as far as possible. Scratches and polishing grooves must be avoided at all costs. For these reasons grain size 120/1000 or 1200 (FEPA-NRM) should be used for wet grinding, if possible one after the other. On changing to a finer paper, the specimen should be cleaned as usual and turned through 90 degrees. After grinding, the specimen should be polished with diamond paste or diamond spray with grain size 7, 2.5 and 1 µm and finally finish polished with 0.25 µm. An intermediate etch has proved valuable for the rapid removal of the deformation layers, but it must not be so strong that a relief remains after polishing.

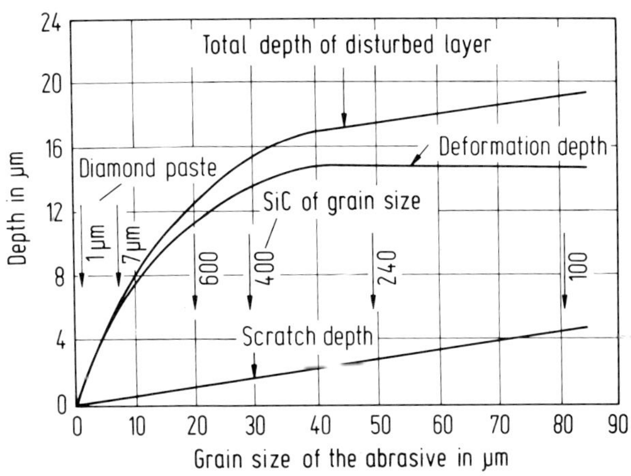

Fig. 36: Influence of the grain size of the abrasive on the depth of the deformed layer 9).

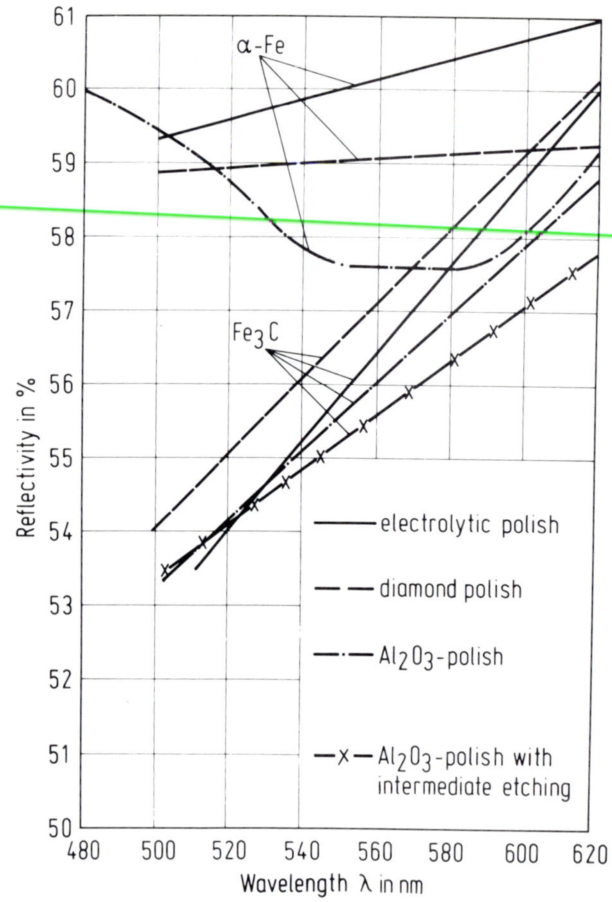

Fig. 37: Influence of the polishing techniques on the reflectivity of α-Fe and cementite 1).

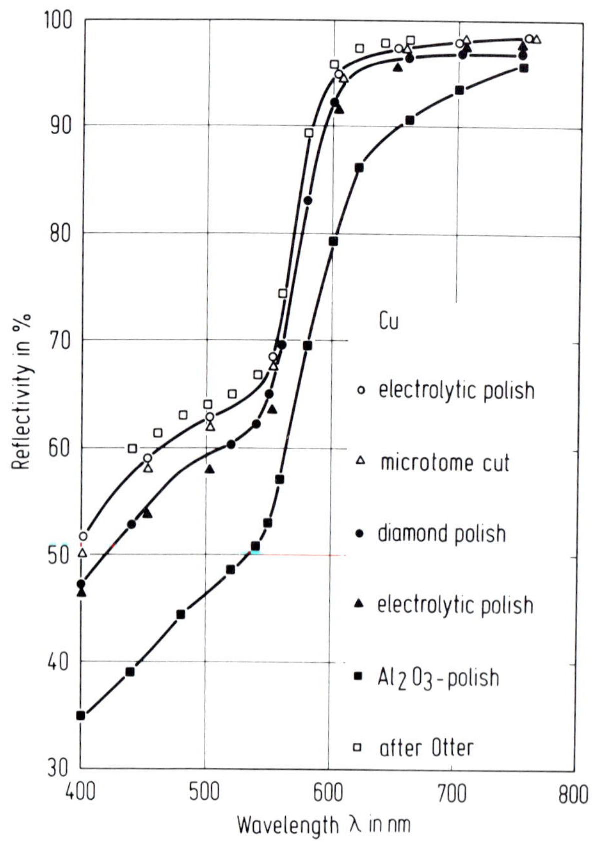

Fig. 38: Influence of the polishing technique on the reflectivity of copper 1).

Following this procedure a surface is produced which is in most cases very suitable for the production of contrast. In the case of some materials such as soft metals like copper, aluminium, brazing alloys or ferritic steels, an additional fine polish with alumina 3 or FF is recommended. Peculiarities in the preparation of individual classes of materials will be treated in more detail in the appropriate chapters.

Another possible preparative technique is electrolytic polishing which also removes the deformation layer produced by grinding. The resulting specimen surface on which a contrast is to be produced may possess large uneven areas and may under certain circumstances have suffered slight electrochemical etching. In general this does not have a deleterious effect on the attainable contrast. On the other hand the very thin oxide layers sometimes produced during electrolytic polishing may affect the contrast. Care must be taken to use only those electrolytes which do not lead to the production of surface layers. Small non-metallic inclusions in metallic materials can easily be dissolved out during electrolytic polishing. In these cases a satisfactory finish can sometimes be achieved by suitable choice of electrolyte.

After polishing, the specimen surfaces must be cleaned carefully with alcohol and dried over a long period. If this rule is not followed, residual moisture from cracks and pores can come to the surface during the evacuation prior to evaporating or sputtering and form patches and rims. These produce strong and distracting contrasts and can lead to an erroneous evaluation of the microstructure.

If the specimens are examined under the optical microscope after polishing it is imperative to ensure that any residues of plasticine and immersion oil are removed before the specimens are placed in the vacuum chamber.

Incompletely hardened mounting media which can result if the proportion of hardener is too low can give off gases during evacuation over long periods and spoil the coating layer.

For the final cleansing and to remove drying stains or organic residues, the specimen can be cleaned before coating by spark discharge in argon. Ion etching by cathodic evaporation is another method of cleaning the surface. Such processes must be carried out with great care, however, otherwise a surface relief can easily be produced and the grain boundaries etched.

It can be seen from Fig. 37 that for microstructures comprising α-iron and cementite, the preparation of specimens by an alumina polish results in a relatively poor reflectivity compared with preparation by a diamond or electrolytic polish. Furthermore the experimental results were scarcely reproducible after an alumina polish. In contrast, the reflectivity values after diamond and electrolytic polishing, although different, were of the same order of magnitude and were easily reproducible. The deformation resulting from grinding and polishing can lead to deviations in the reflectivity values of over 60% (relative).

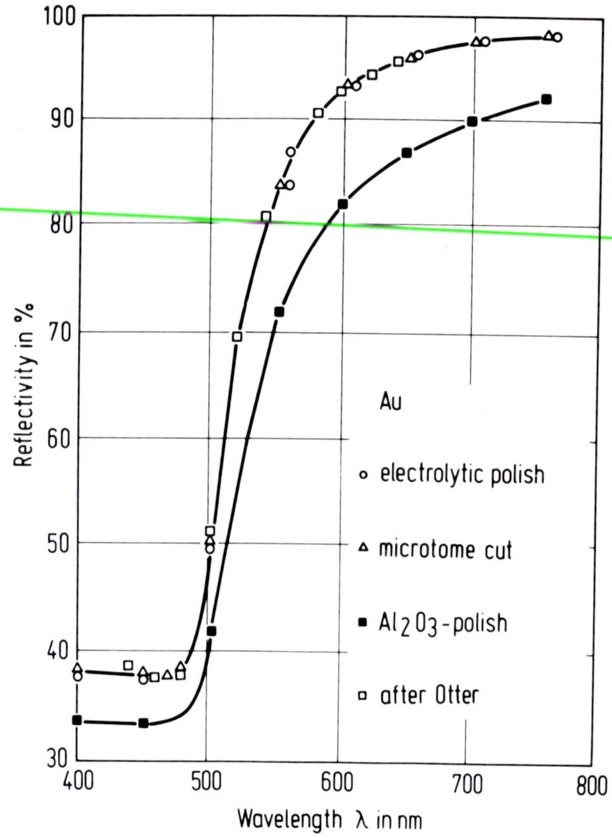

Fig. 39: Influence of the polishing technique on the reflectivity of gold 1).

Figs. 38 and 39 illustrate the influence of different polishing methods on the reflectivity of copper and gold. Electrolytically polished and microtome cut specimens come closest to the ideal reflectivity. Diamond polished surfaces have lower reflectivity values. Alumina polished surfaces exhibit the lowest reflectivity values in this comparative investigation. These results agree with those of other investigations showing that reflectivity values obtained on alumina polished specimens are least reliable 1).

The mean percentage deviations of the reflectivity of polished from ideal surfaces are summarised in Fig. 40. These results from copper, gold, bismuth, nickel and antimony sulphide are also in agreement with earlier findings that an alumina polish commonly yields poor results in

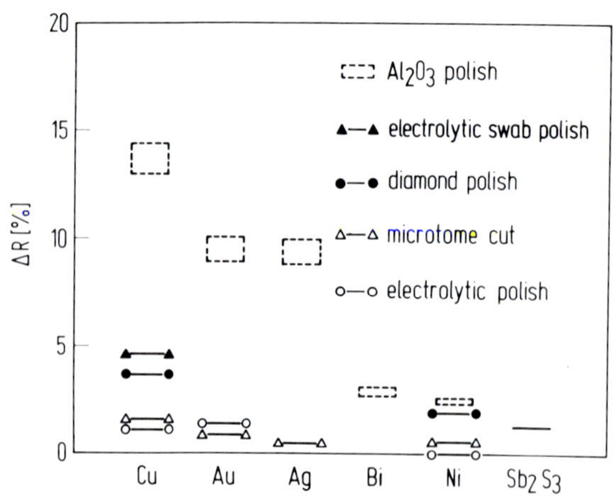

Fig. 40: Influence of the polishing technique on the relative change in the reflectivity ($\Delta R = 0$, ideal surface) 5).

interference metallography. Some values of the reflectivities of metallic and non-metallic phases are given in Table 12 (section 5).

The effect of certain characteristic errors in specimen preparation on the formation of interference layers is illustrated in the following pictures.

1.2.1.2 Examples related to specimen preparation

Scratches on the specimen surface due to polishing and grinding

If the surface of a specimen is coated with a material capable of producing interference, scratches due to grinding and polishing sometimes show up in a particularly awkward fashion. Because of the shadows they cast such preparation defects are particularly noticeable compared with the highly reflecting polished background.

This effect can be reduced by observing the specimen at a wavelength at which the phase with the most scratches has a similar reflectivity to the scratches themselves. In this case the contrast between the scratches and their surroundings is so small that they become invisible. Figs. 41 to 44 show that the scratches stand out more or less clearly depending on the wavelength at which they are observed. The change in brightness of the phases associated with the change in observation wavelength is not so obvious in the pictures because differences in brightness are largely compensated for by the exposure of the film and prints.

Smeared layers, hygroscopic layer materials, deformation layers

Smeared layers can consist of oxidation products which form on the surface of materials with a high affinity for oxygen during specimen preparation. They can also consist of very thin deformed surface layers. In both cases there is a change in the optical properties of the corresponding phase. Figs. 45 to 47 show the same area of the microstructure on a chill cast chromium steel after careful specimen preparation (Fig. 45) and covered with a smeared layer (Fig. 46).

In Chapter 1.2.3 it will be explained that some layer materials react hygroscopically after long exposure to air as a result of which a very fine dispersion of droplets is produced on the surface. An example of this is the evaporated zinc selenide layer shown in Fig. 47. In addition to zinc selenide, zinc telluride is a particularly hygroscopic layer material and should not therefore be stored over long periods, especially in moist air. The microstructure shown in Fig. 47 is that of a specimen exposed to air for one day after coating. Similar phenomena can occur when the vacuum in the evaporating chamber was not good enough or the material to be evaporated was not preannealed.

Too great a pressure on the specimen during polishing can cause such a degree of cold surface deformation that not only are the optical properties changed but also unstable phases undergo transformation. A characteristic example of this is the transformation of residual austenite to deformation martensite in tool steels or the formation of martensite in chemically resistant austenitic chromium-nickel steels. Figs. 48 and 49 show the microstructure of a steel X5 CrNi 18 9 in which the surface regions have been transformed to martensite as a result of excessively heavy deformation (Fig. 48) whereas a correctly polished section is seen after etching to consist exclusively of austenite (Fig. 49). In Fig. 48, the surface is so heavily deformed that the microstructure bears no resemblance to the true structure in Fig. 49. Nor can this be improved by coating.

Drying stains

Drying stains resulting from lack of care in cleaning the specimen surface can critically falsify the structure as seen in the microscope. A poor standard of specimen preparation frequently leads to totally erroneous assessments. Figs. 50 and 51 show the microstructure of an armco iron specimen which was not properly dried before coating with ZnSe.

Cracks and pores

At the end of the grinding and polishing process, cracks and pores in the specimen surface usually contain residual moisture which after drying spreads out over the specimen surface. This "drawing out" of moisture results in areas which on coating ultimately appear in different interference colours from the rest of the specimen surface. Examples of this are shown in Figs. 52 and 53. Fig. 54 shows a hot aluminised sheet mounted in synthetic resin. Shrinkage of the resin resulted in a crevice which led to a ring of moisture. The hot aluminised layer can scarcely be distinguished from the sheet at this point.

Comparison of absorbing and non-absorbing layer materials

The amplitude of the emergent light is attenuated by the selfabsorption of an absorbing layer (c.f. section 1.2.3.2) and the brightness and saturation of the interference colours decreased. For this reason the interference colours are always more highly saturated on a microstructure coated with a non-absorbing material than with an absorbing material of the same refractive index. If the self absorption is very strong this effect leads to a noticeable loss in colour contrast and the layer material ultimately becomes useless. This is particularly true of absorbing layers used at high orders.

Thus if absorbing layers are being used, inadequate colour differentiation should not always be attributed to bad specimen preparation. The differences in colour contrast and hence also in the light-dark contrast are a consequence of the optical properties of the layer material, as shown by the example in Figs. 55 and 56.

Literature to 1.2.1

1) Knosp, H.: Mikrophotometrie, Prakt. Metallographie **7** (1970) 494

2) Wesemeyer, H. and Raether, H.: Über die Reinigung elektrolytisch polierter Metallflächen mittels Kathodenzerstäubung. Naturwissenschaften **39** (1952) 398

3) Jackel, G., Bühler, H.-E. and Robusch, G.: Interferenzschichten-Mikroskopie an opaken und transparenten Werkstoffen. Thyssenforschung **4** (1972) 74

4) Pepperhoff, W. and Ettwig, H.-H.: Interferenzschichten-Mikroskopie. Dr. Steinkopff-Verlag, Darmstadt (1970)

5) Mitsche, R.: Die Anwendung der Mikroreflexionsmessung in der Metallographie. Berg- und Hüttenmännische Monatshefte **107** (1962) 27

6) Pepperhoff, W.: Das optische Verhalten von Metalloberflächen. Handbuch der Mikroskopie in der Technik, Vol. III, Part 2, p. 1. Umschau-Verlag, Frankfurt 1969

7) Petzow, G.: Metallographisches Ätzen. Gebrüder Borntraeger Verlag, Berlin, Stuttgart 1976

8) Semlitsch, M.: Technische Rundschau Sulzer, Forschungsheft 1970

9) Petzow, G. and Exner, H. E.: Recent Developments in Metallographic Preparation Techniques. Microstructural Science, **3** (1975) 291

10) Beraha, E. and Shpigler, B.: Colour Metallography, ASM 1977

11) Petzow, G. and Knosp, H.: Application of Microreflection in Metallography. Metallography **6** (1973) 249

12) Knosp, H., Gahm, I. and Petzow, G.: Quantitative polarisationsmikroskopische Untersuchung von Anisotropieeffekten am Beispiel der (010)-Schnittfläche von Antimonglanz. ZEISS-Mitt. **5** (1971) 331

13) Scheidl, H. and Mitsche, R.: Neue Beispiele aus der Mikroreflexionsmessung. Radex-Rundschau **3/4** (1967) 596

Figs. 41–54: **Preparation artefacts**

Figs. 41–44: **Scratches**

41: Chill cast chromium steel with 2% C and 17% Cr
Coated with ZnSe
Photographed at 540 nm

42: Alloy and preparation as in Fig. 41
Photographed at 560 nm

43: Alloy of 49% Sn, 34% Co, remainder Mn (weight percent)
Coated with ZnSe (c.f. Fig. 56)
Photographed at 530 nm

44: Alloy and preparation as in Fig. 43
Photographed at 560 nm

Figs. 45–47: **Smeared layers, absorption of moisture**

45: Alloy and preparation as in Fig. 41

46: Alloy as in Fig. 45
Specimen covered with a smeared layer
Coated with ZnSe, bright field

47: Alloy and preparation as in Fig. 45
Photographed after one day's exposure to air

Figs. 48 and 49: **Surface deformation**

48: Steel X5 CrNi 18 9
Austenite and martensite
Coated with ZnSe, bright field

49: Steel X5 CrNi 18 9
Austenite
Etched with a solution of 80 g HCl, 5 g HNO_3, 20 g glycerine and 1 g $CuCl_2$

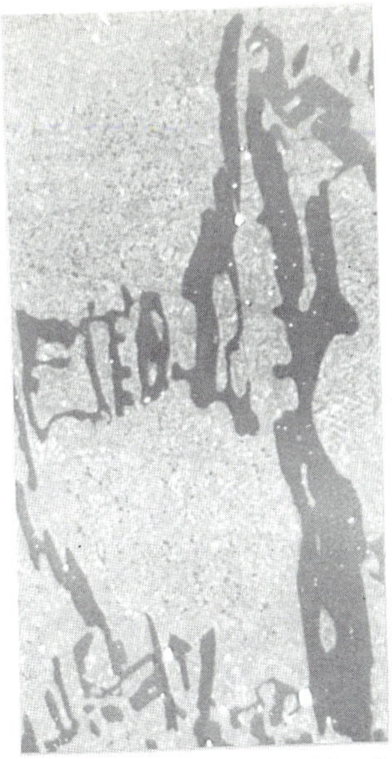

41 500 : 1

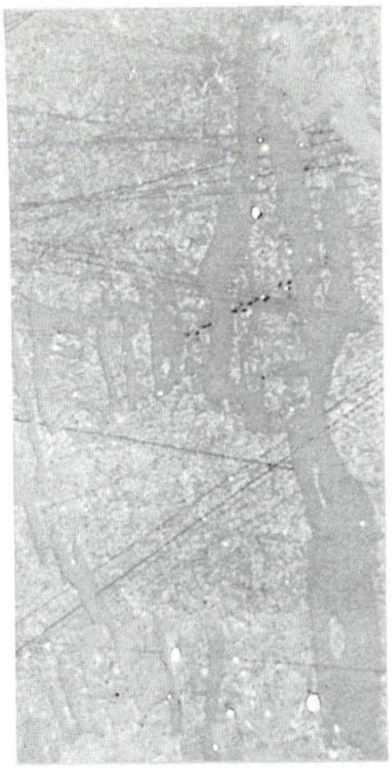

42 500 : 1

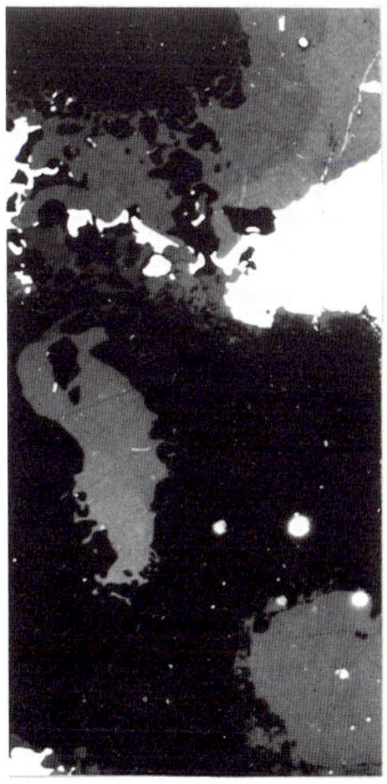

43 200 : 1

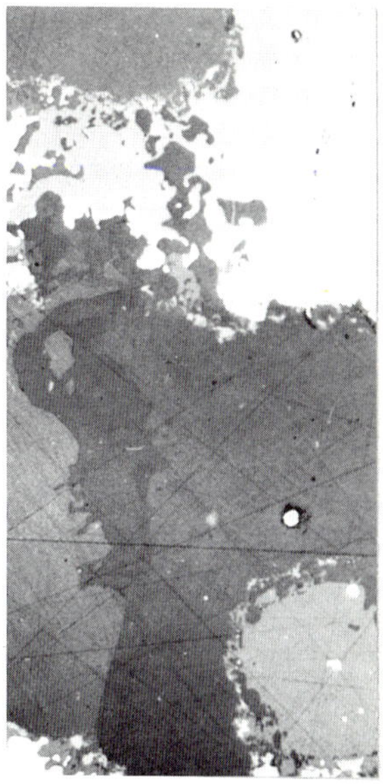

44 200 : 1

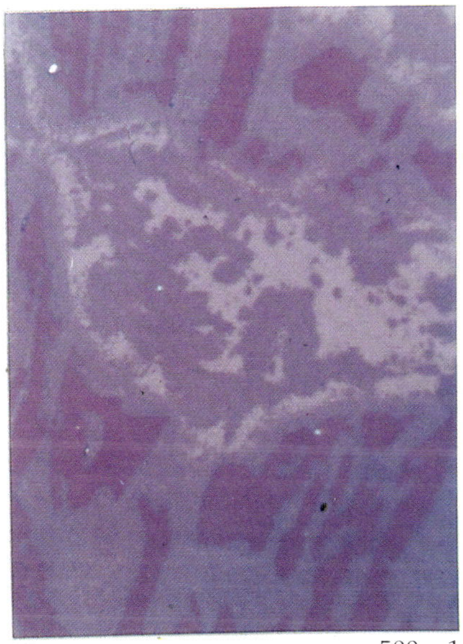

45 500 : 1

46 500 : 1

47 500 : 1

48 200 : 1

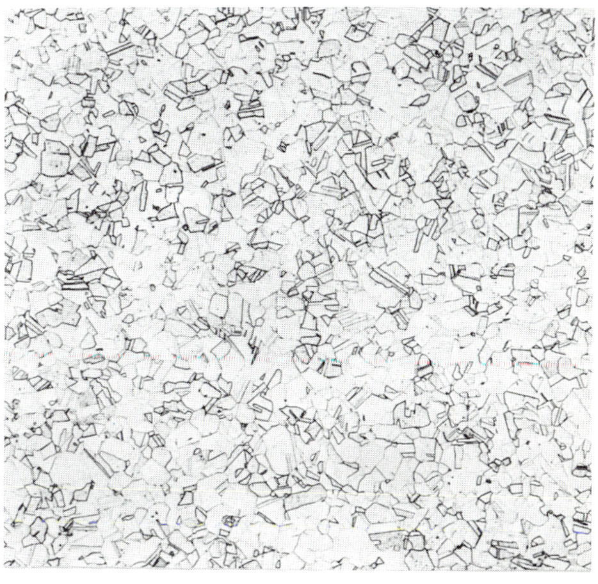

49 200 : 1

Figs. 50–54:	**Drying stains**
50:	Armco-iron with drying stains Coated with ZnSe, bright field
51:	As Fig. 50
52 and 53:	Steel and preparation as in Fig. 41, bright field moisture from cracks
54:	Hot aluminised sheet mounted in synthetic resin Coated with ZnSe, bright field moisture from crack in mount
Figs. 55 and 56:	**Comparison of absorbing and non-absorbing layers**
55:	Alloy as in Fig. 43 Sputtered using an iron cathode and oxygen, bright field
56:	Alloy and preparation as in Fig. 43 but bright field After coating the colours have a greater saturation than in Fig. 55. The phases can be more easily distinguished.

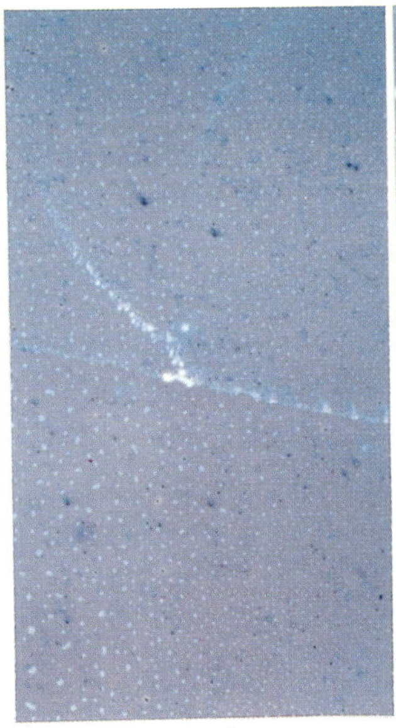

50 100 : 1

51 100 : 1

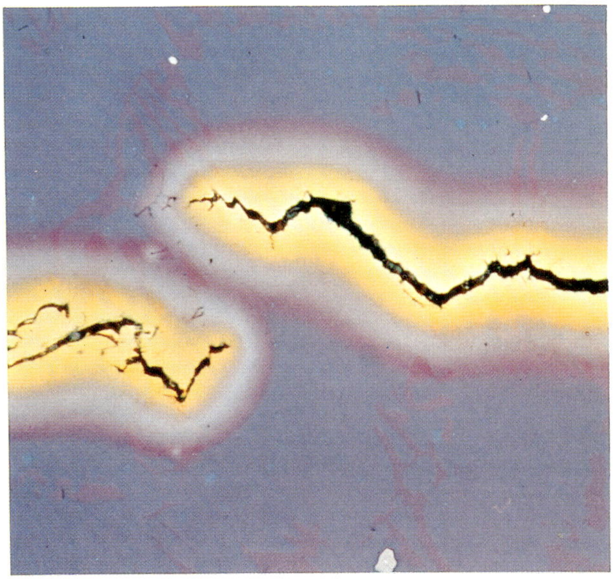

52 200 : 1

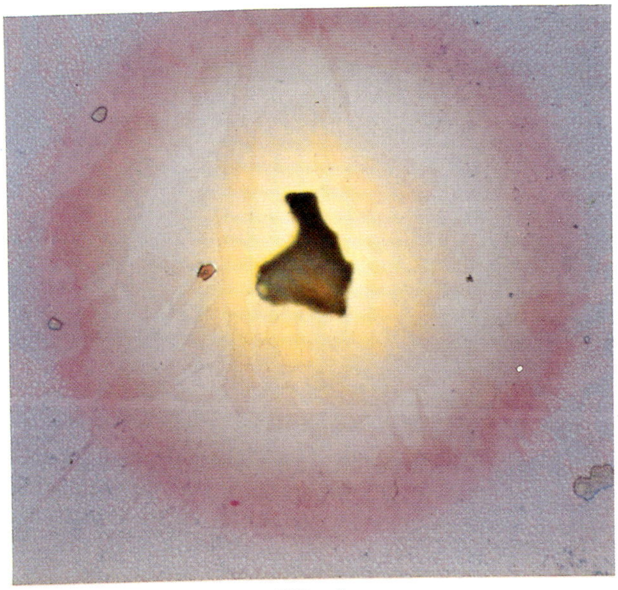

53 500 : 1

54 200 : 1

55 200 : 1

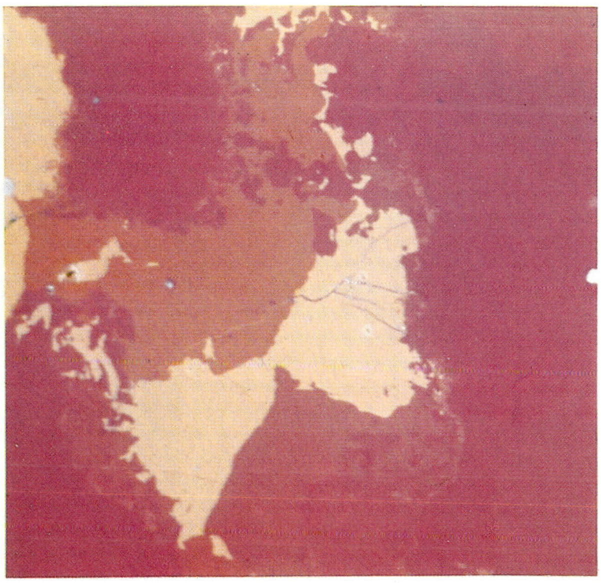

56 200 : 1

1.2.2 Production of interference layers

1.2.2.1 Deposition of layers by evaporation

The deposition of interference layers by evaporation is carried out in vacuum at ca. 10^{-5} mbar. The principle of the apparatus is illustrated in Fig. 57. The vapour source consists of one or more holders for boats or simple sheets of high melting point metals such as tungsten and molybdenum which are heated by the resistance principle.

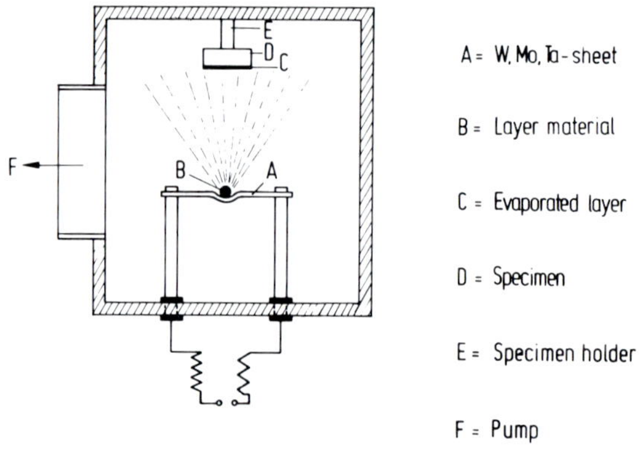

Fig. 57: Schematic representation of an apparatus for coating specimens by evaporation.

The layer material is placed in a depression in the vaporiser sheet or boat (see Fig. 57). A shutter blocks off the vapour source until the desired temperature is reached and the layer material evaporates or sublimes uniformly. The layer materials usually give off gases as they are heated up and this makes the evaporation very erratic. If the specimen were coated during this stage it would be very difficult to check the layer thickness. Furthermore the layer would be inhomogeneous. The shutter can also be used to break off the experiment after a specified time. The rate of evaporation is regulated by altering the heating current. In order to deposit a wedge the specimen is mounted at an angle to the evaporation axis. If the specimen is to be coated as uniformly as possible it is mounted perpendicularly as in Fig. 57. The evaporation process can be terminated after a given time or on a signal from a layer thickness measuring system. If the distance between the vapour source and specimen is sufficiently large, layers of uniform thickness can be deposited on large specimens.

Evaporation Equipment

The equipment, which is supplied by a number of manufacturers, comprises a vacuum chamber, a pumping system, power supply and control unit.

In a small apparatus the vacuum chamber is a glass cylinder of ca. 10 l capacity. It is fitted with several openings into which the flange with the specimen holder, the vapour source or a layer thickness measuring device can be inserted. In large units the vacuum chamber consists of a glass or metal bell with a volume of 100 l or more. In this type of equipment, the vapour source, specimen holder and other attachments are mounted on a base plate. Accessory devices can be built in by means of vacuum seals. It is possible to obtain fittings with one or more interchangeable vapour sources so that several different layers can be deposited successively. One accessory device permits rotation of the specimen during deposition. The power supply units have controls which permit the rate of evaporation to be regulated. In large units, several specimens can be coated simultaneously. In ordinary equipment suitable for metallographical purposes, specimens several centimetres in diameter can be coated uniformly. Layer thickness measuring devices can be supplied with all apparatus (see section 1.2.2.3). In large units the deposition is automatically terminated on achieving the desired layer thickness by swivelling a shutter across the vapour source. In some equipment the specimen can be cleaned before deposition by spark discharge or cathodic evaporation. This is especially important in quantitative assessment where otherwise drying stains could be mistaken for additional "phases" (c.f. Section 1.2.1). The pumping equipment can usually be supplied in serveral capacities. High pumping speeds mean that evacuation is more rapid and thus a greater number of specimens can be processed.

Experimental procedure

The coating unit must always be kept clean. The layers deposited on the equipment during the evaporation process absorb water when the apparatus is opened and this cannot easily be removed by evacuation. It can lead to a deterioration of the vacuum or even to reaction with the layer materials. The apparatus should therefore be held under vacuum when not in use. The specimens should be well dried and free from residues of plasticine and immersion oil which could remain if the specimen had previously been examined under the microscope. If these precautions are not observed the evacuation times are unnecessarily long or stains appear on the specimen (c.f. Figs. 51 and 53, page 51). If possible, therefore, the specimen surface should always be cleaned by spark discharge.

1.2.2.2 Production of layers by sputtering

1.2.2.2.1 Sputtering

Interference layers which enhance the contrast of the microstructure can also be produced by "sputtering". In sputtering, material is atomized by bombarding a surface (target) with high energy particles. Positively charged gas ions produced by electrical discharge are accelerated by a potential difference between the anode (specimen) and cathode (target) so that they bombard the cathode dislodging atoms which then fly out of the surface in all directions. These atoms impinge on the specimen surface, where they form the desired layer. This cathodic evaporation provides a coating source which can easily be controlled by the discharge current and voltage.

If undesirable reactions with the target material are to be avoided, the chamber should be filled with argon or another inert gas. The gas pressure is in the range 0.03 to 0.15 mbar. At the lower limit very few ions are formed because of the very low probability of ionisation. At the upper limit the ions and the atomized target material frequently collide with gas molecules so that the ions are not sufficiently accelerated and the target material atoms do not reach the specimen surface.

Fig. 58 shows a schematic representation of the sputtering apparatus. The specimen which is to be coated with an interference layer is connected as anode. The target is the cathode. Between cathode and anode is a d.c. voltage of 1 to 5 kV.

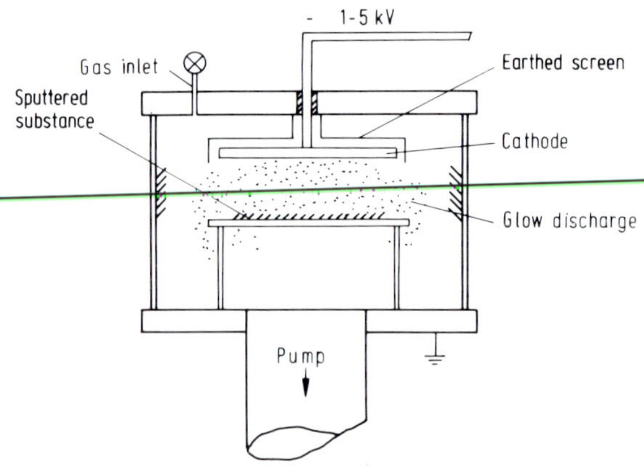

Fig. 58: Schematic representation of a sputtering apparatus.

The material ejected from the cathode consists primarily of neutral atoms. If the impact energy of the ions is low, i.e. at low accelerating voltages, the number of atoms released is negligibly small. At high energies, i.e. accelerated voltages over 500 V, virtually every ion knocks an atom out of the surface of the target. The atomization rate (rate of removal of target material) is dependent primarily upon the target material, the accelerating voltage, the gas pressure and the nature of the atmosphere. The rate of growth of the layer on the specimen depends on the position of the specimen with respect to the target as well as on the rate of atomization. The quantity of material hitting the specimen decreases with the square of the distance between specimen and target. The maximum growth rate is achieved when the specimen lies very close and parallel to the target surface. The deposition rates are of the order of 100 nm/min. If the experimental conditions are chosen so that the growth rate of the layer on the specimen is very rapid the temperature of the specimen may rise to 150^0 C. For details ot layer thickness measurement see section 1.2.2.3.

By reversing the polarity, i.e. by connecting the specimen as target (cathode), the specimen surface can be cleaned before deposition of the layer. Another target must then of course be used for sputtering, otherwise the impurities deposited on the target would be redeposited on the specimen. It is therefore preferable to clean specimens by glow discharge. In some units it is possible to produce a surface relief by ion etching which in some cases facilitates the differentiation of the phases and also shows up the grain boundaries. The procedure for sputtering is the same as for evaporation (c.f. section 1.2.2.1). Additional care must also be taken that the cathode is always absolutely clean and free from grease.

1.2.2.2.2 Reactive Sputtering

In conventional sputtering reactions between the target material and the atmosphere are prevented by the use of inert gases. Such reactions can however deliberately be exploited for producing contrast on metallographic specimens because the layers resulting from the reactions frequently possess outstandingly good optical properties, c.f. section 1.2.3.2. Most commercial sputtering apparatus can be used for the reactive process. The only requirement is a possibility for maintaining the reaction gas at a particular pressure during the reaction, e.g. by a finely adjustable needle valve. Up to now, oxygen (in most cases air is adequate) has been most commonly used, so that the atomized target material is oxidized. Oxidic layers are formed on the specimen, but their optical constants may deviate from those on the compact oxide. Different proportions of reaction gas and unoxidized atoms are incorporated into the layer, depending on the experimental conditions, and reproducible results can be obtained only when these conditions are very closely controlled. The layers are usually not transparent, so the absorption coefficient k_s of the layer is not zero and the equations given in section 1.1.1 are not applicable (c.f. section 1.2.3.2). Up to now the most commonly employed cathode materials have been iron and copper but indium, gold, platinum, palladium, lead and other metals have also been used. Under the conditions used for sputtering even the noble metals are oxidized. Whereas iron oxide is stable, gold and palladium oxide decompose very rapidly at temperatures of 100^0 C and break down if stored at room temperature. Reaction layers composed of the noble metals are of advantage in some cases, however, because of their high transparency.

1.2.2.2.3 Sputtering equipment

Since a vacuum of the order of 0.1 mbar is quite sufficient for sputtering, simple rotary pumps are adequate. The vacuum chamber is usually a small glass cylinder, in large units a bell, in which the specimen lies on a plate. In more sophisticated equipment, the specimens can be rotated and inclined to the target. The cathode, the target, is a disc of the material to be atomized. In large units using noble metal targets the material is deposited as a thin layer on an iron cathode as an economic measure. The targets are easily interchangeable.

The necessary 1000 to 5000 V accelerating voltage is supplied by a power supply unit. This should possess a safety switch to ensure that when the specimen is dismantled the voltage between it and the target is not accidentally left switched on.

The gas pressure for the discharge can be regulated accurately by means of needle valves through which air or another gas can be admitted.

The layer thickness can be measured by the normal techniques but, there are certain limitations which must be taken into account if vibrating quartzes are used (c.f. section 1.2.2.3).

Only one commercial apparatus possesses the possibility of swivelling the specimen under a microscope in the vacuum chamber and so of judging the colour of the layer formed up to that point or the contrast between the phases, in order to follow the formation of the interference layer (c.f. Fig. 60, page 58).

The size of the apparatus is given by the diameter of the largest specimen which can be coated with a uniformly thick layer. Nowadays this diameter ranges in a number of stages from 5 mm to 50 cm. Since evaporation equipment incorporates a vacuum system, a sputtering attachment is frequently supplied as an accessory.

In good equipment, the contamination due to a deposit of hydrocarbons from the pump oil on the specimen is negligible. Such deposits alter the composition and hence the optical constants of the sputtered or evaporated layer in an unknown manner.

1.2.2.3 Devices for the measurement of the layer thickness

1.2.2.3.1 Measurement by means of vibrating quartz

The vibrating quartz device for the measurement of layer thickness operates during deposition by measuring the change in frequency of a simultaneously coated vibrating quartz. The vibrating quartz is mounted next to the specimen in a water cooled holder and coated simultaneously with it. The change in resonant frequency of the quartz is proportional to the mass of the deposited layer. Since the density of the evaporated layer depends on the material used, the change in frequency of the quartz per nm is different for every material. The change in frequency per nm of the evaporated layer can very easily be read off from a nomogram as shown in Fig. 59 which includes the layer materials commonly used in the evaporation method. The nomogram contains three coordinates axes. On the left axis all the curves emerge from one point, a constant which to a first approximation has a value of $K = 0.474 \times 10^{-3}$ for all layer materials. The second axis gives the density of the layer material in g/cm^3. Finally the change in frequency suffered by the vibrating quartz when the coating layer increases in thickness by 1 nm is plotted on the right hand coordinate axis. Starting from the frequency at the beginning of the experiment, the frequency of the desired layer thickness can easily be calculated from these quantities. Newer equipment can be preset for the density of the layer material. The correction based on the nomogram is then carried out by the instrument itself i.e. when the desired layer thickness is reached a signal is given or the coating process is automatically interrupted. Recently instruments have become available which show the layer thickness in nm rather than giving the frequency in hertz. They are capable of making the conversion corresponding to Fig. 59.

When working with vibrating quartzes it must be borne in mind that the instrument indicates only the thickness of the layer coating the quartz. The measuring head must therefore be mounted so that it receives a layer equal in thickness to that on the specimen. If this is not possible, the position of the measuring head with relation to the specimen must be kept constant. A geometrical factor is then included in the measurement which must be determined by calibration. In newer instruments it can be preset and taken account of in the measurement. Under some conditions it is possible to take measurements during sputtering (see section 1.2.2.2) if special measuring heads are used which do not become charged in electric fields. The accuracy of the measurement is however still very uncertain and if the layer thickness is to be checked accurately the sputtering process should be interrupted briefly.

1.2.2.3.2 Optical layer thickness measurement

In principle the layer thickness can be assessed by visual observation of the colour of the specimen surface. It should however be borne in mind that the colours observed depend on the illumination e.g. visible radiation from the vapour source or glow discharge during sputtering. It is therefore advisable to interrupt the coating process for the colour check and to illuminate the specimen with a reproducible source. In the case of unidentified phases or materials it is useful to coat a known reference specimen simultaneously and use its interference colour to set the layer thickness (c.f. Figs. 133 to 138, page 97).

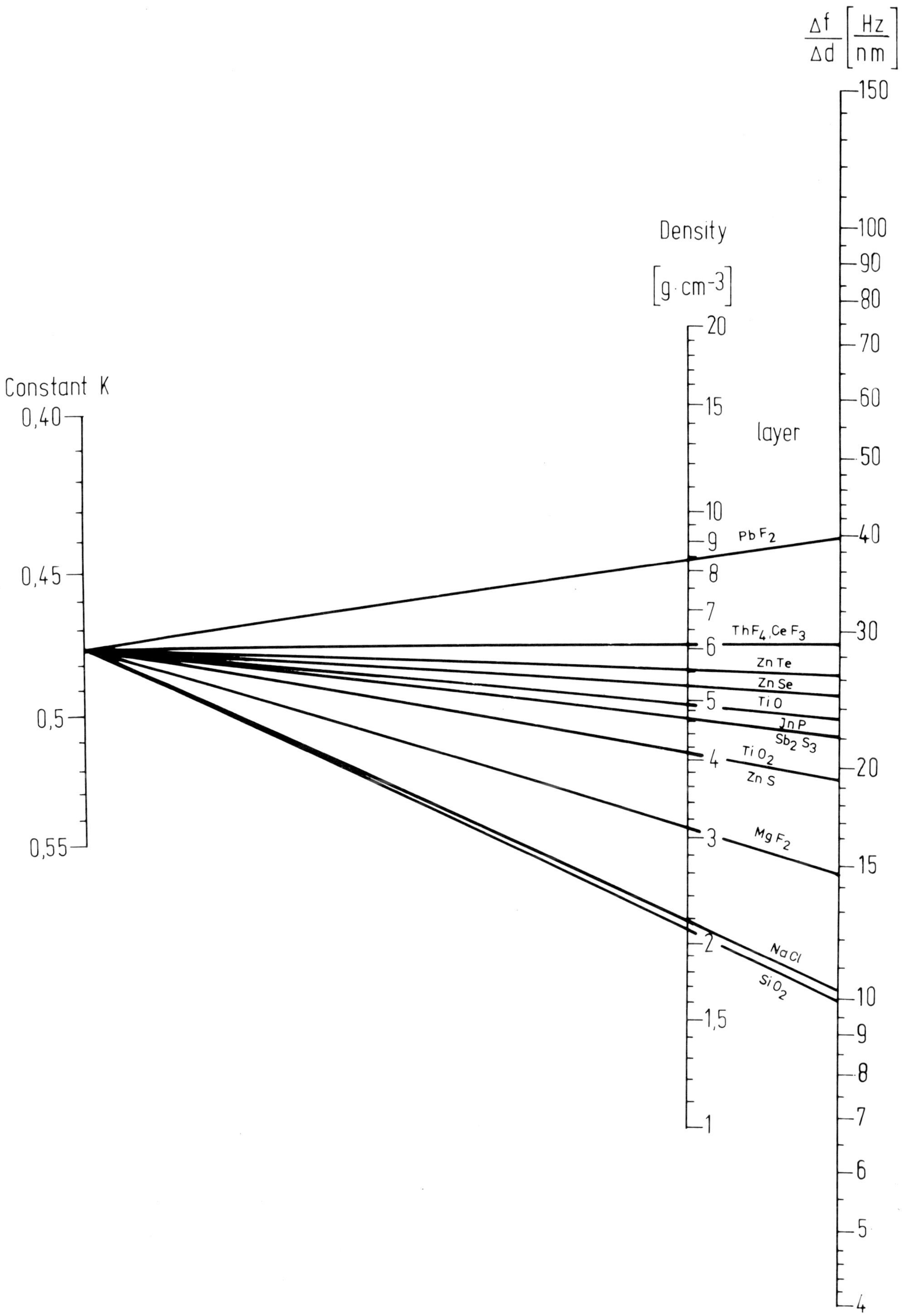

Fig. 59: Nomogram for determining the layer thickness from the indicated quartz oscillator frequency 19). (The curves for the various layer materials have been plotted in the original nomogram).

Fig. 60 shows an experimental set-up for the simple qualitative assessment of the layer thickness. The specimen can be swivelled in front of the objective of an optical microscope as and when desired during the sputtering process. When the desired colour is attained, the coating process is interrupted. It is also possible to attach a microscope photometer and thus approach the reflection minimum more closely.

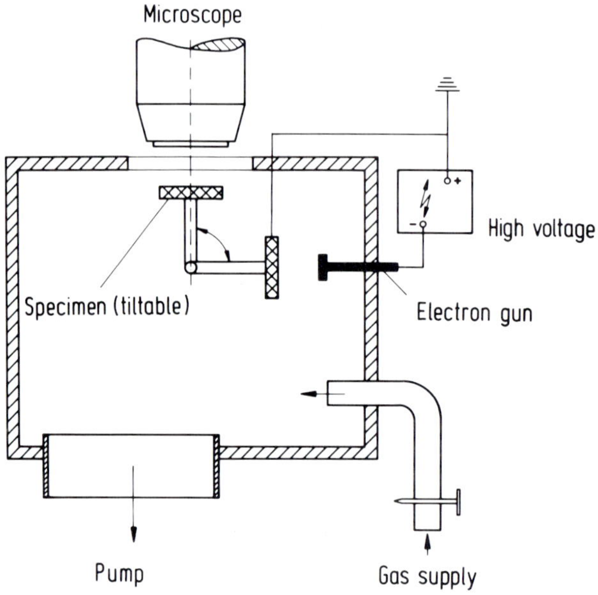

Fig. 60: Experimental arrangement for observing the specimen during sputtering.

In large units the thickness of evaporated layers can be regulated by continuous measurement of the reflectivity of the specimen. A light source emits a modulated beam which passes through a window in the vacuum chamber and is reflected from the specimen surface on to a receiver. The quantity of light reflected is indicated on a control and measuring device.

As a consequence of the coating layer, the reflectivity of the specimen varies with the layer thickness as shown in Fig. 61 (c.f. also Fig. 8, page 15). The shape of the curve itself depends on the optical properties of the layer material and substrate. Thus, in the example below, the reflection minimum due to the highly refracting titanium dioxide is sharper than that due to silicon oxide. This means that the coating process can be broken off at the reflection minimum with greater accuracy in the case of titanium dioxide and zinc sulphide than in the case of silicon oxide.

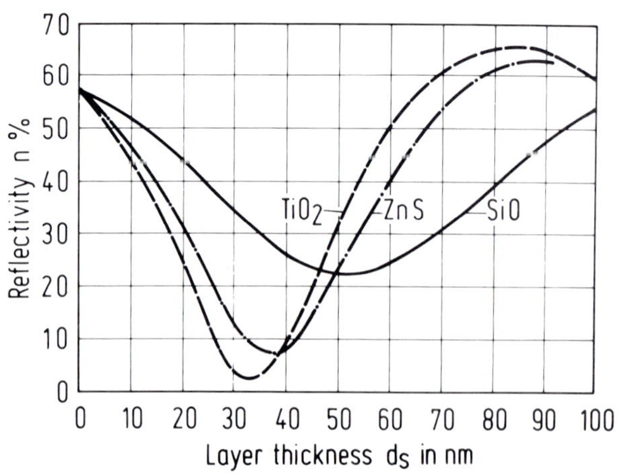

Fig. 61: Variation of the reflectivity with the layer thickness for iron coated by evaporation with titanium dioxide, zinc sulphide and silicon monoxide 18).

1.2.3 Properties of layer materials and of layers

The choice of layer material depends on the optical properties of the material to be coated. The aim is to fulfil the phase and amplitude conditions for one phase thereby reducing its reflectivity to zero. The following rules derive from the basic principles (c.f. section 1.1):

Highly reflecting or strongly absorbing substrates (e.g. metals) necessitate layer materials either with a very high refractive index and largely non-absorbing or with such optical properties that the ratio of refractive index to absorption coefficient results in virtual extinction of one of the phases under investigation in the visible range.

In the case of low reflecting and weakly absorbing materials (e.g. oxides or sulphides) layer materials with a refractive index less than 2.5 find application. This is enough to fulfil the amplitude condition, i.e. total extinction can be achieved by interference. In principle both requirements can be satisfied by layer materials that can be deposited by evaporation or sputtering. In the following, materials suitable for deposition, which have been tested for practical application, are described individually under the heading of the appropriate coating technique.

1.2.3.1 Materials for evaporation and evaporated layers

Table 7 in chapter 5 contains a summary of the most commonly used layer materials. The list shows that there are a number of materials with refractive indices in the range 1.35 to 3.25 (Na_3AlF_6 to ZnTe) which lend themselves well to evaporation. Of the low refractive index materials listed, Na_3AlF_6 (cryolite), MgF_2 (magnesium fluoride), CeF_3 (cerium fluoride) and PbF_2 (lead fluoride) have proved most valuable. Only a few of the compounds mentioned are almost completely non-absorbing. They have an absorption coefficient $k = 0$ in Table 7. In quantitative investigations, for example for the determination of optical constants, the compound zinc sulphide has proved most effective.

The refractive indices and absorption coefficients given in Table 7 for the various layer materials are correct only for a wavelength $\lambda = 550$ nm. Table 8 and Figs. 62 and 63 show the dependence of the refractive index and absorption coefficient on wavelength for some of the compounds in Table 7. The data given are standard values because the optical properties of the evaporated layers are affected within certain limits by the experimental conditions and the exact chemical composition of the material employed.

The optical constants of several phases occurring in steels are summarised in Table 9. Column 7 contains the calculated refractive index for a non-absorbing layer material necessary to satisfy the phase and amplitude conditions. Table 10 contains the optical constants of phases occurring in aluminium alloys.

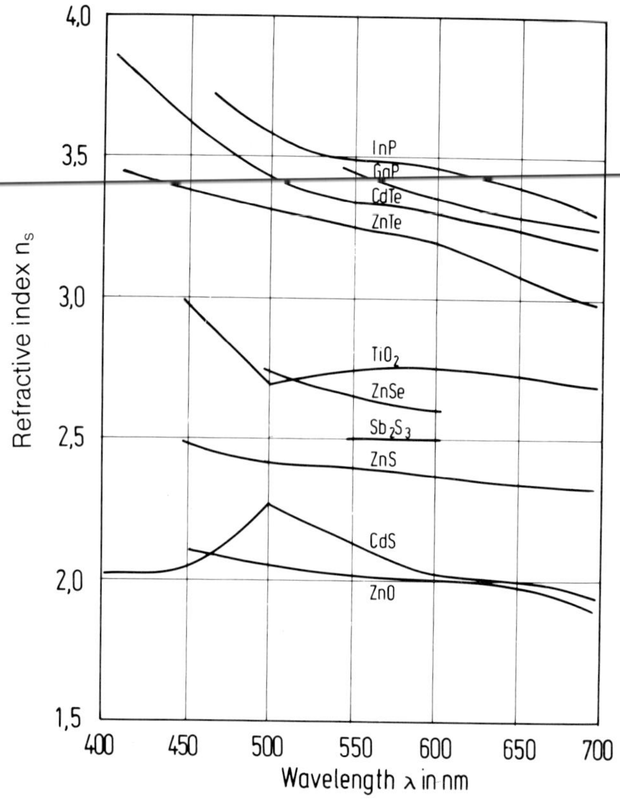

Fig. 62: Refractive index n_s of evaporated layers as a function of the wavelength (cf. Table 8, Chapter 5).

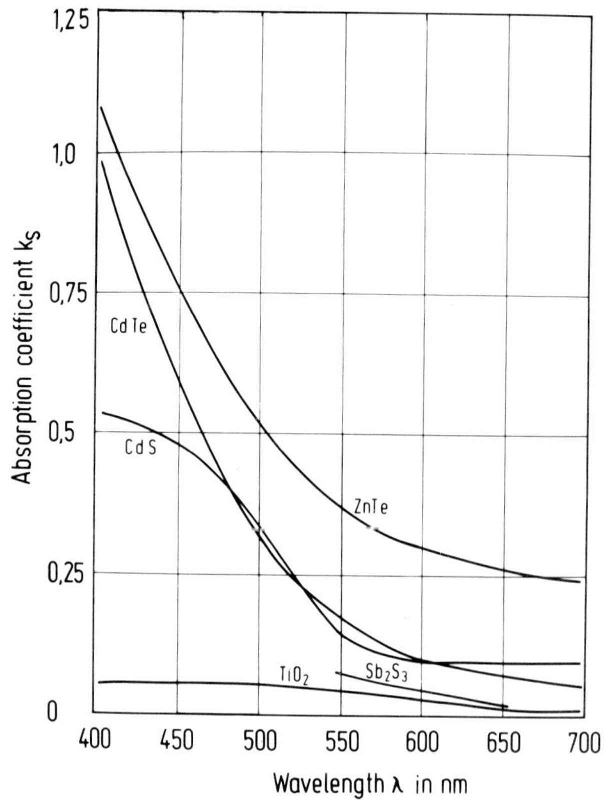

Fig. 63: Absorption coefficient k_s of evaporated layers as a function of the wavelength (cf. Table 8, Chapter 5).

The calculation of the values in column 7 of Table 9 will be illustrated by means of an example. Let α-iron with optical constants $n_p = 2.05$ and $k_p = 3.20$ serve as the example of a metal and Fe_3O_4 with $n_p = 2.30$ and $k_p = 0.70$ as example of an oxide. At 58% the reflectivity of α-iron is high and that of Fe_3O_4 at 19% is low. If a coating layer is required to satisfy both the phase and the amplitude condition, equation 5 gives the required refractive index for a non-absorbing layer ($k_s = 0$) on iron as $n_s = 3.44$ and on Fe_3O_4 as $n_s = 1.64$. For the use of absorbing layer materials, see Fig. 66.

Based on these values, the compounds Na_3AlF_6, MgF_2, CeF_3 and PbF_2 with $n_s = 1.35$ to 1.75 as given by Table 7 are recommended for weakly reflecting materials.

ThF_4 is well suited to the evaporation method and yields good results, but because this substance is slightly radioactive the safety regulations governing the use of radioactive materials must be observed.

ZnS, ZnSe, TiO_2 and ZnTe with $n_s = 2.4$ to 3.25 can be used for coating highly reflecting materials. Of these, ZnS, ZnSe and ZnTe are recommended.

As regards the use of ZnSe and ZnTe, it must be noted that both compounds are slightly hygroscopic, and the layer becomes permeated with fine droplets of moisture if stored for any length of time. TiO_2 is also very suitable but needs a relatively high temperature for evaporation. The vacuum should not be any higher than 10^{-4} torr otherwise there is the danger of decomposition to TiO.

The compounds CdS, CdSe, CdTe or InP are less suitable because they undergo partial decomposition on evaporation and yield very inhomogeneous layers. Specimens have been coated successfully with InP but this substance is not suitable for routine use.

Two very interesting compounds for the evaporation technique are Sb_2S_5 or Sb_2S_3, which show a marked self-absorption at a refractive index of 2.3 and which evaporate with great ease at low temperatures. The disadvantage is that evaporated layers of these materials have less reproducible optical properties than, for example, ZnS. They are therefore not recommended for quantiative investigations.

Tables 7 and 8 are incomplete. In principle any compound which satisfies the following conditions can be used for evaporation:

a) straightforward evaporation procedure (layer material does not decompose, lowest possible evaporation temperature, material preferably not hygroscopic, if so only slightly).

b) interference layers should be homogeneous and durable.

1.2.3.2 Materials for sputtering and sputtered layers

It has already been explained that if sputtering is carried out in the presence of reactive gases, the atomised cathode material reacts with the atmosphere and the stability and optical properties of the compounds making up the deposited layers depend on the type of gas and on the cathode material. Up to now the most commonly used reaction gas has been oxygen. It can be assumed that if this gas is used, the interference layers consist of oxides. The most frequently employed cathode material has been iron, but indium, gold, platinum, palladium, lead, copper, nickel and other metals have also been used.

A list of layer materials produced using oxygen as reaction gas, together with their optical constants (refractive index and absorption coefficient), is given in Table 11 in Chapter 5 for three different wavelengths. The experimental accuracy is also given. By way of further illustration, these relationships are again shown in Figs. 64 and 65. It is characteristic of these oxide layers that marked variations in the refractive index are rarely observed in the visible wavelength range, whereas the absorption coefficient generally shows a noticeable decrease from the short wave to the long wave region.

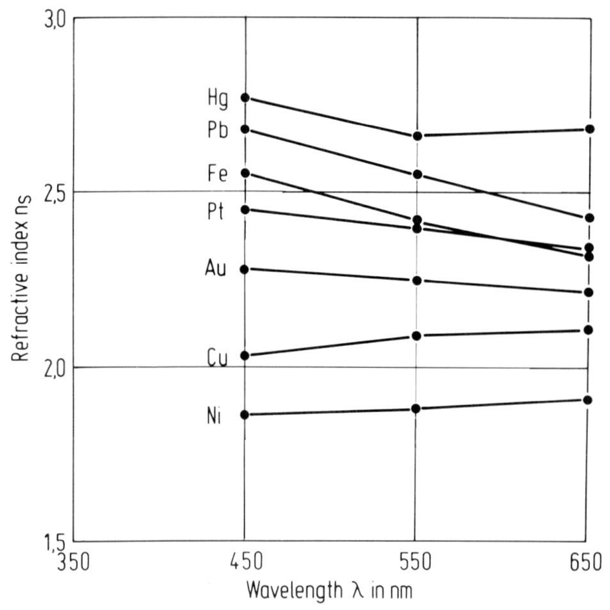

Fig. 64: Refractive index of layers deposited by reactive sputtering with oxygen as a function of the wavelength (cf. Table 11, Chapter 5).

In the case of highly reflecting and strongly absorbing materials (e.g. metals) the amplitude condition can be fulfilled by using non-absorbing layer materials (k_s = 0) with very high refractive indices. Only a very limited number of such materials is available (c.f. Table 7). Instead of non-absorbing layer materials with very high refractive indices, however, it is also possible to use materials with low refractive indices but correspondingly higher self-absorption. This is illustrated by the example of the differentiation of the phases austenite and martensite in a steel with 1% C and 0.6% Mn. (c.f. also section 2.1.1). The optical constants of martensite are given in Fig. 66. The R_{min} value of martensite should be as low as possible in order to achieve good contrast to the austenite. Fig. 66 shows the curves calculated for four values of R_{min} for martensite based on different ratios of refractive index and absorption coefficient, in each case for λ = 540 nm. It can be seen that a reflectivity of 0.1%, for example, can be achieved for martensite using non-absorbing layers (k_s = 0) only if the refractive index is of the order 3.25.

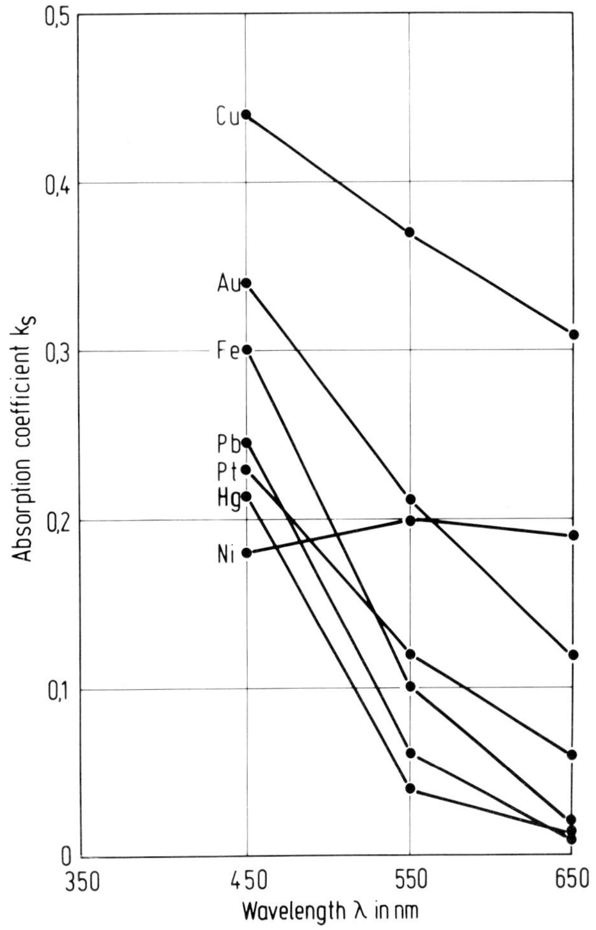

Fig. 65: Aborption coefficient of layers deposited by reactive sputtering with oxygen as a function of the wavelength (cf. Table 11, Chapter 5).

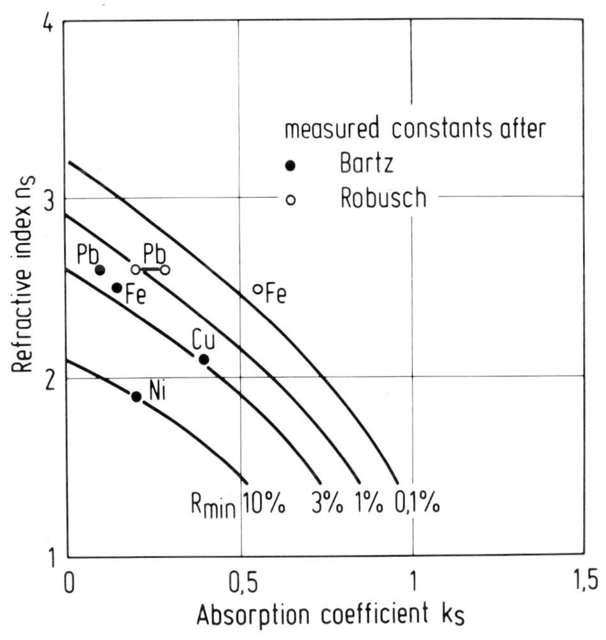

Fig. 66: Realtionship between the refractive index n_s, the absorption coefficient k_s and the reflectivity at the interference minimum R_{min} for α-iron (1% C, 0.6% Mn) sputtered with the cathodes indicated in oxygen. Values for λ = 540 nm.

The optical constants of sputtered layers of the metals lead, iron, copper and nickel are also plotted in Fig. 66. These include values measured on layers sputtered onto transparent substrates 1) as well as those calculated knowing the optical constants of α-iron 3) and the measured reflection minima using an iron and a lead cathode and oxygen as reaction gas 17). The results show that there is still a certain inaccuracy in the values of the optical constants but that lower values for R_{min} can be obtained using absorbing layers than with the non-absorbing materials available at present. One limitation must be pointed out, namely that the colour saturation is lower with absorbing layers than with non-absorbing layers, i.e. the reflection minima are flatter (c.f. Figs. 24 to 35, page 35 and 37).

Literature to 1.2.2 and 1.2.3

1) Bartz, G.: Zur Kontrastierung mikroskopischer Objekte in einer Gasionenreaktionskammer. Prakt. Metallographie **10** (1973) 311

2) Jeglitsch, F. and Mitsche, R.: Die Anwendung optischer Kontrastmethoden in der Metallographie. Radex-Rundschau (1967) 587

3) Stewart, R. L.: Phys. Rev. **45** (1934) 488

4) Bartz, G.: Beiträge elektronenmikroskopischer Direktabbildung. BEDO **7** (1974) 91

5) Bartz, G.: Leitz-Mitteilungen, Sonderhefte Rasterelektronenmikroskopie, Suppl. Vol. (1974) 161

6) Bartz, G. and Holm, R.: Z. f. Metallkunde **67** (1976) 120

7) Hofmann, S. and Exner, H. E.: Z. f. Metallkunde **65** (1974) 778

8) Ondracek, G. and Spieler, K.: Gasätzen – ein neues Kontrastierungsverfahren der Gefügepräparation. Prakt. Metallographie **10** (1973) 324

9) Back, A. and Dietrich, R.: Zur Kontrastierung von Dentalamalgamgefügen. Prakt. Metallographie **11** (1974) 117

10) Petzow, G. and Exner, H. E.: Mikrostructural Science (1975) 291

11) Elssner, G., Pabst, R. and Riedel, S.: Fraktographie und Bruchverlauf in Keramik/Metall-Verbindungen. Prakt. Metallographie **12** (1975) 234

12) Jeglitsch, F. and Novak, H.: Aluminium **51** (1975) 768

13) Ondracek, G. and Spieler, K.: Leitz-Mitteilungen Vol. VI, No. 6 (1976), p. 224

14) Back, A. and Pabst, K.: Leitz-Mitteilungen Vol. VI, No. 6 (1976), p. 32

15) Menzel, D.: Zur Korrosion von Platingeräten. Prakt. Metallographie **13** (1976) 211

16) Beyer, J., Brois, H. J. and Kerkveld, G. E.: Verfahren zur Unterscheidung und quantitativen Beurteilung von Zwischenstufe und Martensit in der Lichtmikroskopie.
Prakt. Metallographie **13** (1976) 324

17) Robusch, G., Engler, G. and Bühler, H.-E.: Möglichkeiten bei der Anwendung absorbierender, insbesondere durch Gasionenbeschichtung aufgebrachter Schichtwerkstoffe in der Metallographie der Eisenlegierungen.
Prakt. Metallographie **14** (1977) 65

18) Pepperhoff, W. and Ettwig, H.-H.: Interfernzschichten-Mikroskopie. Wissenschaftliche Forschungsberichte, Vol. 73. Dr. Steinkopff-Verlag, Darmstadt 1970

19) Catalogue ,,Hochvakuum-Bedampfungstechnik'', Balzers AG, Liechtenstein

2. EXAMPLES OF THE APPLICATION OF THE INTERFERENCE LAYER TECHNIQUE

2.1 DIFFERENTIATION OF PHASES

2.1.1 Differentiation of the three modifications of iron ($\alpha - \gamma - \varepsilon$)

2.1.1.1 General introduction

The three modifications of iron, α (body centred cubic), γ (face centred cubic) and ε (hexagonal), when coated with an interference layer, scarcely differ in colour. They do, however, differ in reflectivity, which decreases in the order ε, γ, α (1–4, 6).

In order to achieve a marked contrast between α, γ and ε, it is expedient to use coating materials with very high refractive indices (cf. text to Fig. 10) or those with medium refractive indices and perceptible self-absorption (cf. section 1.2.3.2).

Quantitative differentiation of α and γ in the form of residual austenite and the transformation microstructure is possible only if the size of the residual austenite particles is greater than the resolution limit of the optical microscope 3). Since in many cases residual austenite is an unstable structural form at room temperature, transformation into martensite can be induced by mechanical grinding and polishing. Particular care must therefore be taken with specimen preparation if residual austenite is to be determined (cf. section 1.2.1 and Fig. 48).

Another metallographic technique for distinguishing ε from α and γ-iron is the use of polarized light, and depends on the hexagonal structure and associated optical anisotropy of ε-iron 4) 6).

Differentiation of the different microstructural forms (martensite, bainite, pearlite) in steel will be considered in section 2.1.2.

Literature to 2.1.1

1) Pepperhoff, W. and Bühler, H.-E.: Metallographischer Nachweis der Ferritbildung in austenitischen Chrom-Nickel-Stählen.
Arch. Eisenhüttenwes. **33** (1962) 711

2) Pepperhoff, W. and Bühler, H.-E.: Lichtmikroskopische Untersuchungen an martensitischen Gefügen.
Arch. Eisenhüttenwes. **34** (1963) 839

3) Bühler, H.-E., Pepperhoff, W. and Schüller, H.-J.: Ermittlung des Restaustenits mit Hilfe des Interferenzschichten-Verfahrens.
Arch. Eisenhüttenwes. **36** (1965) 457

4) Bühler, H.-E., Jackel, G., Thiemann, E. and Baumgartl, S.: Über optische Eigenschaften des hexagonalen ε-Eisens in Eisen-Mangan-Legierungen.
Prakt. Metallographie **6** (1969) 279

5) Ettwig, H.-H., Bühler, H.-E. and Jackel, G.: Absorbierende Schichtwerkstoffe in der Interferenzschichten-Mikroskopie.
Arch. Eisenhüttenwes. **41** (1970) 957

6) Nilson, H., Schüller, H.-J. and Schwaab, P.: Epsilon- und Alpha-Martensit in Mangan- und Chrom-Nickel-Stählen.
Prakt. Metallographie **6** (1969) 269

7) Robusch, G., Engler, G. and Bühler, H.-E.: Möglichkeiten bei der Anwendung absorbierender, insbesondere durch Gasionenbeschichtung aufgebrachter Schichtwerkstoffe in der Metallographie der Eisenlegierungen.
Prakt. Metallographie **14** (1977) 65

8) Bühler, H.-E. and Meyer, L.: Kontraststeigerung in der Auflichtmikroskopie mit Hilfe des Interferenzschichten-Verfahrens und unter Verwendung monochromatischen Lichtes wechselnder Wellenlänge.
Zeiss-Informationen **15** (1967) 118

9) Kohlhaas, E. and Fischer, A.: Kontraststeigerung beim Interferenzschichten-Verfahren durch die Verwendung eines Verlauf-Linienfilters.
Prakt. Metallographie **6** (1969) 339

10) Pepperhoff, W. and Ettwig, H.-H.: Interferenzschichten-Mikroskopie. Dr. Dietrich Steinkopff-Verlag, Darmstadt 1970

2.1.1.2 Differentiation of α, γ and ε using model iron alloys and steels as examples	If the grains are sufficiently large, a perfectly satisfactory light-dark contrast can be produced between ferrite or martensite (both body centred cubic) and austenite. Depending on the conditions the ferrite appears darker (e.g. Figs. 68, 88–91) or lighter (e.g. Figs. 92, 100, 104) than the austenite (c.f. also Figs. 10 and 11).

Fig. 67: **Iron-nickel alloy with 32.5% Ni.**

Spectral reflectivity of austenite (γ) and martensite (α) uncoated and after vacuum deposition of a layer of ZnSe (zero and first-order) and of the absorbing material CdS (zero-order).

In the case of specimens coated with the non-absorbing ZnSe (c.f. Table 7, Chapter 5) the reflectivity at the minimum is independant of the layer thickness. As the interference order goes up, the reflection curves become steeper. Much lower reflection values are achieved using absorbing CdS layers. In this case the reflectivity depends on the order of the deposited layer.

Fig. 68: **Iron-nickel alloy with 32.5% Ni.**

Specimen undercooled to -120^0C, polished electrolytically, vacuum deposited TiO_2 layer, zero-order, bright field, reflection minimum of martensite at 540 nm.

Austenite light, martensite dark. The spectral reflectivity of the austenite and martensite phases coated with a vacuum deposited TiO_2 layer is comparable with that after coating with ZnSe. On the basis of the reflection curves in Fig. 67, therefore, the austenite looks lighter than the martensite.

Figs. 69–71: **Iron-manganese alloy with 14.8% Mn.**

69: Coated with ZnSe, zero-order, bright field, reflection minimum of the martensite at 540 nm.

α-martensite dark, ε-martensite and austenite light to dark grey. The latter two phases cannot be differentiated in this photograph.

70 and 71: Same specimen, preparation and area of the microstructure as in Fig. 69. Photographed in polarized light. Figs. 70 and 71 show the specimen at different positions relative to the polarizing prisms.

The α-martensite is black and the austenite grey in both photographs. Due to its optical anisotropy, the light-dark contrast of the ε-martensite is different for the two positions.

Fig. 72: **Iron-manganese alloy with 14.8% Mn.**

Same specimen as in Figs. 69 to 71. Spectral reflectivity of the α- and ε-martensite after coating with three different thicknesses of ZnSe. The reflectivity of the ε-martensite at the interference minimum is greater than that of the α-martensite at all wavelengths. These results confirm the observation in Fig. 69 that the α-martensite is darker than the ε-martensite.

Fig. 73: **Chemically resistent chromium-nickel steel (X5 CrNi 18 9).**

α- and ε-martensite produced by cooling to low temperatures. Coated with TiO_2, zero-order, interference minimum of the austenite at 540 nm.

The α-martensite is darker than the austenite. Light coloured, fine plates of ε-martensite can be seen between the dark α-martensite plates. These ε-martensite plates are marginally lighter than the austenite and much lighter than the α-martensite (c.f. Fig. 72).

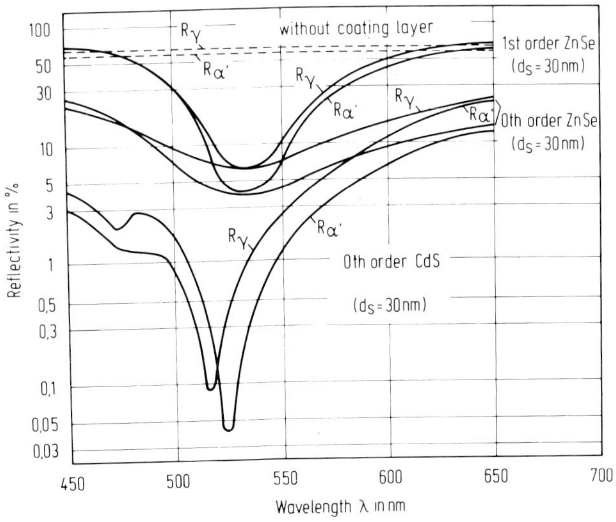

67

68 1000 : 1

69 200 : 1

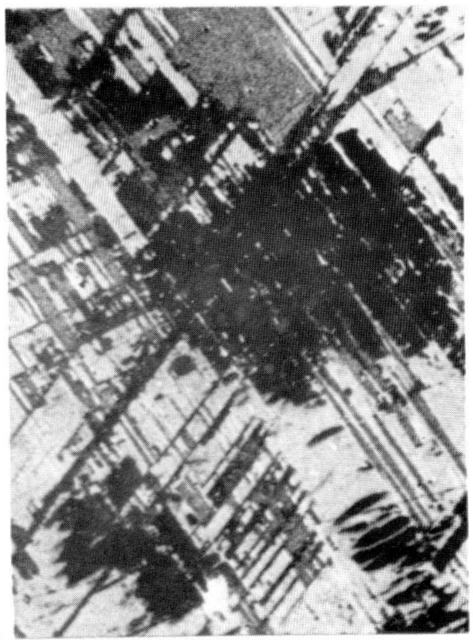

70 200 : 1

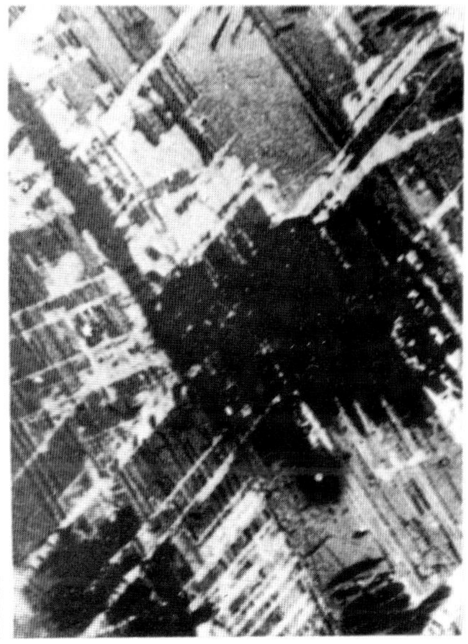

71 200 : 1

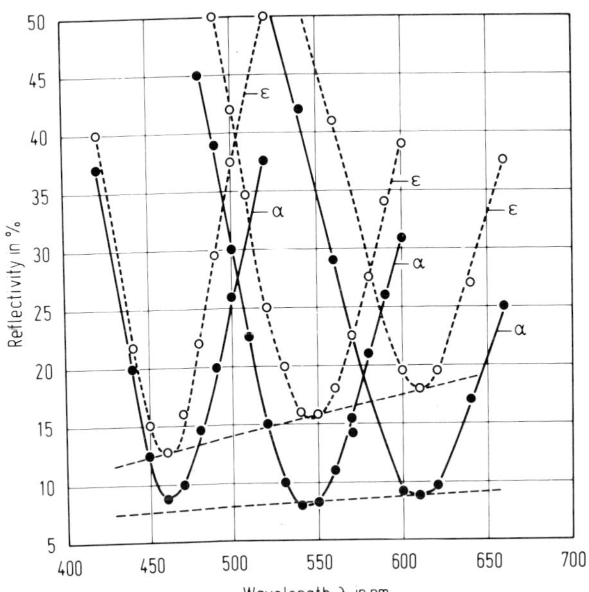

72

73 1000 : 1

2.1.1.3 Differentiation of residual austenite from the transformation microstructure

A particular advantage of the interference layer technique is that the picture it gives of the microstructure is geometrically accurate and the "distortion" resulting from an etching attack is avoided.

This is especially true when the phases present are finely dispersed. In many cases etching does not show up the residual austenite satisfactorily against the transformation microstructure. The extent to which this method permits a quantitative determination of the residual austenite as compared with X-ray or magnetic measurements may be deduced from the following.

Specimens were selected from the time-temperature-transformation diagram for continuous cooling of a steel with 1% C and 2% Mn **(Fig. 74)**, and the residual austenite content determined after various cooling times by X-ray and magnetic methods, after etching in alcoholic nitric acid, and after coating with an interference layer 3). Fig. 75 gives the results of the comparative investigation. All the techniques yielded similar results, except chemical etching, in which case the quantity of residual austenite detected was much lower especially at high rates of cooling.

Fig. 74: **Time-Temperature-Transformation diagram for the continuous cooling of a steel with 1% C and 2% Mn 3).**

Figs. 75–79: **Determination of the residual austenite content in a steel with 1% C and 2% Mn using different methods 3).**

(cf. Time-temperature-transformation diagram in Fig. 74).

75: Summary of the results.

76: Same steel as in Figs. 74 and 75. Time to cool to 500^0 C: 0.3 min, etched in 3% alcoholic nitric acid. The microstructure contains residual austenite and martensite. The patches of residual austenite appear grey and small regions between the white martensite needles appear black.

77: Same material as in Figs. 74 to 76. Specimen etched in 3% alcoholic nitric acid. Time to cool to 500^0 C: 1.5 min. Residual austenite, martensite, bainite and pearlite. The pearlite is very dark. The residual austenite is grey to black. At this magnification the bainite cannot be distinguished as a microstructural constituent.

78: Same material, same specimen, same heat treatment and same area of the microstructure as in Fig. 76. Coated with ZnSe, zero-order. The residual austenite appears white compared with the dark martensite. The resolution of the residual austenite zones within the martensitic regions is considerably better than on the etched specimen in Fig. 76. Also, unlike Fig. 76, the reflectivity of the residual austenite is uniform throughout the microstructure, i.e. has same grey level.

79: Same material, same heat treatment, same area of the microstructure as in Fig. 77. Coated with ZnSe, zero-order. Again the residual austenite appears light against the transformation products which cannot be distinguished from one another. After etching (Fig. 77) the residual austenite zones look smaller than in the coated specimen in this figure. In this case the microstructure is revealed without falsification of the areal proportions (cf. also Fig. 75).

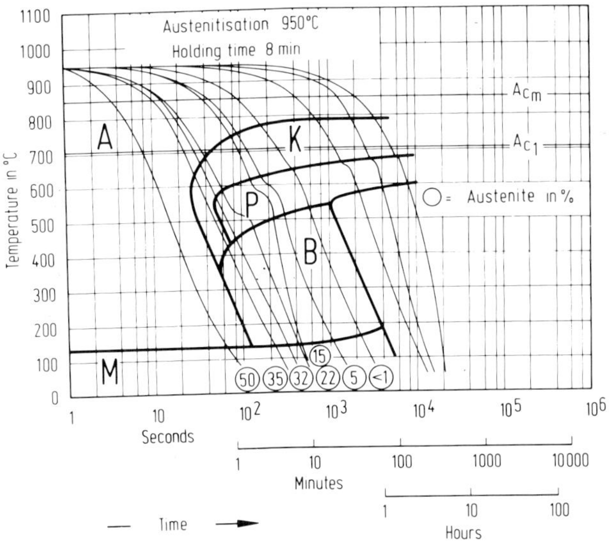

74

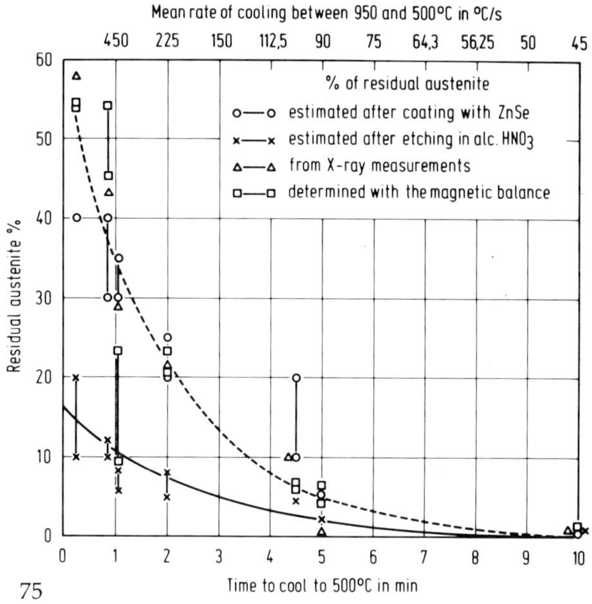

75

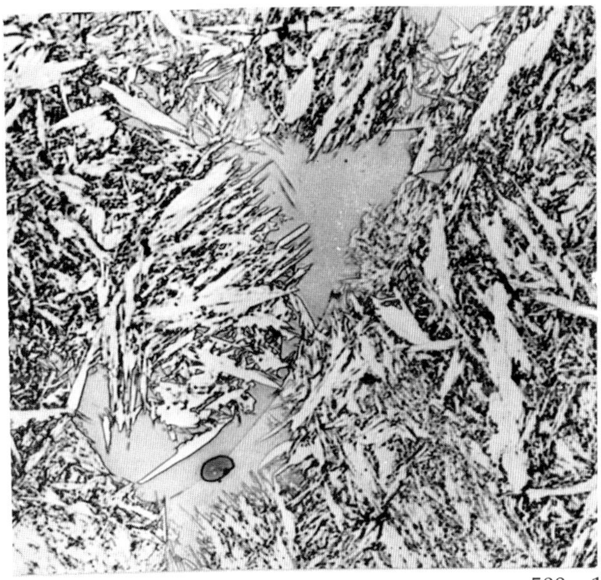

76 500 : 1

77 200 : 1

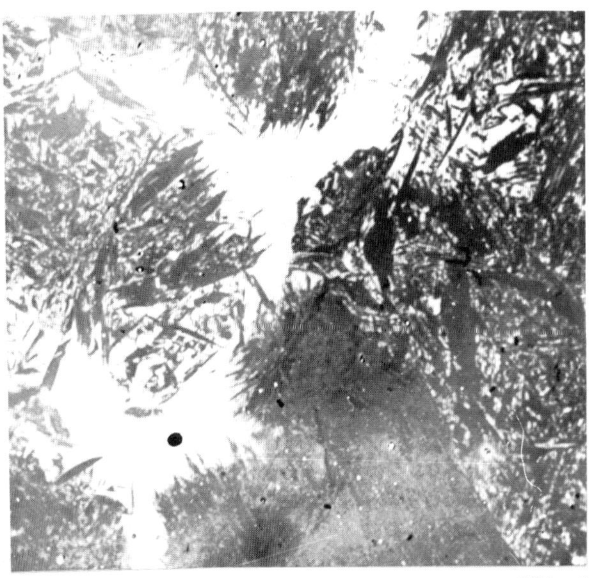

78 500 : 1

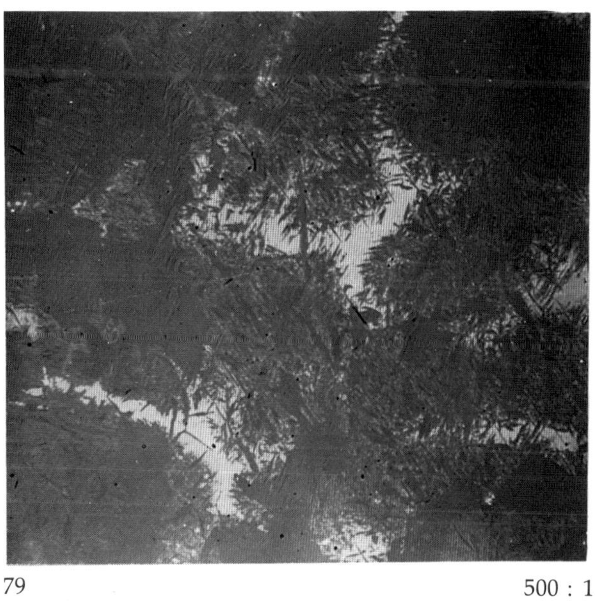

79 500 : 1

2.1.1.4 Influence of the refractive index of the layer on the light-dark contrast between ferrite and austenite

Coating with highly refracting materials reduces the reflectivity of ferrite or austenite from about 60% to values of 2 to 6% (Figs. 67 and 86). The slight difference in reflectivity between the two phases α and γ in the polished state (reflectivity ratio $\gamma : \alpha = 1.05$) explains why the two modifications can scarcely be differentiated visually if at all in the unetched condition. A surface layer, either a vacuum deposit of zinc selenide or a sputtered layer, increases the ratio to infinity or reduces it to zero (cf. Figs. 11 and 87). The contrast can thus be increased to the best possible value of 1 by an interference layer (cf. Fig. 10).

It can further be seen from the reflection curves for the coated condition, Fig. 86, that the interference minima for the two phases lie approximately at the same wavelength. As far as the differentiation of austenite and ferrite is concerned, however, this means that in the coated condition there is virtually no difference in colour (cf. Fig. 95) but only in reflectivity at one wavelength.

The light-dark contrast between the two phases can be enhanced as desired by the choice of a layer with suitable refractive index (Fig. 87). This is described in detail with the aid of Figs. 10 and 11. Since no non-absorbing layer materials with a refractive index >3.7 are available at present, the amplitude condition for austenite cannot be satisfied by this means. This is possible only using absorbing layers (cf. section 1.2.3.2).

Figs. 80 and 81: **Case-hardening steel 14 NiCr 14 carburized to 1.5% C.**

80: Heat treatment: 830^0C/15 min/oil. Vacuum deposited ZnSe layer, zero-order, wavelength of interference minimum of martensite 540 nm, bright field.

Residual austenite white, transformation microstructure and carbides dark grey to black.

81: Same steel, same specimen, same heat treatment and same area of the microstructure as in Fig. 80, etched in 3% alcoholic nitric acid. Martensite white and very dark, carbides light grey.

The martensite looks white or black depending on the carbon content. After this etch, the residual austenite is no longer discernible within the martensitic regions.

Figs. 82 and 83: **Steel with 1.04% C and 1.02% Mn.**

82: Heat treatment: 1150^0C/1 h/ice water. Vacuum deposited TiO_2 layer, bright field, wavelength of the reflection minimum of martensite 540 nm.

Martensite dark grey, residual austenite white.

Even very small areas of residual austenite are faithfully reproduced.

83: Same specimen, same heat treatment and same area of microstructure as in Fig. 82. The coated specimen was repolished and etched with 3% alcoholic nitric acid. Martensite white to dark grey.

After this preparation the residual austenite zones can no longer be seen because, as in Fig. 81, the martensite appears both white and dark grey.

Figs. 84 and 85: **Steel 100 Cr 6.**

84: Heat treatment: 1050^0C/2 h/air. Vacuum deposited ZnSe layer, zero-order, reflection minimum of matrix 540 nm. Residual austenite white, transformation microstructure dark grey.

The very finely dispersed residual austenite is clearly visible.

Fig. 85: See page 76

80 500 : 1

81 500 : 1

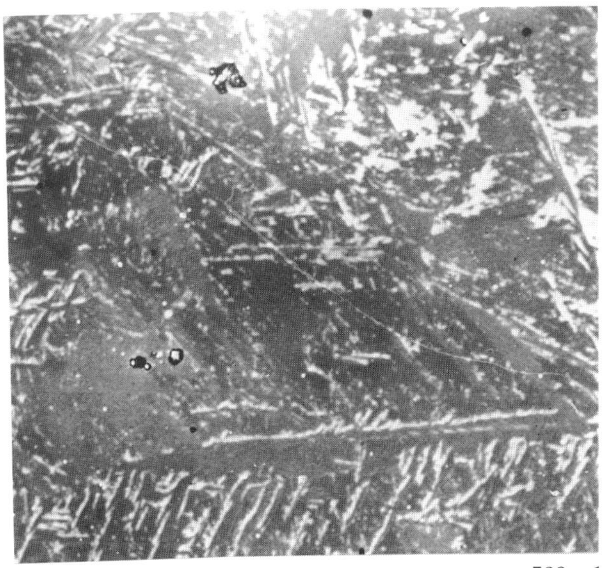

82 500 : 1

83 500 : 1

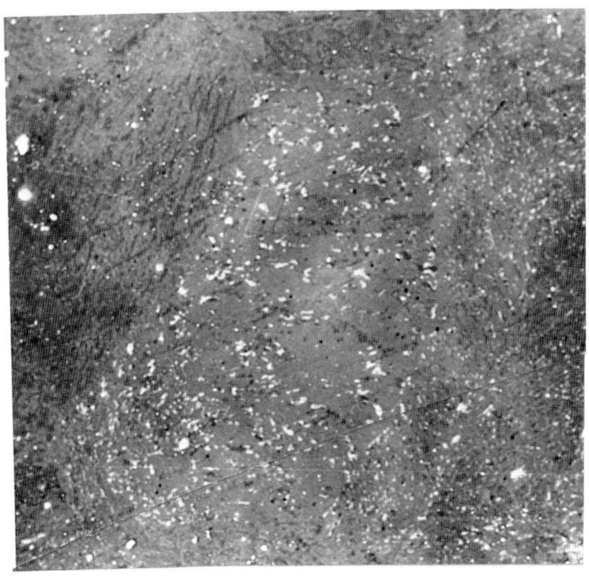

84 500 : 1

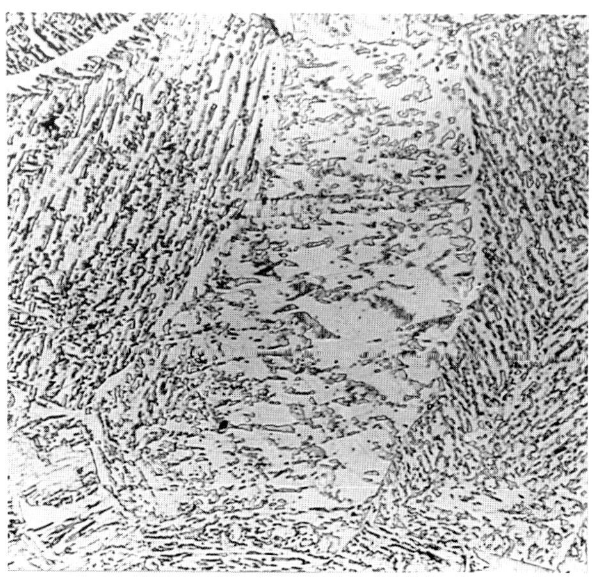

85 500 : 1

Fig. 85: Same specimen and same area of microstructure as in Fig. 84. The coated specimen was repolished and etched with 3% alcoholic nitric acid.

Bainite, ferrite light grey, carbides and martensite dark grey.

The carbides and martensite cannot be clearly differentiated in all parts of the figure at this magnification. The residual austenite is not distinguishable from the transformation microstructure.

Fig. 86: **Carbon steel with 1.6% C.**

Reflectivity as a function of observation wavelength in the uncoated condition, coated with a vacuum deposited ZnSe layer of zero-order and a sputtered layer (gaseous ion coating), produced using iron or lead cathodes under oxygen. The numerical table on the top right gives the values of the reflectivity at the minimum in the uncoated condition, after vacuum deposition of ZnSe at 540 nm and after sputtering at 560 nm.

Fig. 87: **Carbon steel with 1.6% C and iron-nickel alloy with 32.5% Ni.**

Ratio of the reflectivities of the γ- and α-phases as a function of the refractive index n_s of the layer material for non-absorbing layers (cf. also Figs. 10 and 11).

Figs. 88–91: **Iron-nickel alloy with 32.5% Ni, as cast.**

The four pictures show the same area of the specimen after deposition of different vacuum deposited layer materials. The contrast increases with increasing refractive index corresponding to Fig. 87. The photometrically determined increase in the contrast is not fully realised in the photographic reproduction. Apparently a reflectivity ratio of about 1.5 increases the contrast to such a level that no further improvement is possible on a photograph. The contrast in Fig. 91 looks somewhat poorer because although indium phosphide has the highest refractive index it has also a noticeable absorption. This makes the colours appear less saturated and the contrast in a black and white photograph does not seem as good.

88: Vacuum deposited zinc sulphide layer, zero-order, reflection minimum of martensite at 540 nm. Austenite white, martensite black.

89: Vacuum deposited ZnSe layer, zero-order, reflection minimum of martensite at 540 nm. Austenite white, martensite black.

90: Vacuum deposited ZnTe layer, zero-order, reflection minimum of martensite at 540 nm. Austenite white, martensite black.

91: Vacuum deposited InP layer, zero-order, reflection minimum of martensite at 540 nm. Austenite white, martensite black.

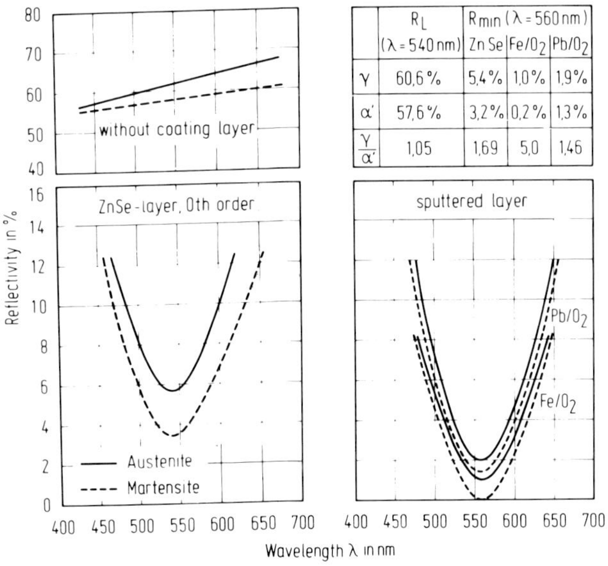

86

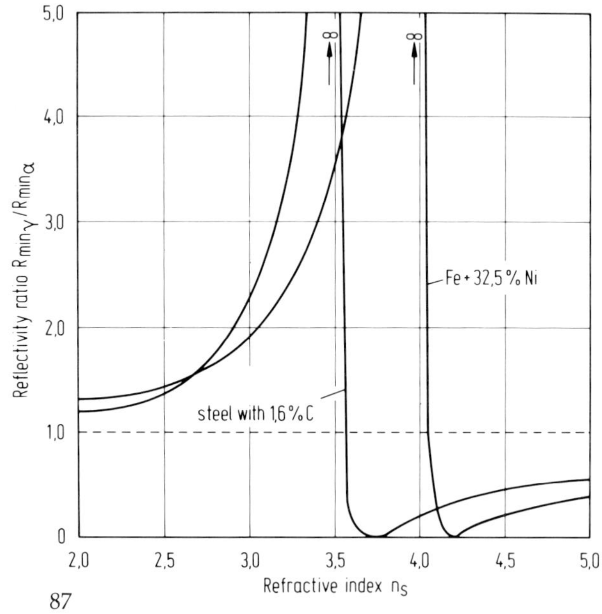

87

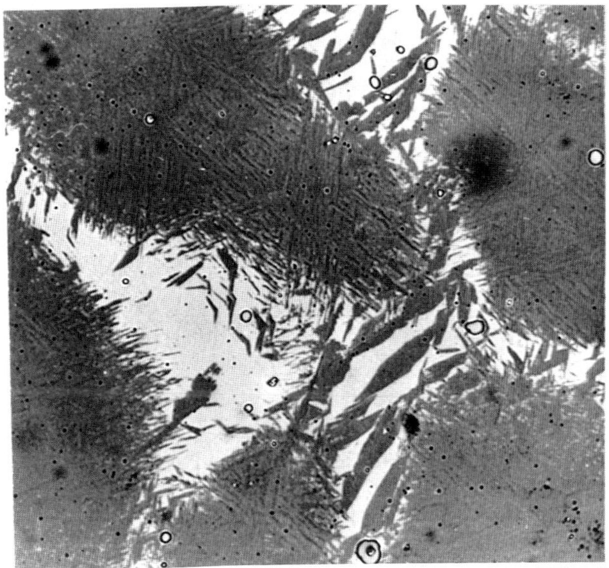

88 200 : 1

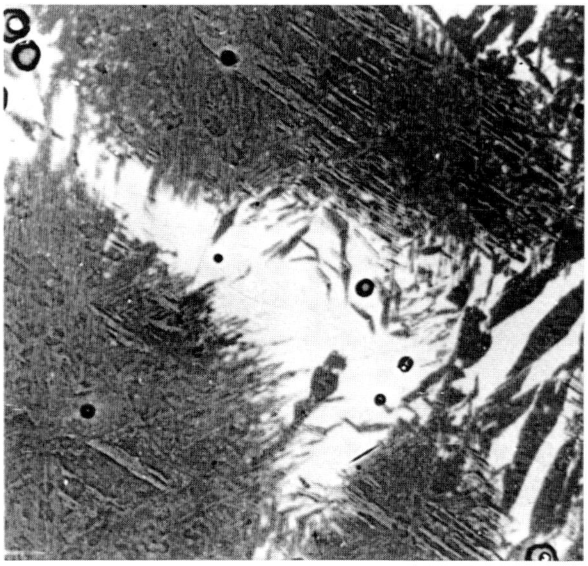

89 200 : 1

90 200 : 1

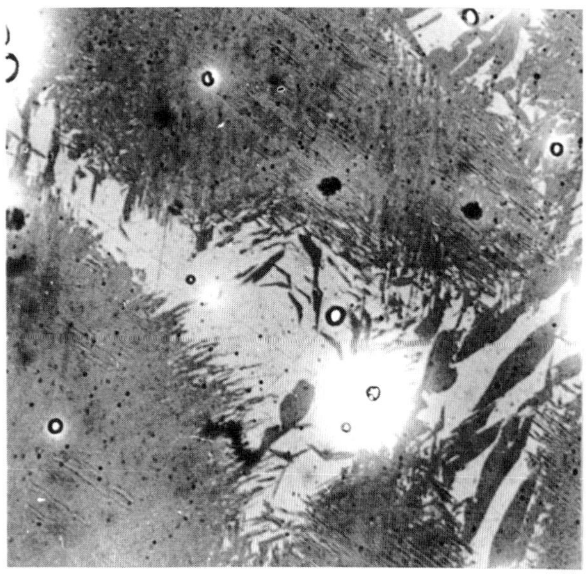

91 200 : 1

2.1.1.5 Influence of observation wavelength on the light-dark contrast between α- and γ-iron

It can be seen from Fig. 66 that because of the very slight self-absorption of the layer the amplitude condition can be satisfied even at low refractive indices and for highly reflecting materials. It has further been shown that both the refractive index and the absorption coefficient of the layer material in the range of the visible spectrum can exhibit a noticeable dispersion (cf. Figs. 62–65 and Tables 8 and 11). According to the ratio of the refractive index to the absorption coefficient, therefore, it should be possible to produce maximum light-dark contrasts between two phases even in cases where the reflection minima lie virtually at one wavelength. As an example, we shall consider the case of the two allotropic modifications of iron, ferrite and austenite. If a steel with 1% C and 0.6% Mn is coated by sputtering (iron cathode, oxygen) the residual austenite appears dark against the transformation microstructure at an observation wavelength of 510 nm (Fig. 92). At longer wavelengths (560 nm, Fig. 93), however, the contrast is reversed. The reflectivity ratio for the two phases has been measured for the above steel, Fig. 94. In agreement with the two micrographs, the austenite has a lower reflectivity than the transformation microstructure at short wavelengths. At 550 nm, maximum contrast between the two microstructural constituents is achieved for the coating selected. At longer wavelengths still, the residual austenite appears lighter than the martensite or bainite on the photograph on account of its higher reflectivity.

The difference in reflectivity in the colour reproduction is shown once again in Fig. 95. From this photograph it is again clear that if the spectral location of the interference minima is almost the same there is hardly any difference in the colour of the two allotropic modifications of iron.

An indium phosphide coating yields a curve very similar to that in Fig. 94. In the short wave range indium phosphide has a high refractive index and a relatively high self-absorption. Both parameters decrease as the wavelength increases (cf. Fig. 62).

According to the picture sequence Figs. 96 to 100, the light-dark contrast between austenite and martensite coated with indium phosphide can be reversed by observation at different wavelengths.

Figs. 92–100: **Steel with 1.0% C and 0.6% Mn.**

92: Heat treatment: 1150°C/water, sputtered layer produced with an iron cathode under oxygen, reflection minimum of martensite at 540 nm, photographed at 510 nm. Martensite grey, residual austenite black.

93: Same specimen, heat treatment and surface layer as in Fig. 92, photographed at 560 nm. Martensite dark grey, residual austenite white.

The contrast between the martensite and austenite is reversed compared with Fig. 92.

94: Steel with 1% C and 0.6% Mn, sputtered layer produced using an iron cathode under oxygen. Ratio of the reflectivities of the γ and α-phases as a function of the observation wavelength.

Figs. 95–100: See page 80

92 500 : 1

93 500 : 1

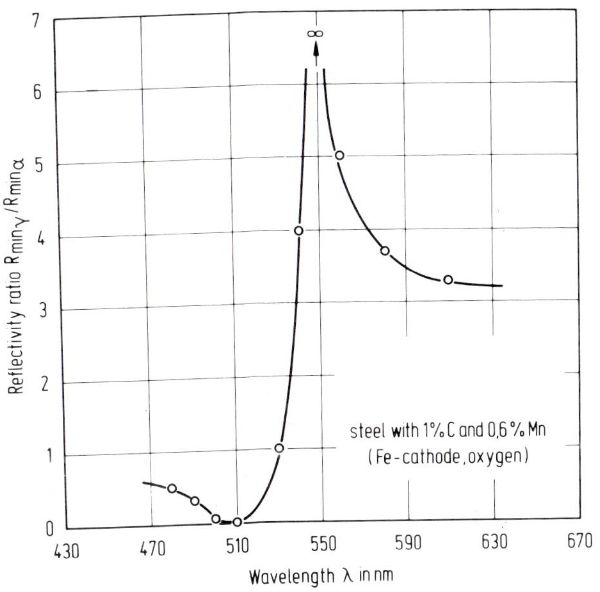

94

95 500 : 1

96 500 : 1

97 500 : 1

98 500 : 1

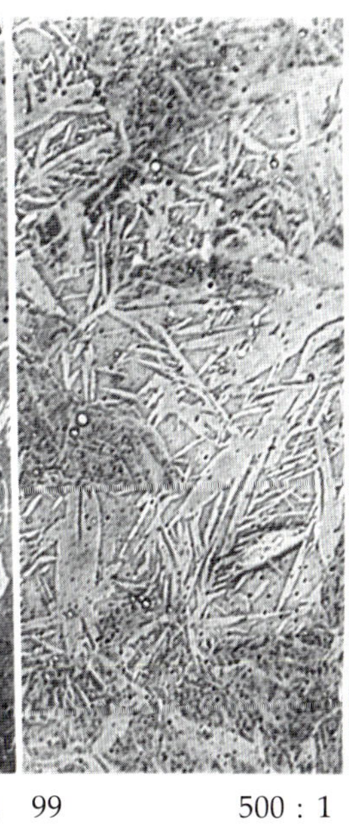

99 500 : 1

100 500 : 1

Fig. 95:	Steel, heat treatment as in Fig. 92. Coated with ZnSe, zero-order. Reflection minimum of martensite at 490 nm.

Martensite: red (9 : 2 : 5)

Austenite: yellow (4 : 3 : 3)

The shade equivalent wavelength of the complementary colours (cf. section 1.1.3) for martensite lies at 493 nm (cf. Fig. 21), the shade equivalent wavelength for austenite lies at 485 nm. The wavelengths of the reflection minima for the two phases differ only slightly and this is in agreement with the measurements in Fig. 86. This demonstrates the reliability of a colour assessment according to the colour plates, although colour differences due to printing must be expected.

Figs. 96–100: Material and heat treatment as in Fig. 92, vacuum deposited indium phosphide layer, zero-order. Reflection minimum of martensite at 540 nm. The photographs taken at different wavelengths demonstrate the reversal in contrast between the martensite and austenite phases.

96: Photographed at 680 nm. Martensite dark, residual austenite white.

97: Photographed at 520 nm. Martensite dark grey, residual austenite white.

98: Photographed at 485 nm. Martensite dark grey, residual austenite white.

99: Photographed at 470 nm. Martensite and austenite cannot be differentiated because they have the same reflectivity.

100: Photographed at 450 nm. Martensite light grey, austenite black. The contrast between martensite and austenite is the reverse of that in Fig. 96.

The reversal of the contrast between the martensite and austenite can also be achieved using contrast enhancement by absorbing layers produced by sputtering. The following set of pictures illustrates this possibility and also the effect of the layer thickness on the contrast between several phases.

Fig. 101:	**Steel with 1% C and 0.6% Mn.** Sputtered layer produced with an iron cathode under oxygen. The left hand side of the diagram shows the change in reflectivity with wavelength for various layer thicknesses. The location of the reflection minimum is marked by the heavy dashed curve. The amplitude condition for martensite is satisfied at 550 nm. The reflectivity has decreased to zero. The heavy dashed curve from the left hand diagram is plotted in the right hand diagram. Also shown are the reflection curves for the residual austenite while the heavy full curve marks the positions of the reflection minima for different layer thicknesses. The amplitude condition for residual austenite is satisfied at ca. 510 nm.
102:	Sputtered layer produced with an iron cathode under oxygen. Ratio of reflectivities of austenite and martensite according to Fig. 101. Greatest possible contrast is achieved at 510 nm and at 550 nm. The corresponding curves for a chromium steel chill casting and steel X10 CrNi 30 9 show that the observation wavelength giving optimum contrast depends on the composition of the steel, cf. Fig. 162.
Figs. 103–111:	**Chill cast chromium steel with 2.5% C and 17.8% Cr.** Coated as cast by sputtering with an iron cathode under oxygen, zero-order. The picture sequence shows how optimum contrast can be achieved between austenite, martensite and carbide phases by changing the layer thickness and observation wavelength, cf. Figs. 157–161.
Figs. 103–105:	The layer thickness was chosen so that the reflection minimum of martensite lay at 470 nm.
103:	Reflection curves for the three phases. The wavelengths at which Figs. 104 and 105 were photographed are marked at 490 and 520 nm.
104:	Coating as in Fig. 103, photographed at 490 nm. Carbide black, austenite dark grey patches, martensite light grey background. As shown by the reflection curves in Fig. 103, the carbide is the darkest phase. The next lightest phase is the austenite and the lightest phase of all is the martensite.
105:	Coating as in Fig. 103, photographed at 520 nm. Carbide black, martensite and austenite white. As shown by the reflection curves in Fig. 105 the reflectivities of austenite and martensite are equal. They cannot therefore be distinguished at this wavelength.

Figs. 106–108: The coating was applied so that the reflection minimum of martensite lay at 530 nm.

106: Reflection curves for the three phases. The wavelengths 530 nm and 600 nm at which Figs. 107 and 108 were photographed are marked on the diagram.

107: Coating as in Fig. 106, photographed at 520 nm. Martensite dark grey, austenite light grey to white, carbide white.

108: Coating as in Fig. 106, photographed at 600 nm.

Martensite dark grey, austenite white. Due to the different slopes of the reflection curves in Fig. 106 a contrast reversal for the carbide is possible simply by changing the observation wavelength.

Figs. 109–111: The coating was applied so that the reflection minimum for martensite lay at 570 nm.

109: Reflection curves for the three phases.

The wavelengths 570 nm and 610 nm at which Figs. 110 and 111 were photographed are marked on the diagram.

110: Coating as in Fig. 109, photographed at 570 nm. Martensite black to dark grey. In this photograph the martensite can be distinguished from the remaining transformation microstructures containing ferrite by a very small difference in grey level. Austenite light grey, carbide white.

111: Coating as in Fig. 109, photographed at 610 nm. Carbide and martensite light grey to dark grey, austenite white. As shown by Fig. 109 the martensite and carbide phases cannot be distinguished because they have practically the same reflectivity.

These examples show clearly that any degree of light-dark contrast can be produced using absorbing layer materials and observing at different wavelengths. They also show, however, that the conditions must be matched to the particular substrate material because the optical properties of the individual phases vary considerably e.g. from steel to steel. In this context, Fig. 102 shows that keeping the observation wavelength and all other experimental conditions constant, different reflectivity ratios are obtained for austenite and martensite for three different steels.

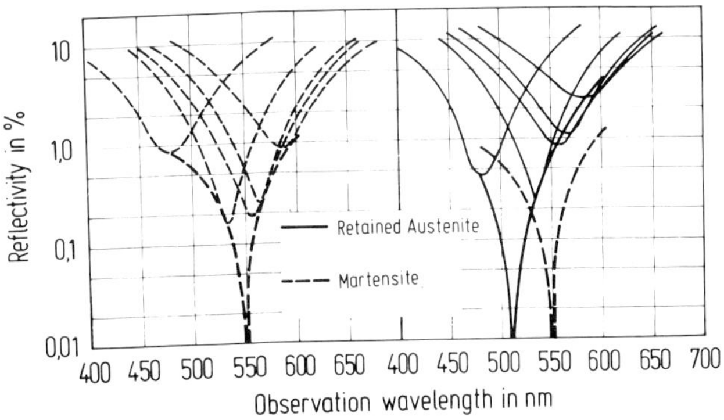

101

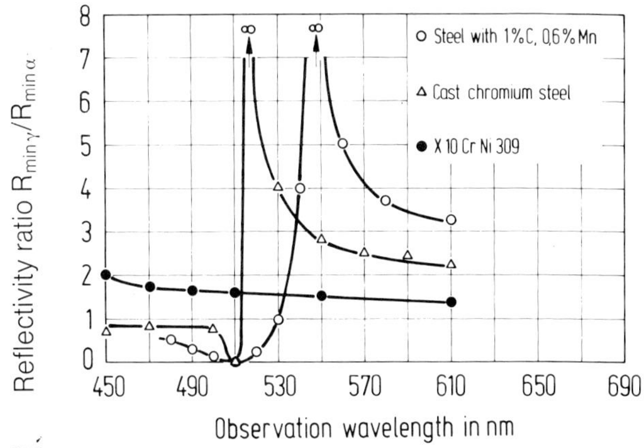

102

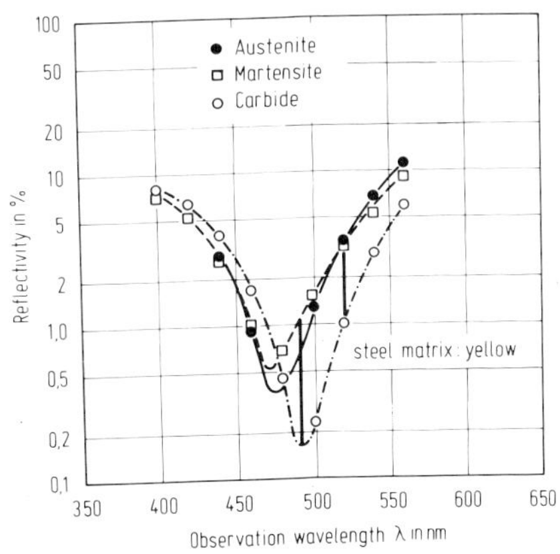

103

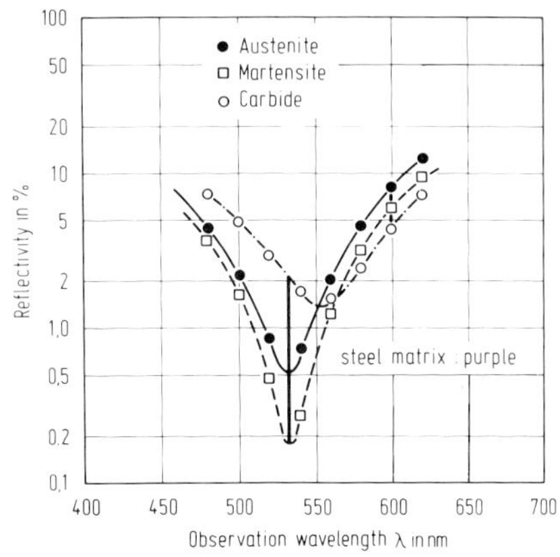

106

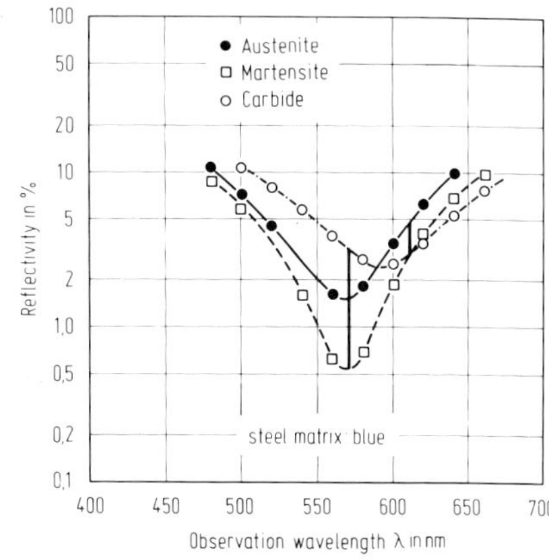

109

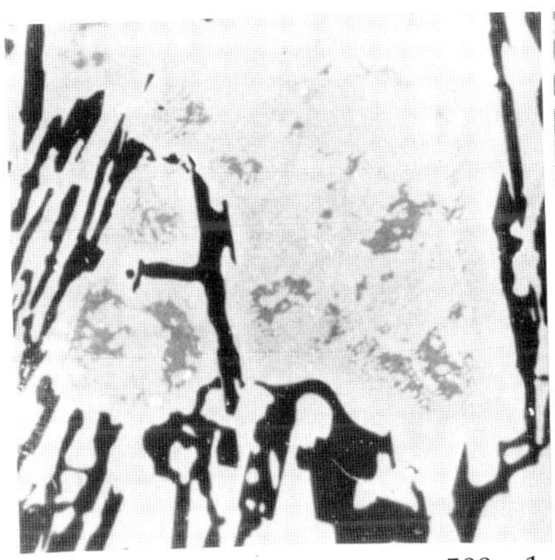

104 500 : 1

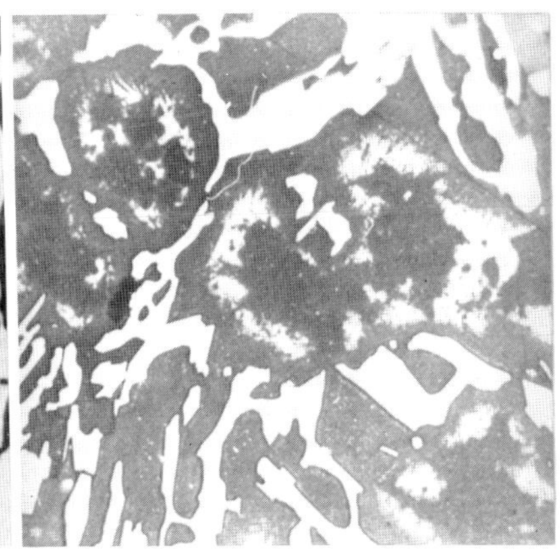

107 500 : 1

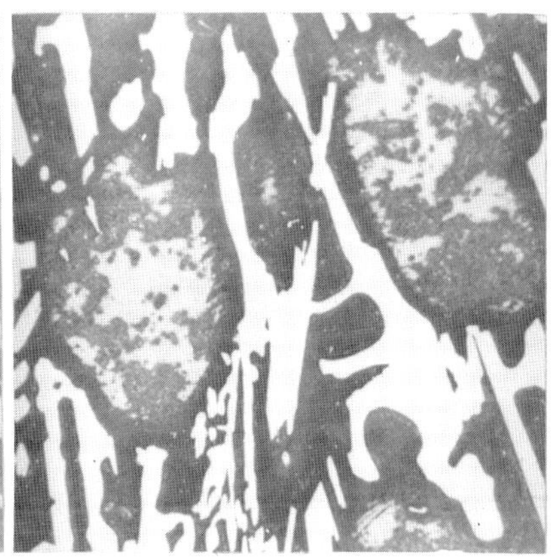

111 500 : 1

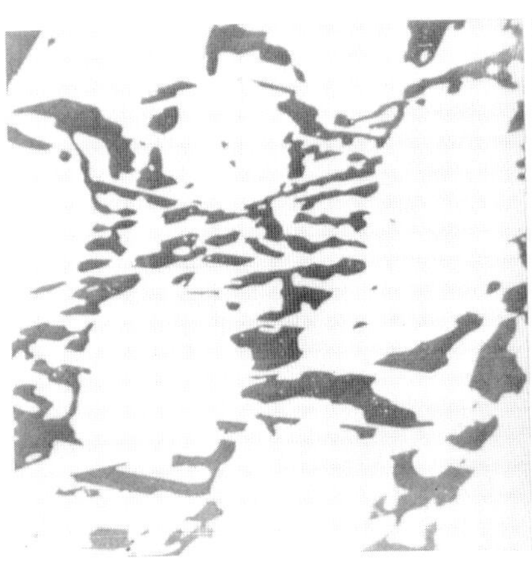

105 500 : 1

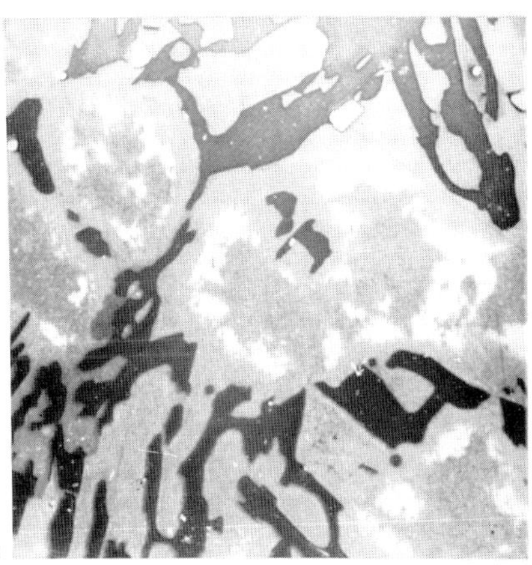

108 500 : 1

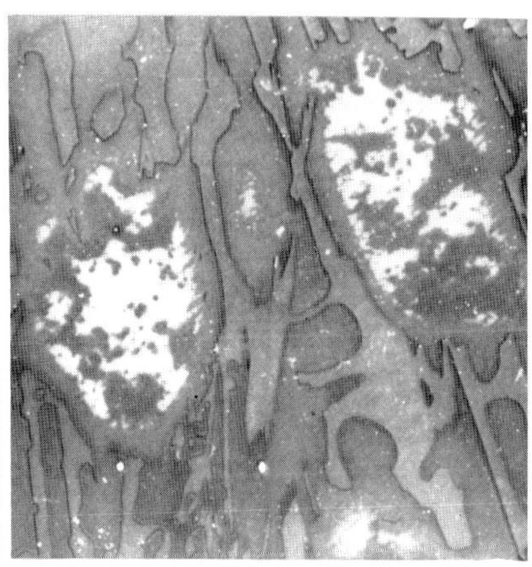

112 500 : 1

2.1.2 Iron Alloys and Steels

2.1.2.1 General Introduction

In describing the microstructure of a steel it is important not only to assess the form of the microstructure but also to estimate the undissolved carbides, non-metallic inclusions and intermetallic phases and to recognize concentration variations in the alloying elements. By the form of the microstructure we understand the arrangement of individual phases such as grain boundary cementite or ferrite in Widmannstätten structures, characteristic phase mixtures such as ferrite and carbide in pearlite or bainite, or martensitic structures. All such microstructural forms can occur individually or together.

Carbides are not generally classified with the "non-metallic inclusions". This term is usually reserved for oxides, suphides, nitrides or borides. In every case it is desirable to know the type, quantity, form and arrangement of the phases. Variations in the concentration of alloying elements can arise due to segregations or diffusion treatment. They can be shown up by interference layers only if there is a significant difference in the optical constants of the regions rich in and deficient in the alloying element.

In the case of **unalloyed or microalloy steels,** not only non-metallic inclusions but also the formation of ferrite and the formation, quantity and configuration of pearlite are of primary interest. For neither of the latter does the deposition of an interference layer afford any advantage over normal metallographic methods. Pearlite can be contrasted very well by the usual etching techniques. The important determination of the grain size of ferrite is possible only by methods reacting strongly of differences in orientation or which show up grain boundaries. On the other hand the evaluation of the distribution, size and shape of carbides in soft annealed steels, e.g. some deep drawing steels, is facilitated by the use of interference layers for contrast enhancement since errors due to relief formation resulting from chemical etching are particularly likely at low concentrations (c.f. Figs. 112 to 116).

In low and medium alloy steels, a group which includes high carbon steels, residual austenite may persist after certain heat treatments in addition to the transformation microstructures described above. As long as the areas of residual austenite are sufficiently large, they can be shown up by the deposition of interference layers, whereas they are generally undetectable by conventional metallographic techniques (see Section 2.1.1). Essential to the description of a microstructure are the proportions of ferrite, pearlite, bainite and martensite. Since interference layers do not usually provide any definite information about pearlite, bainite and martensite, normal methods of contrast production are preferred.

In the case of high alloy steels, austenite, ferrite and carbide and in some cases inter-metallic phases occur together in various proportions dependent on the steel. In this case a satisfactory differentiation of the phases and of the different carbide types is scarcely possible by chemical etching. Examples of the application of the interference layer method to this group of steels are provided by Figs. 157 to 180.

Non-metallic inclusions found in steel include the sulphides, oxides, nitrides and borides. The most important oxide phases besides FeO are Al_2O_3 (including spinels) and SiO_2. Phase mixtures can also occur. The most important sulphide inclusions are FeS and MnS. Both phases form solid solutions and can also take up other elements. Of the nitrides, only TiN and ZrN form inclusions large enough to be visible in the light microscope. AlN, NbN and VN are usually so finely dispersed that they are optically undetectable. Nitrogen can also exist in carbides in the form of carbonitrides.

Non-metallic inclusions, especially nitrides, borides and sulphides can be made to stand out very clearly using the interference layer technique (c.f. Figs. 118 to 150).

A metallographic assessment of tool steels should involve a description of the type, quantity, form and arrangement of the carbides. Here too interference layer metallography can make a valuable contribution (c.f. Figs. 151 to 156). Any residual austenite is generally optically undetectable because it is so finely dispersed. Similarly, carbides precipitated from martensite during annealing are beyond the resolution limit of the light microscope.

Literature to 2.1.2

1) Pepperhoff, W. and Bühler, H.-E.: Metallographischer Nachweis der Ferritbildung in austenitischen Chrom-Nickel-Stählen.
Arch. Eisenhüttenwes. **33** (1962) 711

2) Pepperhoff, W. and Bühler, H.-E.: Lichtmikroskopische Untersuchung an martensitischen Gefügen.
Arch. Eisenhüttenwes. **34** (1963) 839

3) Schüller, H.-J. and Bühler, H.-E.: Gefügeentwicklung an austenitischen Chrom-Nickel-Stählen durch Interferenz-Aufdampfschichten und Ätzmittel.
Berg- und Hüttenm. Monatshefte **109** (1964) 105

4) Bühler, H.-E., Pepperhoff, W. and Schüller, H.-J.: Ermittlung des Restaustenits mit Hilfe des Interferenzschichten-Verfahrens.
Arch. Eisenhüttenwes. **36** (1965) 457

5) Kessler, H. and Pitsch, W.: Die kristallographischen Eigenschaften des im Martensit bei schnellem Aufheizen rückumgewandelten Austenits in einer Fe-Ni-Legierung mit 32 Gew.-% Ni.
Arch. Eisenhüttenwes. **38** (1967) 321

6) Peter, W., Müller, H. and Kohlhaas, E.: Metallkundliche Untersuchungen über den Einfluß von Molybdän auf das Ausscheidungsverhalten hochwarmfester Nickel-Chrom und Nickel-Chrom-Kobalt-Legierungen.
Arch. Eisenhüttenwes. **38** (1967) 329

7) Pepperhoff, W.: Entwicklung des Gefüges von Ferrolegierungen durch Interferenz-Aufdampfschichten.
Arch. Eisenhüttenwes. **32** (1961) 651

8) Robusch, G. and Engler, G.: Nachweis von Restaustenit mit Hilfe aufgedampfter Interferenzschichten.
Prakt. Metallographie **9** (1972) 553

9) Robusch, G., Engler, G. and Bühler, H.-E.: Möglichkeiten bei der Anwendung absorbierender, insbesondere durch Gasionenbeschichtung aufgebrachter Schichtwerkstoffe in der Metallographie der Eisenlegierungen.
Prakt. Metallographie **14** (1977) 65

10) Bühler, H.-E., Baumgartl, S., Meyer, L., Mörck, G., Nikoleizig, A. and Robusch, G.: Metallographie oberflächenveredelter Stähle – metallische Schichten –.
Special Vol. 7 of Prakt. Metallographie. Edited by Bühler H.-E., Hillnhagen E. and Horn, E., Dr. Riederer-Verlag GmbH, Stuttgart (1977), 5

11) Bühler, H.-E. and Jackel, G.: Tafeln zur quantitativen Metallographie mit aufgedampften Interferenzschichten.
Arch. Eisenhüttenwes. **41** (1970) 859

12) Jackel, G. and Bühler, H.-E.: Interferenzschichten-Mikroskopie in der Metallographie in: Fortschritte in der Metallographie, Special Vol. 3 of Prakt. Metallographie, Dr. Riederer-Verlag GmbH, Stuttgart (1972), 107

13) Silber, F. A. and Judtmann, H.: Aufgedampfte Interferenzschichten als Hilfsmittel zur Identifizierung von Einschlüssen.
Prakt. Metallographie **9** (1972) 318

14) Semlitsch, M.: Interferenz-Aufdampfschichten bei Mikrogefügeanalysen von Werkstoffen mit der Elektronenstrahl-Mikrosonde.
Prakt. Metallographie **8** (1971) 511

15) Scheidl, H. and Mitsche, R.: Neue Beispiele aus der Reflexionsmessung. Radex-Rundschau, **M 3/4** (1967) 596

16) Naumann, F. K.: Beitrag zum Nachweis der σ-Phase und zur Kinetik ihrer Bildung und Auflösung in Eisen-Chrom und Eisen-Chrom-Nickel-Legierungen. Arch. Eisenhüttenwes. **34** (1963) 187

17) Pepperhoff, W., Bühler, H.-E. and Dautzenberg, N.: Beitrag zum Aufbau des Systems Eisen-Chrom-Kohlenstoff im Bereich der technischen Ferrochrom-Legierungen. Arch. Eisenhüttenwes. **33** (1962) 611

18) Unpublished work of Thyssen AG and Mannesmann-Forschungsinstitut GmbH.

Differentiation of ferrite, martensite and cementite

Figs. 112–116: **Iron-carbon alloy containing 0.5% C.**

112: Heat treatment: 850°C 15 min/water + 715°C 96 h/air + 730°C 250 s/water.

Etched for 25 s in 1% alcoholic nitric acid.

Photographed through a green filter.

ferrite: white to light grey [1]

martensite: dark grey [2]

cementite (Fe_3C): white specks [3]

Coarse carbides are formed during the pretreatment at 850 and 715° C. They are partially dissolved during the anneal at 730° C. The austenite present at this temperature is transformed to martensite on quenching. After the normal etch in nitric acid it is difficult to distinguish the cementite from the ferrite.

113: Same specimen and heat treatment as in Fig. 112.

Etched for 5 s in 1% alcoholic nitric acid.

Sputtered with a copper cathode in air, zero order.

Photographed at 550 nm.

ferrite: light grey [1]

martensite: dark grey [2]

cementite: black [3]

The interference layer enhances the contrast between the carbides and the ferrite. Ferrite and martensite cannot be differentiated at this wavelength.

Figs. 114–117: See page 88

114: Same specimen, same heat treatment and same preparation as in Fig. 113.

Photographed at 550 nm.

ferrite and martensite: light green (24 : 6 : 3)

cementite: dark green (22 : 6 : 4)

The colour photograph shows clearly the differences in saturation and brightness of the carbides and of the ferrite and martensite.

115: Same specimen and same heat treatment as in Fig. 113.

Etched for 7 s in 1% alcoholic hydrochloric acid.

Sputtered with a copper cathode in air, zero-order.

Photographed at 600 nm.

ferrite: light grey

martensite: black

cementite: medium grey

By changing the pre-etching conditions and observation wavelength the martensite can clearly be distinguished from the ferrite. This is not possible simply by the deposition of a layer. In contrast to Fig. 112, the pre-etch produces such a minimal relief that falsification of the proportions of the phases need not be feared in a quantitative assessment.

116: Same specimen, same heat treatment and same preparation as in Fig. 115.

Photographed at 600 nm.

ferrite: yellowish red (4 : 4 : 2) (original 6 : 2 : 2)

martensite: reddish brown (8 : 4 : 3) (original 7 : 3 : 4)

cementite: pink (10 : 1 : 2) (original 6 : 3 : 2..3)

The photograph shows the differences in brightness and saturation of the colours of the ferrite and martensite. There are no differences in shade in the original because of the use of colour filters. In printing, however, the large differences in brightness are converted into differences in shade.

Fig. 117: **Phosphide eutectic in a basic Bessemer pig iron**

Etched for 5 s in 1% alcoholic nitric acid. Coated with ZnSe, zero order.

Photographed at 540 nm.

cementite: light grey

Fe_3P: dark grey [1]

graphite: black [2]

The pearlite [3] is not completely resolved and cannot therefore be identified clearly. Within the phosphide [1] differences in the phosphorus concentration can be recognised by light and dark patches.

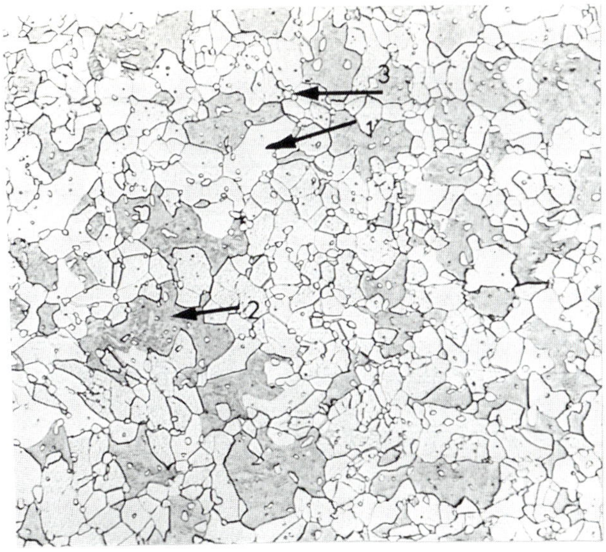

112 500 : 1

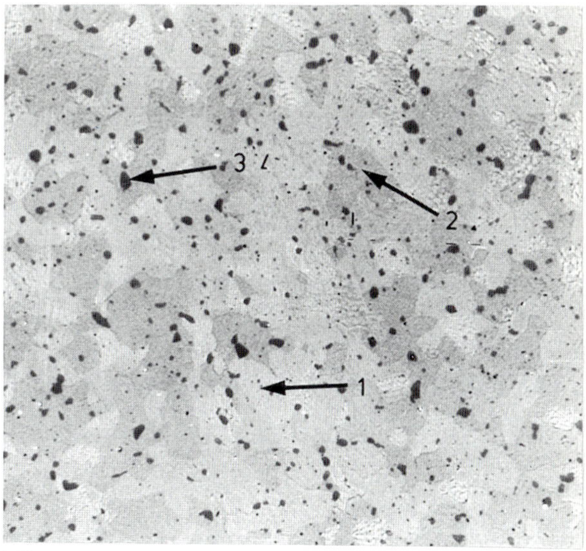

113 500 : 1

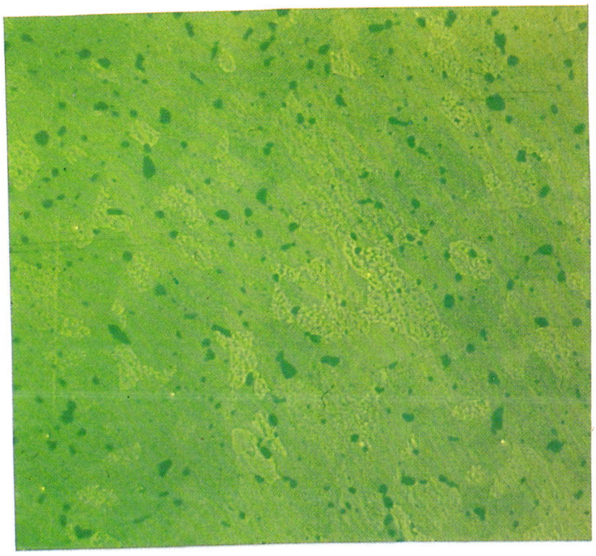

114 500 : 1

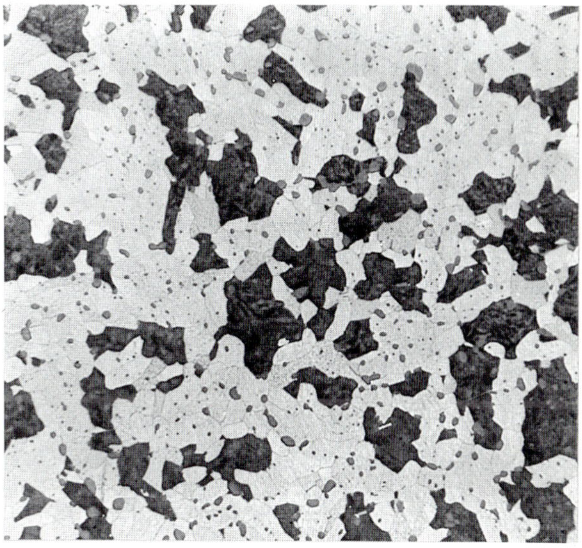

115 500 : 1

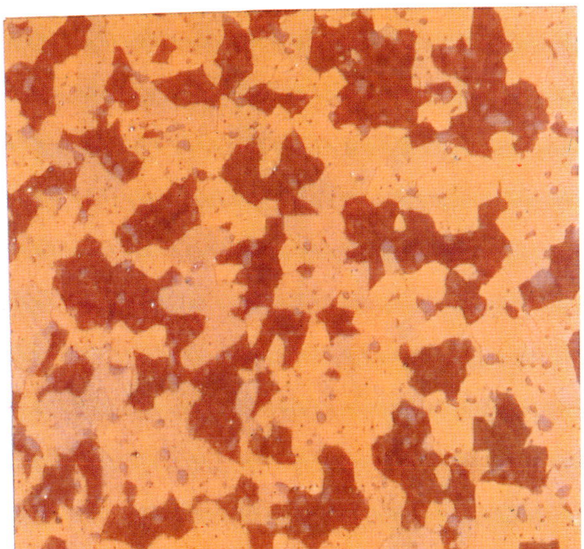

116 500 : 1

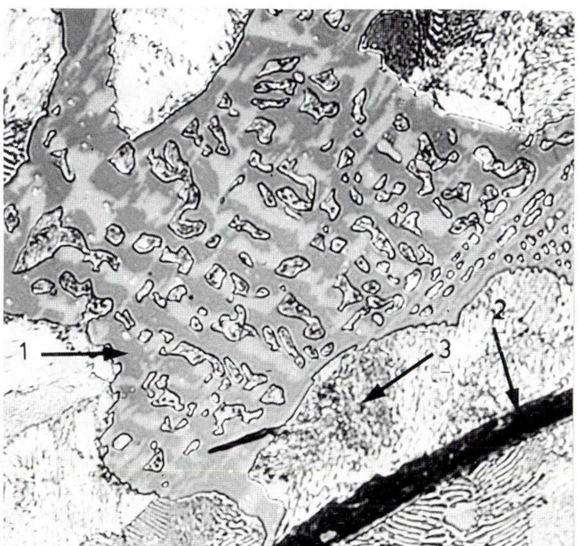

117 500 : 1

Oxides and sulphides in steel

The estimation of oxide and sulphide inclusions is of vital importance in the assessment of the purity of steels. Oxidic phases must also be identified and evaluated in investigations of oxidation and scale layers. The differentiation of these phases is scarcely possible using conventional metallographic methods. Reflectivity curves for these phases are given in the following. The practical applications are illustrated by micrographs.

Fig. 118: **Reflectivity of the oxides Fe_2O_3 and Fe_3O_4**

In the uncoated condition Fe_2O_3 is somewhat brighter than Fe_3O_4. The contrast between these phases is of the order of 0.3. After vapour phase deposition of ZnS, the contrast at 475 nm has risen to ca. 0.5. By changing the wavelength of the illumination the contrast between these two phases can be reversed. This makes it easier to identify them.

Fig. 119: **Reflectivity of the sulphides FeS and MnS**

In the uncoated condition the contrast between FeS and MnS is ca. 0.3. By coating with ZnS, this contrast can be increased to 0.9 at 575 nm. By varying the observation wavelength the MnS and then the FeS can be made to appear brighter in turn (c.f. Figs. 123 and 133 to 142).

Fig. 120: **Influence of the refractive index of the layer on the reflectivity of Fe_2O_3 and Fe_3O_4 at the interference minimum**

By varying the layer material either the Fe_2O_3 or the Fe_3O_4 can be made to appear darker at the interference minimum. By coating with ThF_4 and PbF_2 the reflectivity of one of the phases can be reduced virtually to zero, with the result that the contrast between the two phases is a maximum. The dependence of the reflectivity on the wavelength must be taken into account for each refractive index (Fig. 118).

Fig. 121: **Influence of the refractive index of the layer on the reflectivity of MnS and FeS at the interference minimum, (calculated)**

The curves for FeS and MnS diverge more than those for Fe_2O_3 and Fe_3O_4, Fig. 120. Once again, the MnS and then the FeS can be made to appear the darker of the two phases at the reflection minimum by a suitable choice of layer refractive index. The arrows show which reflectivity values can actually be realised with the available layer materials (c.f. Figs. 62 and 64 and also chapter 5). Fig. 119 illustrates the wavelength dependence for a ZnS coating.

Fig. 122: **Difference in wavelength of the reflection minimum of Fe_3O_4 and Fe_2O_3 dependent on the refractive index of the layer**

The difference in the wavelength of the reflection minima (c.f. Fig. 118) increases as the refractive index of the layer increases. The individual experimental values show a marked scatter because the reflection minima are not always very pronounced (c.f. Fig. 118). As well as the absolute value of the reflection minimum, an increase in the difference between the wavelengths of the minima also leads to an increase in the contrast (c.f. Figs. 118 and 119). Observed visually, a difference in the wavelength of the reflection minima appears as a difference in colour (c.f. Figs. 123 and 134).

Fig. 123: **Reflectivity of Fe, FeS and MnS as a function of wavelength.**

Sputtered with a copper cathode in air.

The data for the curve were obtained using the specimen illustrated in Fig. 134. The practical significance of the shape of the curve is discussed with reference to Figs. 133 to 144.

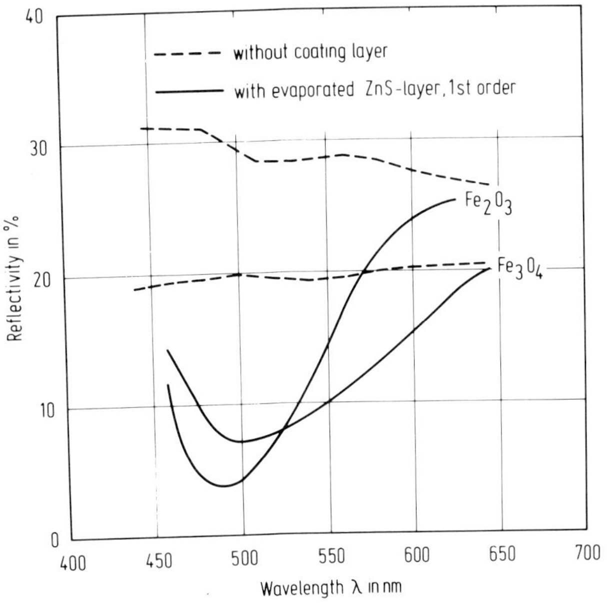

118

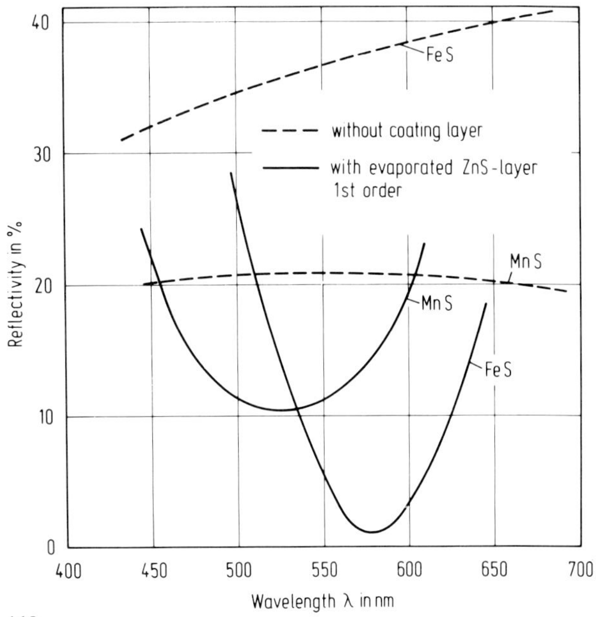

119

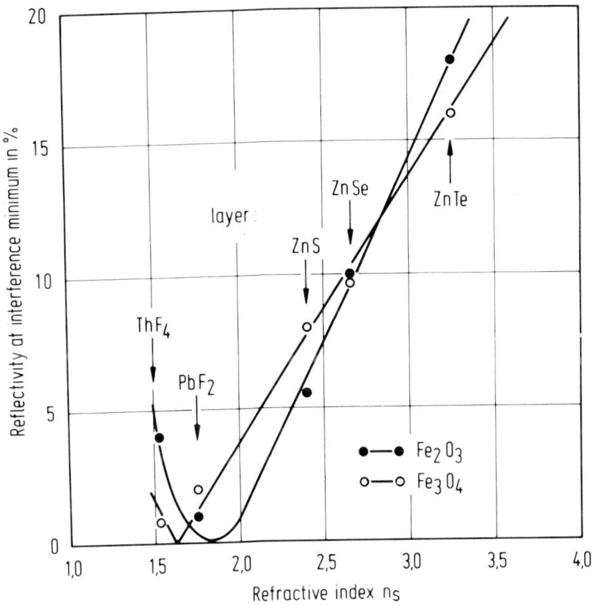

120

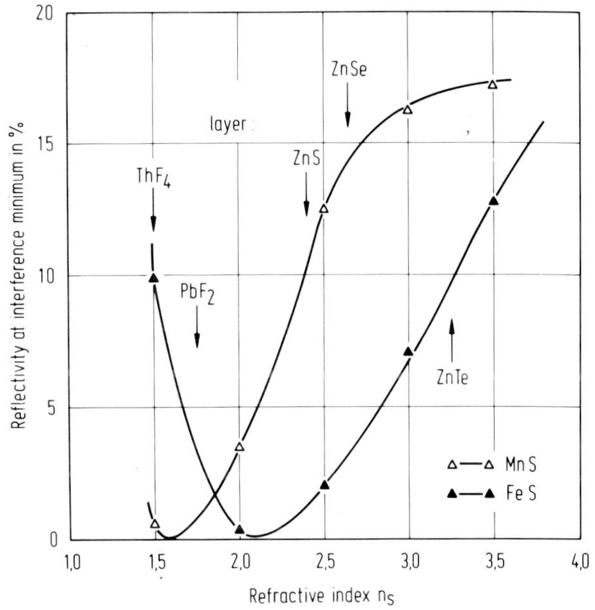

121

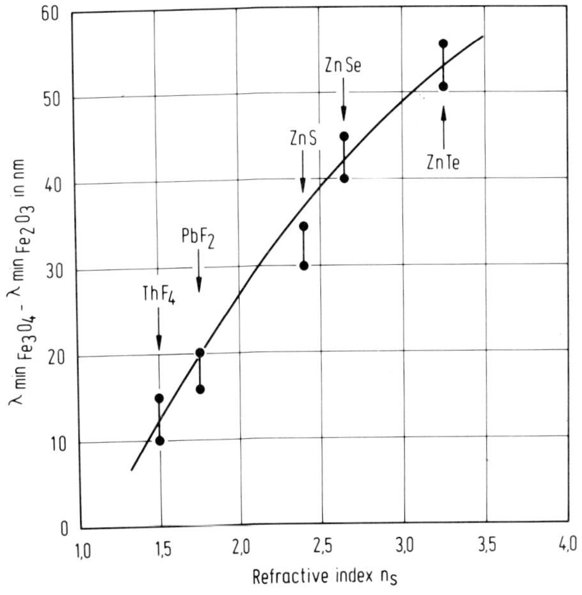

122

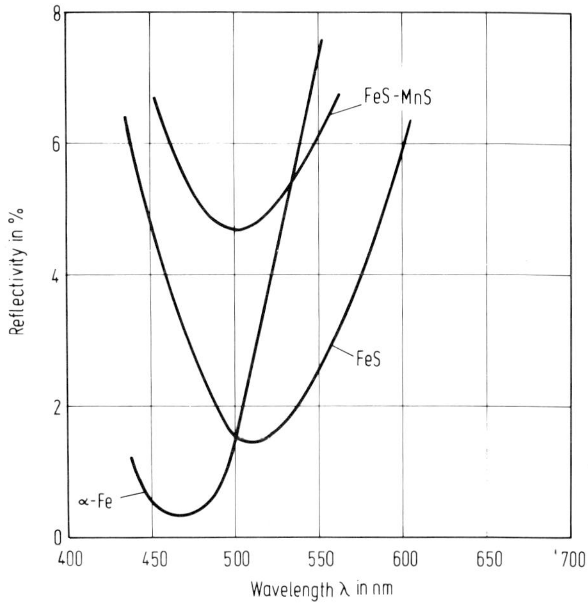

123

Figs. 124–131: **Differentiation between FeO and FeS**

124: **Reflectivity of MnS, FeS and FeO as a function of wavelength.**

Coated with ZnS, first order.

Under these conditions the wavelengths of the reflection minima of the three phases are widely separated. By suitable choice of observation wavelength all possible light-dark contrasts between the phases can be produced. The differences in colour of the phases are well marked. Similar results are obtained by sputtering.

Figs. 125–128: **Free cutting steel containing 0.09% C, 0.6% Mn and 0.2% S.**

125: Heat treatment: 1050^0 C 5 h in dry oxygen/air.

Sputtered with a copper cathode in air, zero order.

iron matrix: red (8..9 : 2 : 3)

FeO: pale violet (11 : 0...1 : 3)

FeS: violet (12 : 3 : 4)

For the disposition of the phases see also Fig. 127. A eutectic of FeO and FeS has formed near the FeO. All the phases can easily be distinguished by their colours.

126: Same material and same heat treatment as in Fig. 125.

Polished specimen, bright field.

Once again the FeO can clearly be recognised by its dark colouration. The contrast between FeS and iron is adequate for a visual assessment but at the lower limit for a quantitative determination.

127: Schematic representation of the arrangement of phases in Figs. 125–128.

128: Same specimen, same heat treatment, same preparation and same area of the microstructure as in Fig. 125.

Photographed at 560 nm.

The FeO appears white and the FeS black. The contrasts are considerably better than in Fig. 126, the eutectic in particular can be resolved very easily.

Figs. 129–131: **Free cutting steel containing 0.09% C, 0.7% Mn and 0.2% S.**

129: Heat treatment: 1050^0C 30 min + 1200^0C 60 min in dry oxygen.

Polished specimen, bright field.

In some cases the individual phases cannot be identified definitely on the basis of brightness alone.

Figs. 130 and 131: See page 94

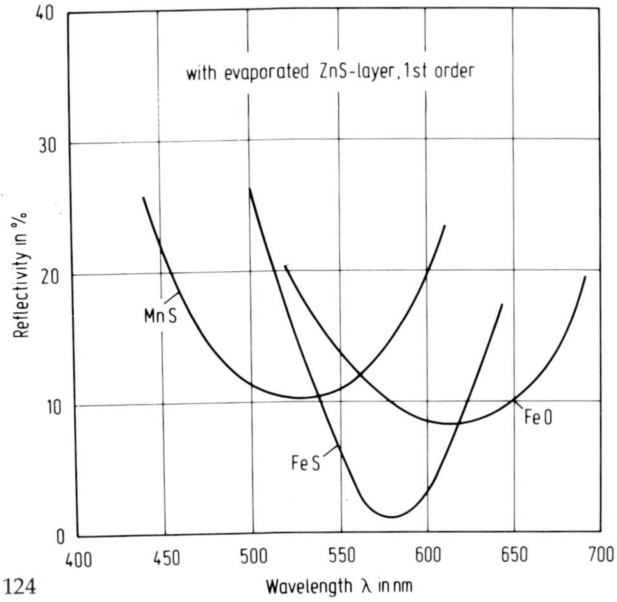

124

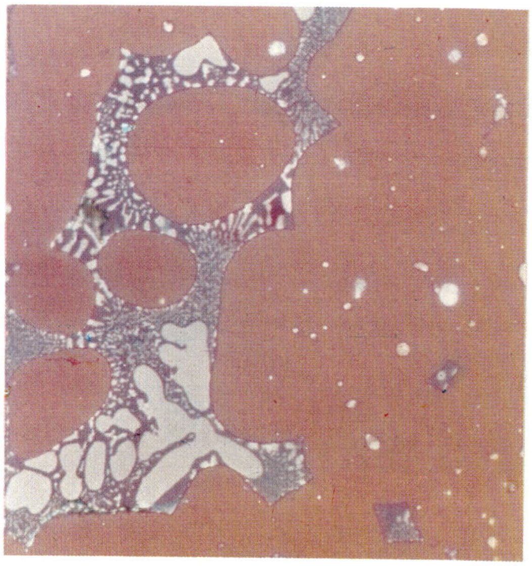

125 500 : 1

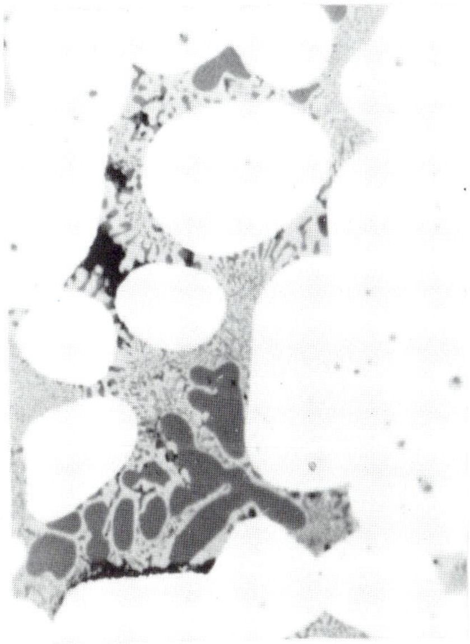

126 500 : 1

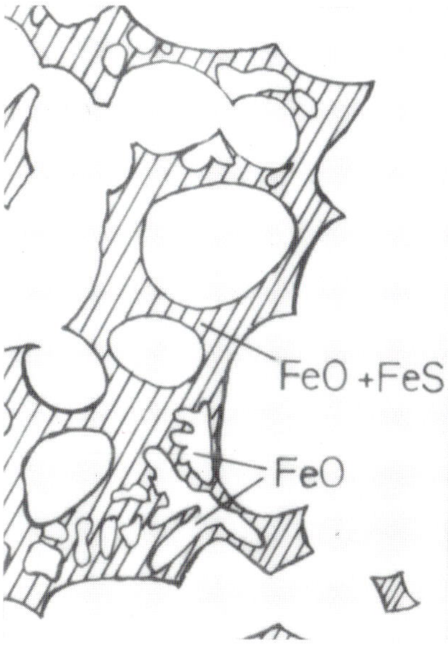

127 500 : 1

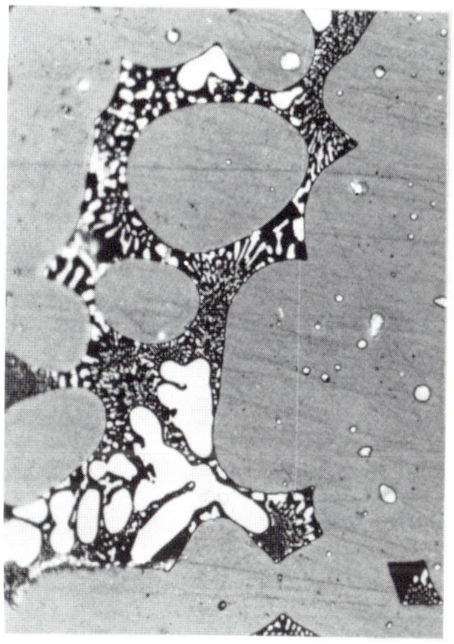

128 500 : 1

129 500 : 1

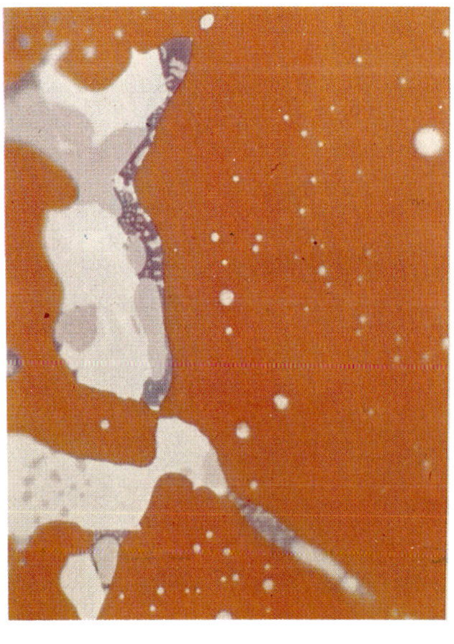

130 500 : 1

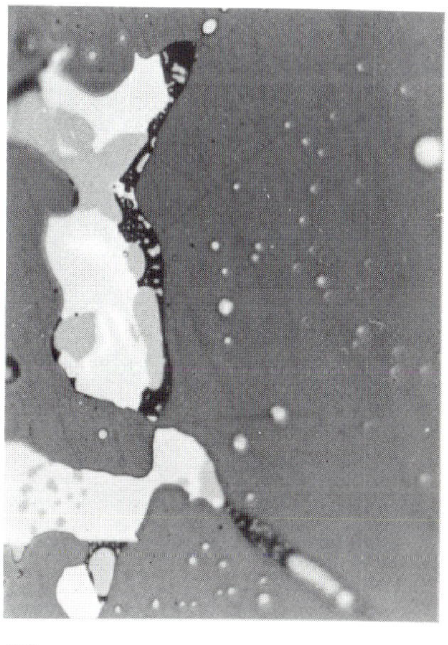

131 500 : 1

130: Same specimen, same heat treatment, same preparation and same area of the microstructure as in Fig. 129.

Sputtered with a copper cathode in air, zero order.

matrix: yellowish red (5 : 5 : 3)

FeO: grey (– : 0 : 1.5)

FeS (in eutectic): violet (12 : 2 : 3)

MnS: pale violet (11 : 0..1 : 2)

The individual phases can be identified by their colour.

131: Same specimen, same heat treatment, same preparation and same area of the microstructure as in Fig. 130.

Photographed at 580 nm.

The phases can be distinguished on the black and white photograph because of the good contrast.

Figs. 132–144: Differentiation between FeS and MnS

In many cases, quantitative data concerning the composition of solid solutions in binary and ternary systems can be obtained by reflectivity measurements (c.f. section 2.2.1).

Figure 132 illustrates this for the binary system FeS-MnS.

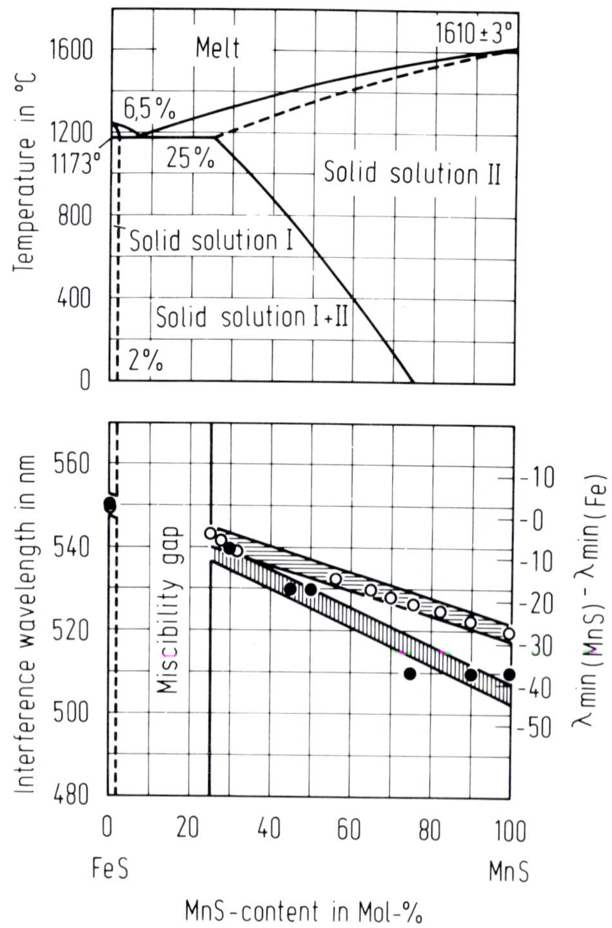

Fig. 132: Change in wavelength of the reflection minimum of FeS-MnS solid solutions with the change in MnS content after deposition of ZnS as compared with the phase diagram 18).

The ordinates of the lower diagram give the difference between the wavelengths of the reflection minima for the FeS-MnS solid solutions and the iron matrix. Calculations from the two series of measurements have given different values but the correlation between the iron content in the MnS and the optical constants of manganese sulphide can be considered reliably established.

Figs. 133–142: **Specimen containing 80% FeS and 20% MnS.**

Figs. 133 to 142, showing a specimen consisting of FeS and MnS, demonstrate the change in contrast and in the saturation of the colours with increasing thickness of the sputtered layer (c.f. also Figs. 30 to 35). The thickness of the layer is indicated by the sputtering time. The specimen involved consisted only of FeS and MnS. For purposes of comparison, the iron mounting of the specimen was also coated and photographed. This photograph appears on the left of the FeS-MnS micrograph in each case.

133: On cooling, large plates of FeS have precipitated from an FeS-MnS solid solution.

Sputtered with a copper cathode in air, zero order.

Sputtering time: 35 s

iron: pale yellow (2 : 3 : 1)

FeS: yellowish brown (4 : 4 : 2)

MnS (matrix): light brown (4 : 3 : 1)

134: Same specimen and same heat treatment as in Fig. 133.

Sputtered with a copper cathode in air, zero order.

Sputtering time: 55 s.

iron: dark yellow (4 : 6 : 1)

FeS: dark orange (5 : 5 : 2)

MnS: bright yellowish red (5 : 3 : 2)

The shade of the iron corresponds to the complementary colour of the reflection minimum in Fig. 123. In the case of FeS and MnS the printing process has caused colour shifts.

Figs. 135–138: See page 96

135: Same specimen, same heat treatment and same area of the microstructure as in Fig. 133.

Sputtered with a copper cathode in air, zero order.

Sputtering time: 65 s.

iron: dark yellow (4 : 6 : 1)

FeS: red (7 : 4 : 4)

MnS: bright red (6 : 3 : 2)

136: Same specimen, same heat treatment and same area of the microstructure as in Fig. 133.

Sputtered with a copper cathode in air, zero order.

Sputtering time: 75 s.

iron: bright red (6 : 4 : 2)

FeS: blue (15 : 2 : 4)

MnS: red (9 : 2 : 2..3)

137: Same specimen, same heat treatment and same area of the microstructure as in Fig. 133.

Sputtered with a copper cathode in air, zero order.

Sputtering time: 85 s.

iron: red (9 : 3 : 3)

FeS: blue (18 : 5 : 3)

MnS: violet blue (14 : 1 : 4)

138: Same specimen, same heat treatment and same area of the microstructure as in Fig. 133.

Sputtered with a copper cathode in air, zero order.

Sputtering time: 95 s.

iron: reddish (10 : 2 : 4)

FeS: turquoise (18 : 5 : 3)

MnS: blue (16 : 1 : 4)

Comparison ot the colour data shows that as the sputtering time is prolonged, that is as the layer thickness increases, so the saturation of the colours decreases. This is also valid in the originals for FeS. At the same time the photographs show that the contrasts are best in Fig. 135, becoming poorer for both thinner and thicker layers. The differences in brightness shown by the original specimens are cancelled out by the photographic process and cannot therefore be reproduced.

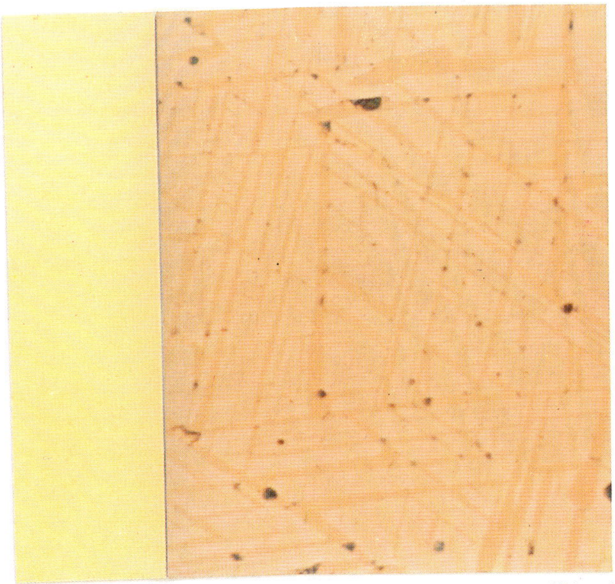

133 650 : 1

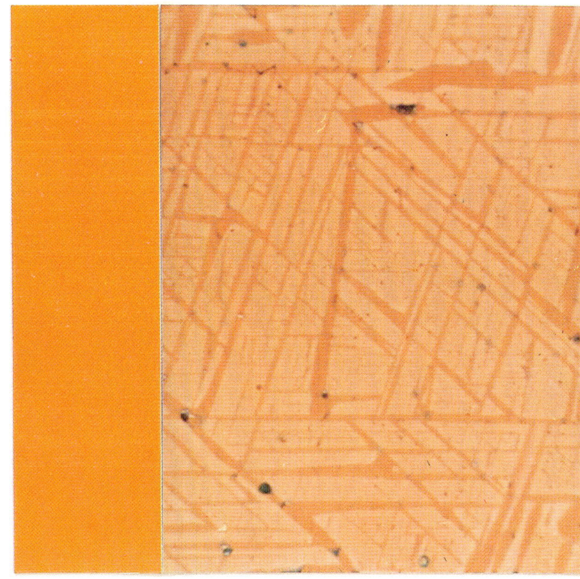

134 650 : 1

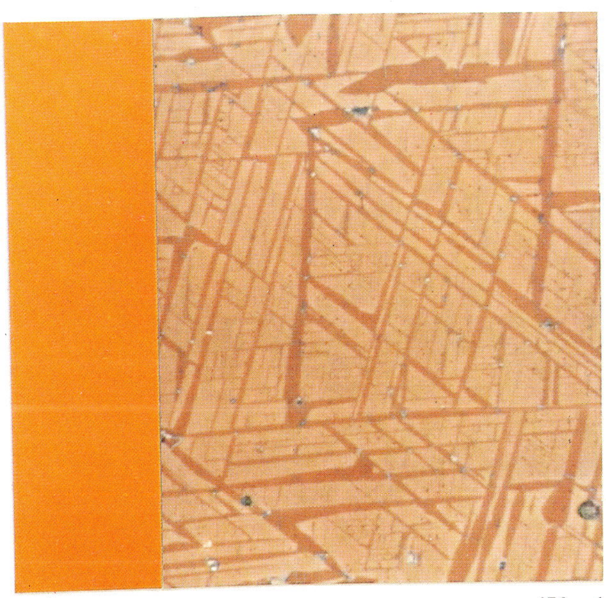

135 650 : 1

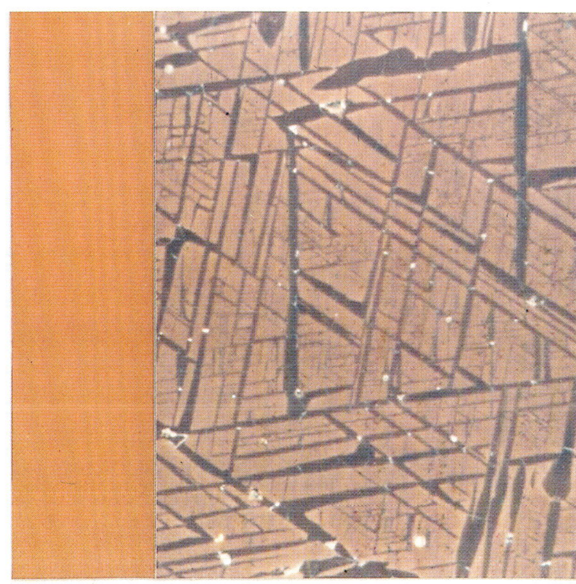

136 650 : 1

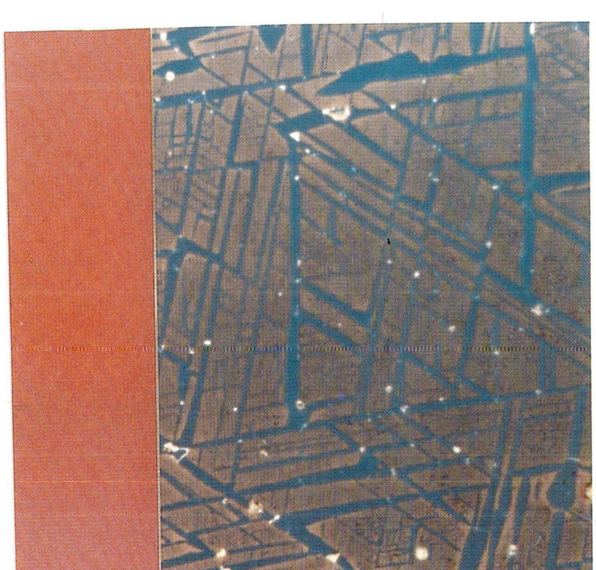

137 650 : 1

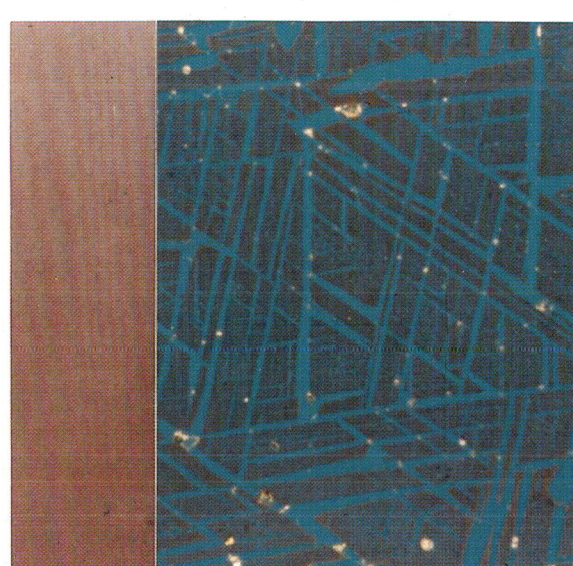

138 650 : 1

Figs. 139–142: Figures 139 to 142 are photographs taken at different wavelengths of the specimen shown in colour in Fig. 134. The reflectivities of the three phases in this specimen were measured as a function of wavelength and are plotted in Fig. 123. This series of photographs shows the change in relative brightness of the individual phases with the wavelength of the illumination.

139: Same specimen, same heat treatment and same preparation as in Fig. 134.

Photographed at 480 nm.

According to Fig. 123, iron has the lowest reflectivity and MnS the highest reflectivity at this wavelength. FeS is intermediate. The photograph shows that the darkness of the individual phases follows an order corresponding to their reflectivities: iron is black, FeS is dark grey, MnS is light grey.

140: Same specimen, same heat treatment and same preparation as in Fig. 134.

Photographed at 500 nm.

At this wavelength the reflectivities of iron and FeS are equal. There is consequently no difference in brightness between iron and FeS on this photograph.

141: Same specimen, same heat treatment and same preparation as in Fig. 134.

Photographed at 520 nm.

At this wavelength, FeS has the lowest reflectivity and MnS still has the highest. Consequently the FeS plates appear darker than the iron and MnS is the brightest phase.

142: Same specimen, same heat treatment and same preparation as in Fig. 134.

Photographed at 540 nm.

At this wavelength, iron has the greatest reflectivity and so appears brighter than the MnS. FeS is the darkest phase.

In Fig. 139, iron is the darkest phase, in Fig. 142 the lightest. This illustrates the potential use of monochromatic light. In Figs. 139 to 142 the relative brightness of the individual phases is correctly reproduced but the absolute difference in brightness as expressed in the reflectivity values in Fig. 123 cannot be reproduced in black and white photographs. The range of brightness that can be represented by printing and reproductions on paper is far from adequate to represent the actual large differences in reflectivity.

Figs. 143 and 144: **Steel X 15 CrNiSi 20 12.**

Recuperator sheet, region near to the surface.

Heat treatment: service specimen, exact history unkown.

Coated with ZnSe, zero order, bright field.

143: FeS: red (8 : 4 : 4)

MnS: yellow (3 : 5 : 2)

FeS and MnS have precipitated at the surface.

144: Same steel, same heat treatment and same preparation as in Fig. 143, bright field.

The contrast between the phases is good on a black and white bright field photograph.

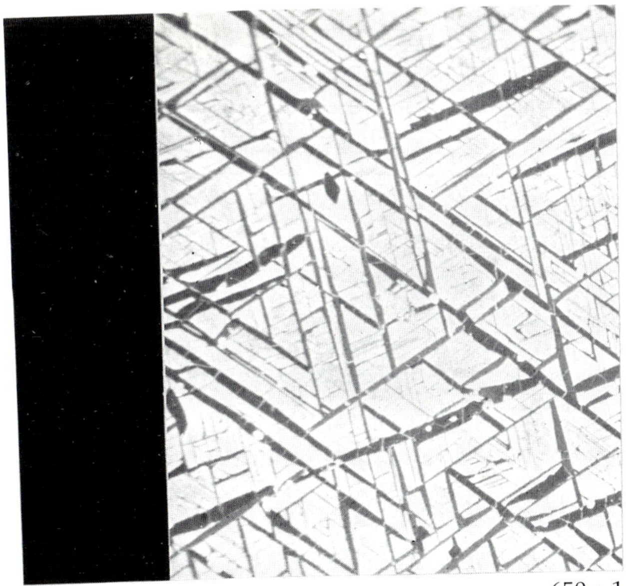
139 650 : 1

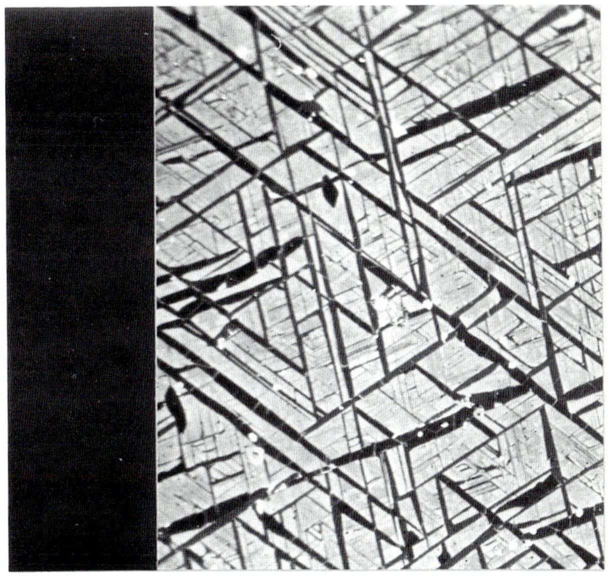
140 650 : 1

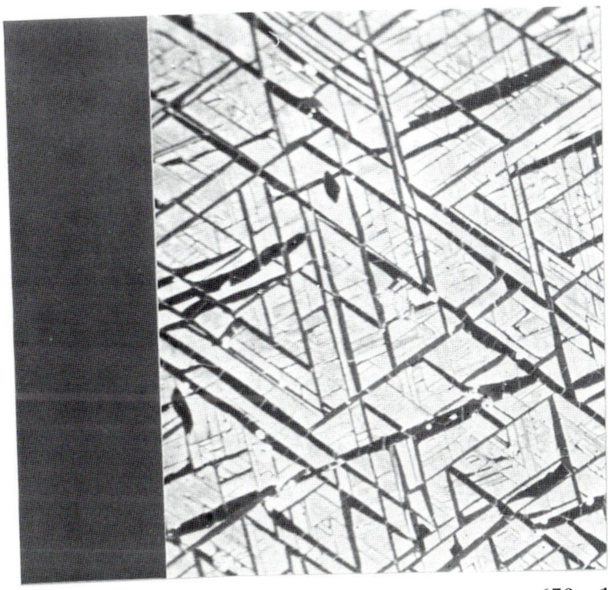
141 650 : 1

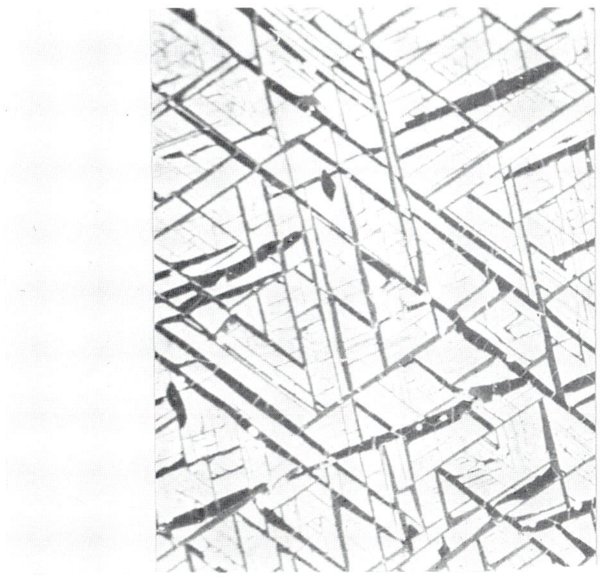
142 650 : 1

143 1000 : 1

144 1000 : 1

Sulphide inclusions and carbonitrides in steel

Fig. 145: **Free-cutting steel with 0.09% C, 0.9% Mn and 0.22% S.**

Sputtered with a copper cathode in air, zero order.

matrix: red

Photographed at 540 nm.

The sulphides show up in great contrast to the matrix and so can easily be measured, for example with automatic instruments for image analysis.

Figs. 146–150: **Precipitates in a steel like St 52 additionally alloyed with 0.15% Ti.**

The specimens come from a section of a rough slab in the rolled condition. They are coated with a zero-order layer of ZnSe. The colour of the matrix is usually set at golden yellow (λ_{min} = 480 nm).

146: **Distribution of the elements carbon, nitrogen and titanium in a titanium carbonitride.**

This example shows how the chemical composition of phases can be determined using a microprobe. As with the particle illustrated, the minimum required dimensions are approximately 5.5 μm. The particle which has been examined, shown in the upper picture, is similar to that shown in Fig. 148.

147: matrix: α-iron, yellow (2 : 4 : 1)

MnS: violet blue (12 : 0..1 : 4) [1]

titanium carbosulphide: red (6 : 3 : 3) [2]

titanium carbonitride: dark red (6 : 4 : 4) [3]

Titanium carbosulphide accumulates on the MnS. A titanium carbonitride is visible at the upper left edge [3].

148: matrix: α-iron, yellowish green (1 : 3 : 1)

MnS: blue (elongated) (16 : 0..1 : 4) [1]

titanium carbonitride: large inclusion, yellow to red (2 : 3 : 2) to (8 : 5 : 4) [2]

titanium carbosulphide: yellow (4 : 3 : 3) [3]

The change in colour within the titanium carbonitride inclusion from the edge to the middle indicates a variable composition corresponding to the measurements in Fig. 146.

149: matrix: α-iron, yellow (2..3 : 4..5 : 1..2)

MnS: greyish blue (elongated) (14 : 0..1 : 2) [1]

TiO_2: blue (17 : 3 : 1) [2]
(single inclusion in MnS)

titanium carbosulphide: violet (11 : 3 : 4) [3]

The colour of the matrix is different from that in Fig. 147 because the thickness of the layer is different. The MnS is also lighter for the same reason. This causes the other phases to stand out clearly.

150: matrix: α-iron, greenish yellow (1 : 2 : 2)

MnS: elongated bluish green inclusions (18 : 0..1 : 4)

titanium carbonitride: large inclusion: edge (6 : 3 : 3), centre (2 : 3 : 2)

titanium carbosulphide: small inclusions (3 : 5 : 3) [1]

The variable colouration of the large titanium carbonitride inclusion is a clear indication of the variable concentration as measured for a titanium carbonitride in Fig. 146.

145 500 : 1

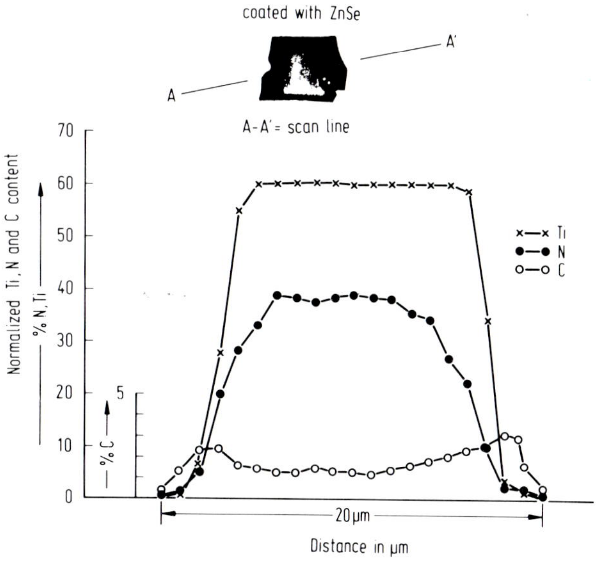

146

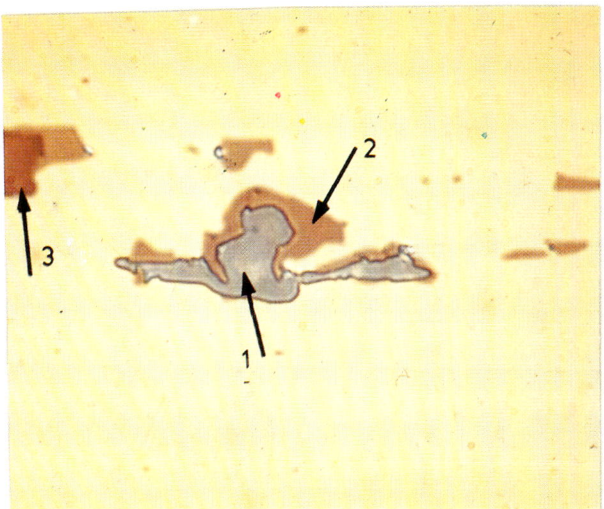

147 1600 : 1

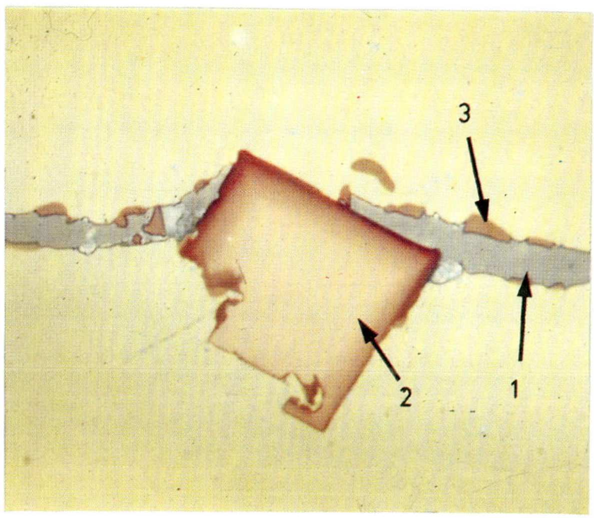

148 1600 : 1

149 500 : 1

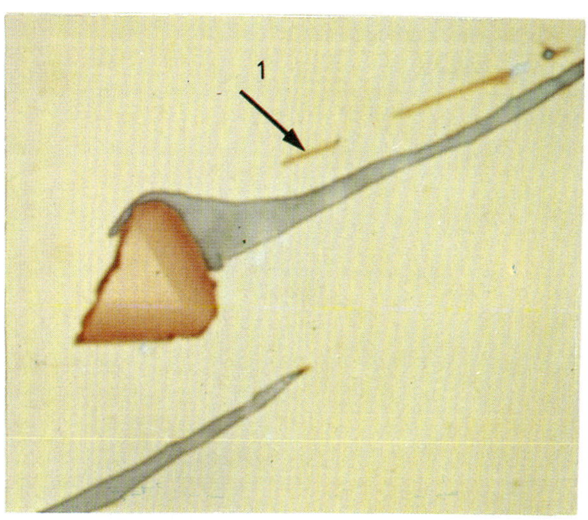

150 1600 : 1

Carbides in high-speed tool steels

The high-speed tool steels belong to the ledeburitic group. The so-called ledeburite eutectic, which forms on solidification, consists of austenite and the carbides MC, M_6C and M_2C. The formation of M_6C carbide is promoted by tungsten, silicon and nitrogen. It thus occurs in steels containing 18% tungsten and in high-speed tool steels with a high silicon and nitrogen content. The M_2C carbide, which contains mainly molybdenum, iron and vanadium forms predominantly in the region of crystalline segregations during solidification. Its formation is favoured by an increase in the carbon, molybdenum and/or vanadium content. The carbide M_2C is stable to soft annealing (about 700^0C). It decomposes by reaction with the surrounding austenite to M_6C and MC when heated to the hot shaping temperature. These carbide reactions can be demonstrated particularly well by coating and observation in monochromatic light.

Figs. 151–153: **Steel S 6-5-2 (0.9% C, 4% Cr, 5% Mo, 1.8% V, 6.5% W).**

151: As cast, soft annealed + 1125^0C 15 min/water.

Coated with ZnSe, zero-order, bright field.

The martensitic background is red (8 : 0..1 : 3).

The ledeburite consists essentially of large brownish-red M_2C carbides (6 : 4 : 3) [1]. Only isolated patches of M_6C (yellow) (3..5 : 4 : 3) [3] are visible. In the background, finely dispersed, dark red MC (7 : 3 : 4) [2] is visible together with one or two larger particles.

152: Same specimen, same heat treatment, same preparation and same area of the microstructure as in Fig. 151.

Photographed at 570 nm.

MC: black [2]

M_2C: grey [1]

M_6C: white [3]

M_2C and M_6C can be differentiated better in this photograph than in the colour photograph in Fig. 151.

153: Same specimen, same heat treatment and same preparation as in Fig. 151.

Photographed at 550 nm.

MC and M_2C: black

M_6C: light grey

MC and M_2C cannot be differentiated at this wavelength. Only M_6C stands out clearly against the matrix.

Figs. 154–156: **Steel S 6-5-2.**

154: As cast, soft annealed + 1125^0C 7 h/water.

Coated with ZnSe, zero-order, bright field.

matrix: martensite, red (8 : 0..1 : 3)

M_2C: orange red (6 : 4 : 2) [1]

M_6C: yellow (4 : 5 : 1) [2]

MC: dark red (7 : 3 : 4) [3]

Due to the prolonged heat treatment as compared with the specimens shown in Figs. 151 to 153 the M_2C carbides have decomposed almost completely to M_6C and MC.

Figs. 155 and 156: See page 104

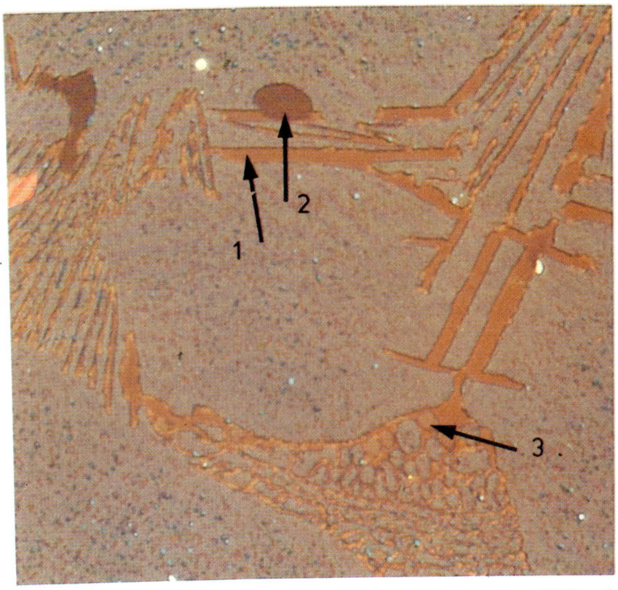

151 500 : 1

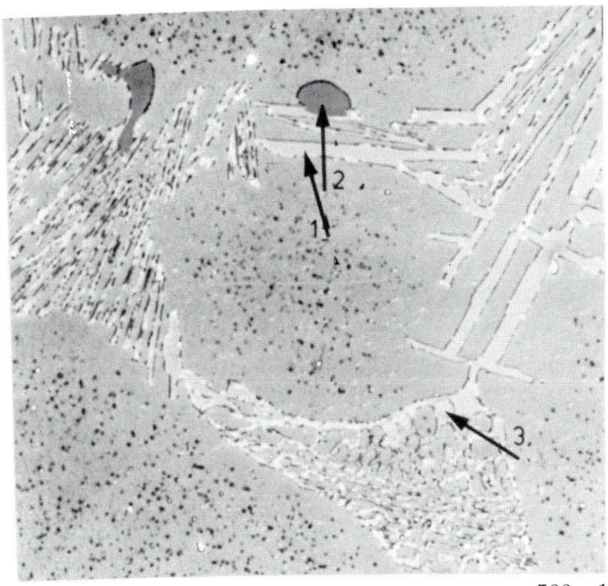

152 500 : 1

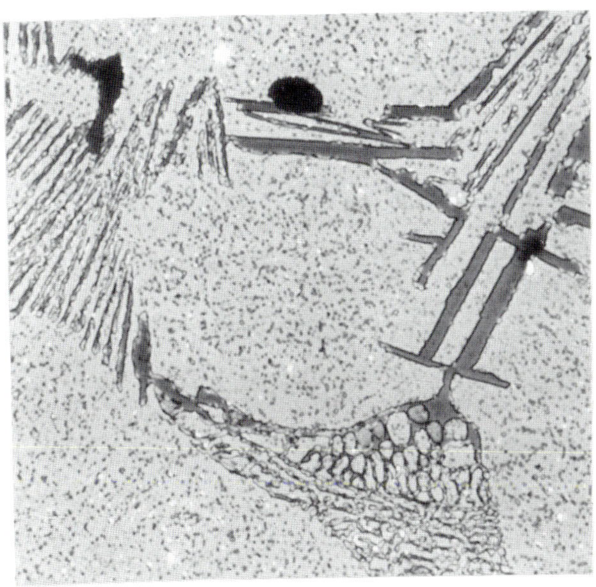

153 500 : 1

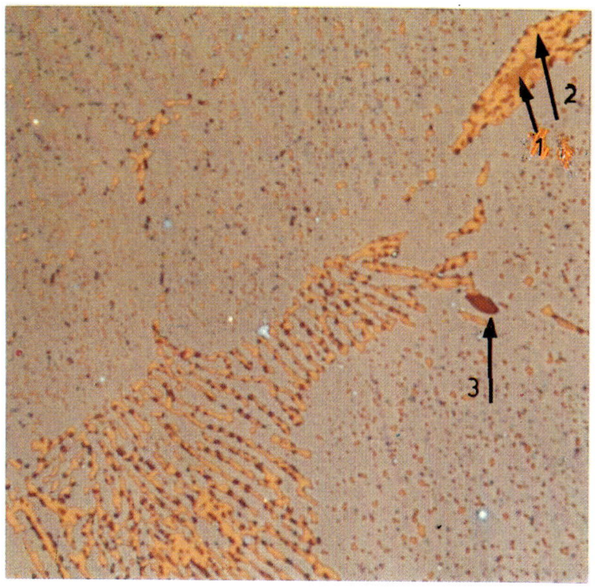

154 500 : 1

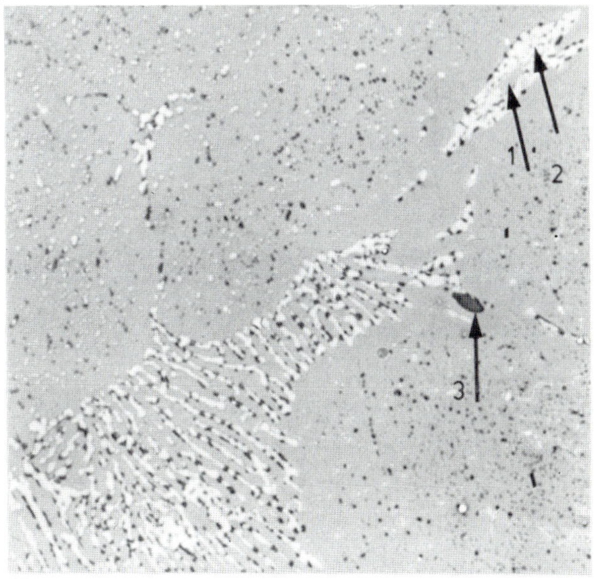

155 500 : 1

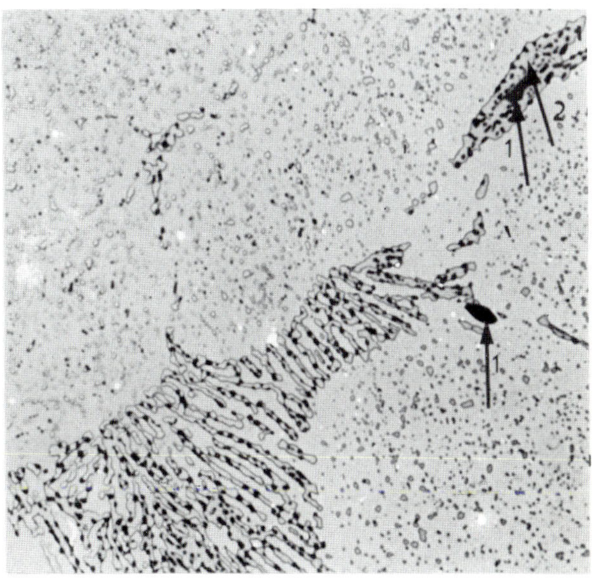

156 500 : 1

155: Same specimen, same heat treatment, same preparation and same area of the microstructure as in Fig. 154.

Photographed at 580 nm.

M_2C: grey [1]

MC: black [3]

M_6C: white [2]

At this wavelength the MC and M_6C carbides can be more easily differentiated than in the colour photograph Fig. 154.

156: Same specimen, same heat treatment and same preparation as in Fig. 154.

Photographed at 540 nm.

MC and M_2C: black [1]

M_6C: dark grey [2]

At this wavelength the carbide M_6C is somewhat darker than the matrix; the carbides MC and M_2C cannot be distinguished from one another.

These examples show that the use of monochromatic light sometimes yields better contrasts than are obtained in colour reproductions. Above all, the choice of wavelength determines which phase shall appear dark as compared with the others.

Martensite, residual austenite and carbides

Figures 157 to 162 illustrate examples of the differentiation between martensite, residual austenite and carbides, in some cases supplementary to section 2.1.1.

Figs. 157–161: **Chilled cast chromium steel containing 2.5% C, 17% Cr, 1% Ni and 0.5% Mo.**

As cast.

Same specimen as in Figs. 103 to 111.

157: Etched with a mixture of hydrochloric, acetic and picric acids.

Three phases can be distinguished: the matrix, a light precipitate and a dark precipitate. The microstructure is very difficult to interpret.

158: Same material, same heat treatment and same area of the microstructure as in Fig. 157.

Coated with ZnSe, zero-order.

Colour of the matrix: purple

Photographed at 520 nm.

martensite matrix: light grey

residual austenite: white [1]

eutectic carbides: black [2]

eutectic microstructure: dark grey [3]

After deposition of a coating layer it is possible to give a clear interpretation of the structure in Fig. 157.

Figs. 159–165: See pages 106 and 107

157 100 : 1

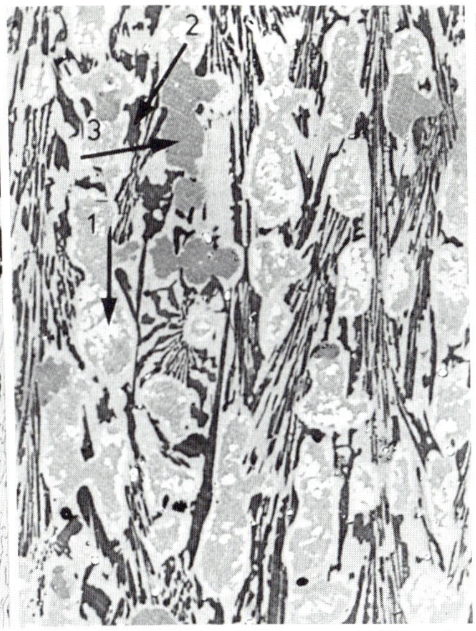

158 100 : 1

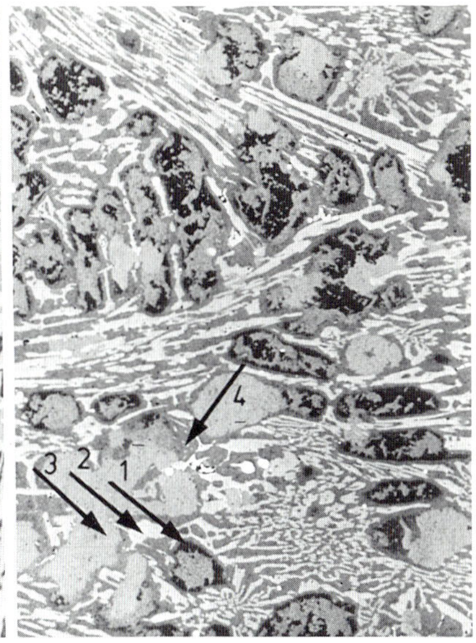

159 100 : 1

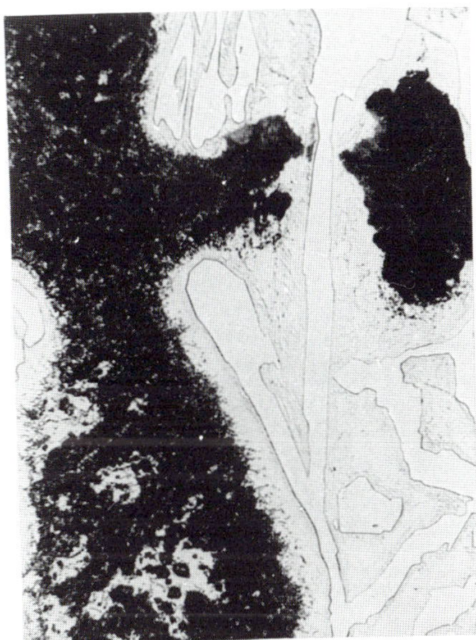

160 500 : 1

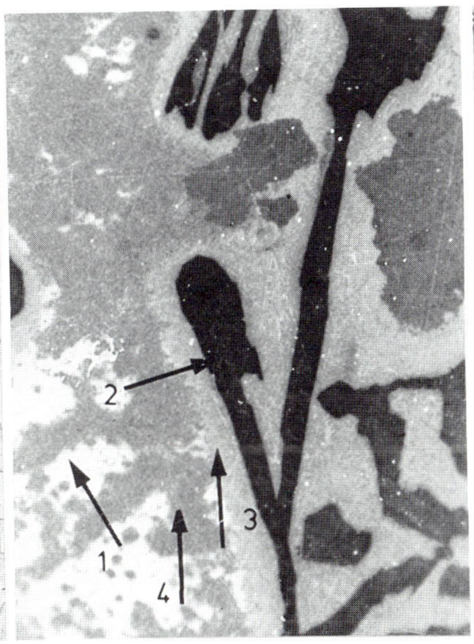

161 500 : 1

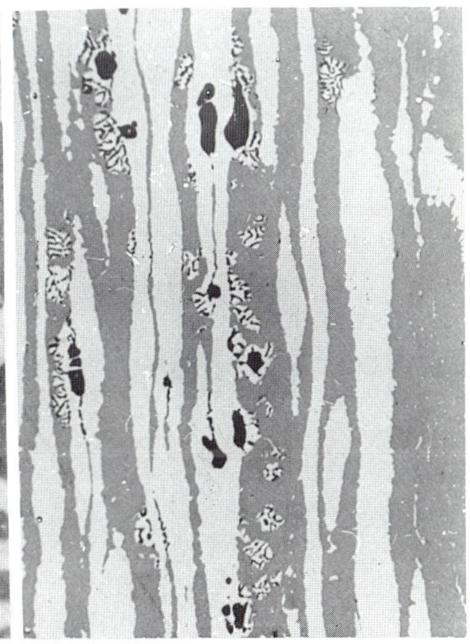

162 500 : 1

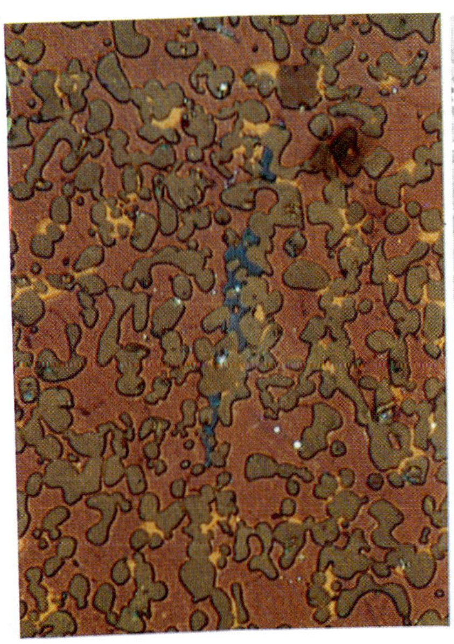

163 500 : 1

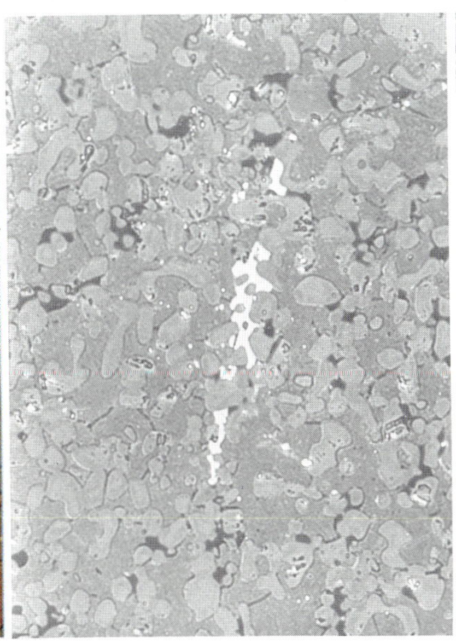

164 500 : 1

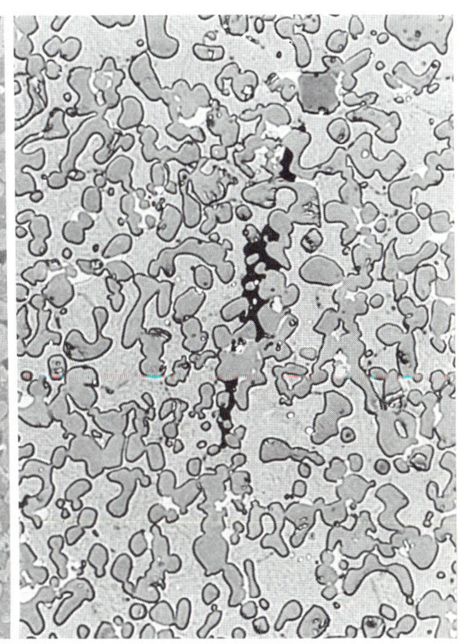

165 500 : 1

159: Same material and same heat treatment as in Fig. 157.

Coated with Sb_2S_3.

Colour of matrix: purple

Photographed at 520 nm.

martensite matrix: dark grey [4]

residual austenite: black [1]

eutectic carbides: white [2]

eutectic microstructure: light grey [3]

In comparison with Fig. 158 this example shows how the choice of layer material can be used to vary the reflectivity of the individual phases within wide limits. Again, with this coating layer, differentiation of phases and resolution are considerably better than in Fig. 157.

160: Same material, same heat treatment and same preparation as in Fig. 157.

Even at high magnifications the individual phases cannot clearly be differentiated after etching.

161: Same material, same heat treatment, same preparation and same photographic conditions as in Fig. 158.

Same area of the microstructure as in Fig. 160.

martensitic matrix: dark grey [4]

residual austenite: white [1]

eutectic carbide: black [2]

eutectic microstructure: light grey [3]

The individual phases can be distinguished clearly after deposition of a coating layer and the resolution is markedly improved.

Fig. 162: **Steel X 10 CrNi 30 9** (c.f. Fig. 102)

Sputtered with an iron cathode in oxygen, zero-order.

Colour of matrix: purple

Photographed at 510 nm.

austenite: light grey

ferrite: dark grey

σ-phase: black

This photograph again shows the good contrast that can be produced between austenite and ferrite. At the same time it is a first example of how satisfactorily coating layers can be used to contrast σ-phase in steels (see also Figs. 169 to 174).

Figs. 163–165: **Ferrotitanide**

Ferrotitanide is a material produced by powder metallurgy and containing about 50 vol.% of TiC, used mainly for parts liable to severe wear. Depending on the requirements, e.g. high temperature strength, corrosion resistance etc., the TiC is sintered with steel powder of the appropriate chemical composition, whereby the two can react to form other carbides. In addition, e.g. during subsequent heat treatments, carbides can form in the steel matrix. The microstructure of the high temperature material is a good example of the use of layers in the differentiation of several carbides. It has the composition (% by mass):

34.5% TiC, 0.75% C, 10.0% Cr, 3.0% Mo, 0.5% V, 0.8% Cu, 1.0% Mn, remainder iron.

The material is soft annealed and hardened: 1000^0C 2 h/furnace + 720^0C 6 h/furnace + 1050^0 C/air.

163: Material and heat treatment as above, coated with ZnSe, zero-order, bright field.

matrix: purple (6 : 3 : 5)

TiC: greenish (1 : 3 : 5)

M_7C_3: yellow (2 : 5 : 1)

M_3C: blue (17 : 4 : 4)

After coating, the individual phases in the material can clearly be distinguished by the colour.

164: Same specimen, same heat treatment, same preparation and same area of the microstructure as in Fig. 163.

Photographed at 470 nm.

The individual phases can also be distinguished easily in the black and white photograph.

165: Same specimen, same heat treatment, same preparation and same area of the microstructure as in Fig. 163.

Photographed at 590 nm.

A comparison of Figs. 163, 164 and 165 shows that the phases can be distinguished just as well in a black and white as in a colour photograph and that by choosing illumination of a suitable wavelength the relative brightnesses of the individual phases can be varied over a wide range.

Carbonitrides in austenitic cast alloys

Figs. 166–168: **Steel X 45 CrNiSi 25 30.**

Heat treatment: 45000 hours service in nitriding atmosphere.

166: Sputtered with an iron cathode in oxygen, bright field.

austenite matrix: light brown (4 : 3 : 3)

$M_{23}(C,N)_6$: yellowish red (6 : 4 : 3) [1]

$M_2(C,N)$: red (9 : 3 : 4) [2]

$M_6(C,N)$: violet red (9 : 3 : 5) [3]

During the anneal, the carbides have taken up nitrogen. Other carbides containing nitrogen have precipitated in the vicinity of the $M_{23}C_6$ carbides. In the colour photograph, $M_6(C,N)$ can scarcely be distinguished from $M_2(C,N)$.

167: Same material and same heat treatment as in Fig. 166.

Coated with ZnSe, zero-order.

Photographed at 535 nm.

austenite: white

$M_{23}(C,N)_6$: light grey [1]

$M_2(C,N)$: medium grey [2]

$M_6(C,N)$: dark grey [3]

Using a different type of layer and monochromatic light, very good contrasts are obtained, in some cases better than in the colour photograph in Fig. 166.

168: Same matrial and same heat treatment as in Fig. 166.

Sputtered with a copper cathode in argon.

Photographed at 570 nm.

The individual phases have the same grey tone as in Fig. 167. A comparison of the photographs shows that similar contrasts can be achieved between individual phases using quite different coating techniques.

σ-phases in steel

σ-phase can appear after prolonged annealing, particularly in steels with a high chromium content. It is generally undesirable on account of its brittleness and its predisposition to precipitate at grain boundaries. σ-phase can be revealed by etching with NaOH 16). A simple method that simultaneously reveals other phases is the production of a surface layer. An example of this has already been seen in Fig. 162. σ-phase has a tetragonal lattice and should be optically anisotropic although this cannot be observed under normal conditions. Only after the deposition of a layer did this anisotropy first became observable 17).

Fig. 169: **Iron-chromium alloy containing 2.6% Cr.**

Heat treatment: 500^0C 250 h/air.

Coated with TiO_2, zero-order, bright field.

iron-chromium solid solution: white

$M_{23}C_6$: grey

σ-phase: black

σ-phase precipitates are coarse and occur preferentially at grain boundaries. They stand out in good contrast to the matrix and carbides.

Figs. 170–177: See page 110

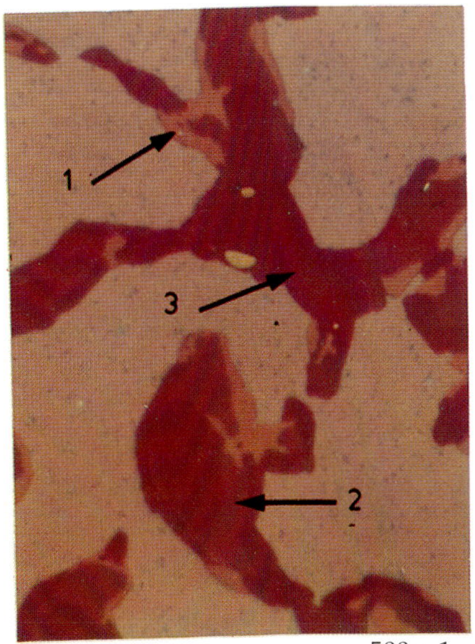

166 500 : 1

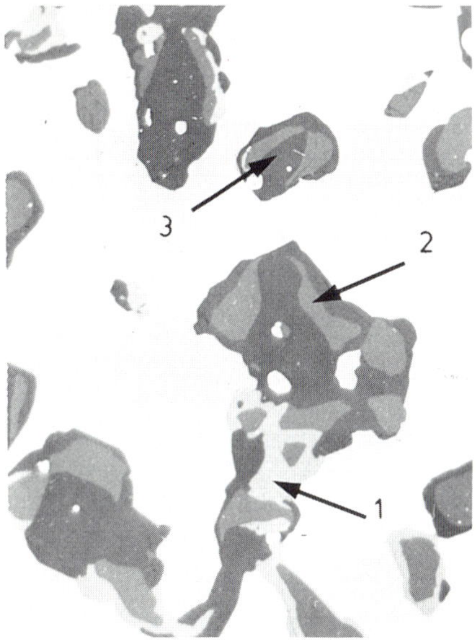

167 500 : 1

168 500 : 1

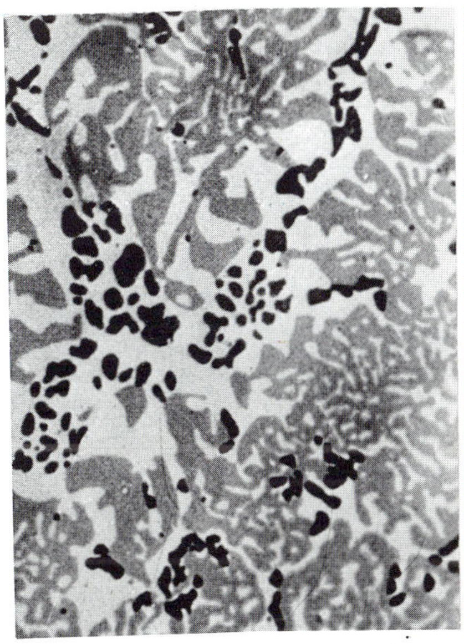

169 100 : 1

170 100 : 1

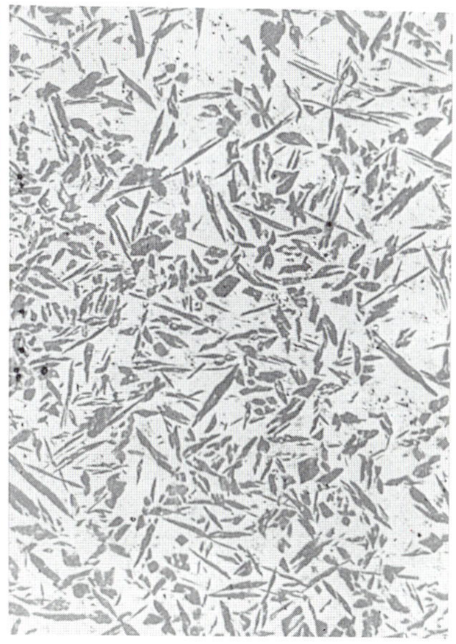

171 200 : 1

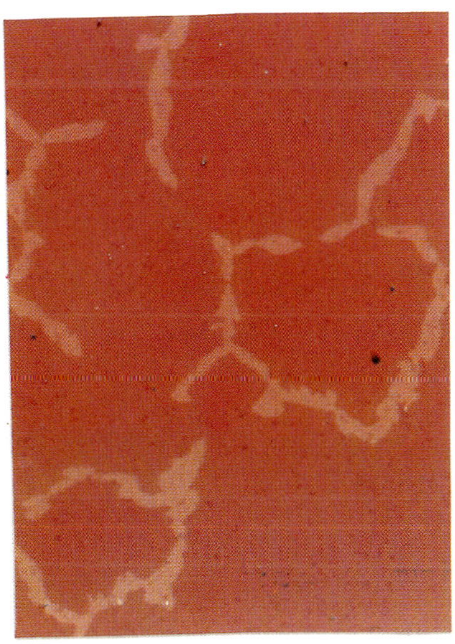

172 200 : 1

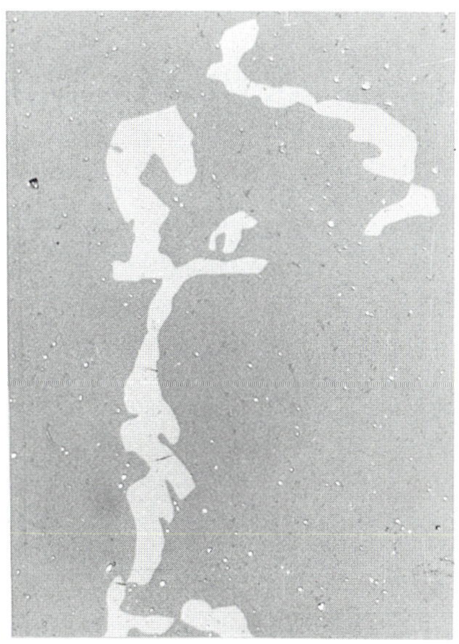

173 500 : 1

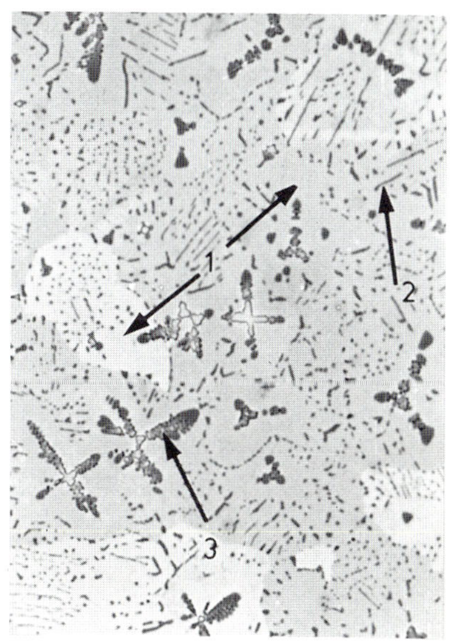

174 500 : 1

Figs. 170 and 171: **Iron-chromium alloy with 30% Cr and 3% Si.**

170: Heat treatment: 1300°C 20 h/air + 800°C 5 h/air under cyclic stress.

Sputtered with a titanium cathode in air.

matrix: iron chromium solid solution: orange (6 : 3 : 3)

σ-phase: red (8 : 4 : 4)

By deforming the specimen while precipitation takes place, the σ-phase can be made to precipitate uniformly within the grains.

171: Same specimen, same heat treatment and same preparation as in Fig. 170.

Photographed at 550 nm.

In this photograph a very good contrast between the σ-phase and the matrix has been achieved by the use of monochromatic light.

Figs. 172 and 173: **Model alloy containing 70% Fe and 30% V.**

Heat treatment: 1010°C 5 h/water.

172: Sputtered with a copper cathode in air, zero-order.

matrix: iron-vanadium solid solution: red (8 : 3 : 3)

σ-phase: bright red (7 : 2 : 3)

σ-phase appears in several iron systems. In addition to the technically most important iron-chromium system, the iron-vanadium system has been selected as an example. During the anneal at 1010°C, σ-phase has precipitated at the grain boundaries.

173: Same material, same heat treatment and same preparation as in Fig. 172.

Photographed at 550 nm.

The rather poor colour contrast in Fig. 172 can be improved to a good black and white contrast by use of a monochromatic filter.

174: **Iron-vanadium alloy containing 60% V.**

Heat treatment: 900°C 15 min./water

matrix: iron-vanadium solid solution: dark grey to light grey [1]

σ-phase: fine black precipitates [2]

slag inclusions: large, interlaced, dendritic, black to white precipitates [3].

After sputtering with a copper cathode, the σ-phase was not visible. The precipitates are so fine that they can no longer be resolved. This photograph shows the limits of the interference layer technique. In this case etching would be preferable although it results in an exaggeration of the proportion of σ-phase in the microstructure.

2.1.3 High temperature nickel base alloys

High temperature nickel alloys are hardened by the γ′-phase which is coherent with the face centred cubic nickel solution (γ). As the proportion by volume of γ′-phase increases, a corresponding increase in the high temperature fatigue strength is also observed provided that the precipitates do not exceed a certain size. Cast alloys for highly stressed turbine blades operating at around 900°C therefore contain more than 50 vol.% of γ′-phase. In addition, these alloys usually contain carbides.

A high temperature cast nickel alloy has been selected as an example. As well as γ and γ′-phases, the primary carbides MC and M_6C are present in the as cast microstructure. During heat treatment or stressing at high temperatures, the MC is converted by a reaction with the nickel solid solution to γ′-phase, M_6C and $M_{23}C_6$.

Literature to 2.1.3

1) „The Superalloys", ed. by C. T. Sims and W. C. Hagel, John Wiley & Sons, New York (1972)

2) Schubert, F.: Untersuchungen der das Zeitstandverhalten oberhalb 900°C von technischen hochwarmfesten Nickelgußlegierungen bestimmenden Einflußgrößen. Thesis, Technische Hochschule Aachen (1974)

3) Jackson, J. J., Donachie, M. I., Henricks, R. J. and Gell, M.: Einfluß des prozentualen Volumens feinster γ′-Ausscheidungen auf das Kriechen in gerichtet erstarrtem (DS) Mar – M 200 + Hf. Metallurg. Trans. **8A** (1977) 1615

4) von den Steinen, A. and Schubert, F.: Hochwarmfeste Nickel- und Kobaltlegierungen.
in: Werkstoffkunde der gebräuchlichen Stähle, Part 2.
Verlag Stahl und Eisen, Düsseldorf (1977)

5) Hammond, C. and Nutting, J.: Die physikalische Metallurgie von Superlegierungen und Titanlegierungen. Metal. Sci. **11** (1977) 474

6) Schubert, F.: Hochwarmfeste Legierungen.
in: Festigkeitsverhalten bei höheren Temperaturen.
Beitrag zum Kontaktstudium Werkstoffkunde Eisen und Stahl IV.
Verlag Stahleisen (1978)

Figs. 175–177: **Cast nickel alloy containing 10.2% Co, 7.8% Cr, 13.8% W, 5.3% Al, 5.3% Ta, 0.13% C, 0.05% Zr, 0.017% B and 0.34% Fe.**

175: Heat treatment: 1000^0C, 1000 h/air.

Coated with ZnSe, first order.

After this heat treatment, the specimen is overaged. An area has been selected in which the carbide decomposition is not too far advanced. After coating with ZnSe, the background γ solid solution appears flesh-coloured and the γ'-phase in the background is bright red. The large carbide particle in the centre of the picture, yellow (4 : 5 : 3), is M_6C. The elongated dark carbides (8 : 3 : 4) have the MC structure. Precipitates of $M_{23}C_6$ (yellow specks) can be seen at the top left. It is easier to distinguish the phases in the monochrome photographs (Figs. 176 and 177).

176: Same specimen, same preparation, same area of the microstructure.

Photographed at 470 nm.

The γ and γ'-phases are scarcely distinguishable. M_6C and $M_{23}C_6$ are black, MC is dark grey. As a consequence of the decomposition described above the $M_{23}C_6$ carbides are located mainly in the vicinity of the MC carbides.

177: Same specimen, same preparation and same area of the microstructure as in Fig. 175.

Photographed at 560 nm.

The γ-solid solution and the $M_{23}C_6$ carbides are white. M_6C and γ'-phase are grey. MC is black. The ends of the MC carbides are not deep black because a change in the chemical composition of the carbides has begun there as a result of reaction with the γ solid solution.

Figs. 178–180: **NIMOCAST Alloy 739 (0.15% C, 22.4% Cr, 19.0% Co, 2.0% W, 1.0% Nb, 1.4% Ta, 3.7% Ti, 1.9% Al, 0.1% Zr, 0.009% B, remainder Ni)**

178: Heat treatment: 1160^0C, 4 h/air $+1000^0$C 6 h/air $+900^0$C 24 h/air $+700^0$C 16 h/air.

Electrolytic etch, 30 s in H_3PO_4; 3 V; 0.16 A.

The roughening of the background corresponds to the dispositions of γ and γ'-phases. The γ'-phase is resolved in only a few places as fine bright dots. The black patches are particles which dissolved out during etching. In addition to the matrix, light coloured phases are visible, but it is difficult to distinguish between them.

179: Same specimen, same area of the specimen as in Fig. 178.

Unetched, sputtered for 1.5 min with a Cu-cathode in air, bright field.

The γ'-phase is too finely dispersed to be visible. This indicates that in Fig. 178, the particles appeared bigger than they really were due to the effects of etching. The other phases can be distinguished considerably better than in Fig. 178, but they still cannot be identified individually.

180: Same specimen, same preparation and same area of the microstructure as in Fig. 179.

Photographed with an oil immersion objective.

Magnified section of Fig. 179. The colours have been changed by the oil immersion objective. The phases can be distinguished very clearly.

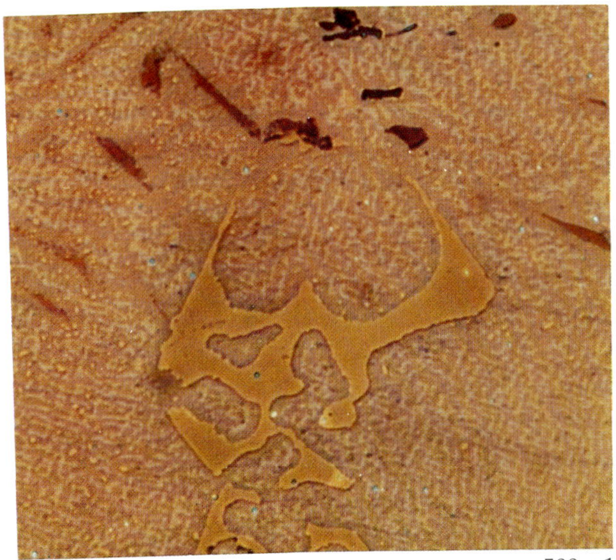

175 500 : 1

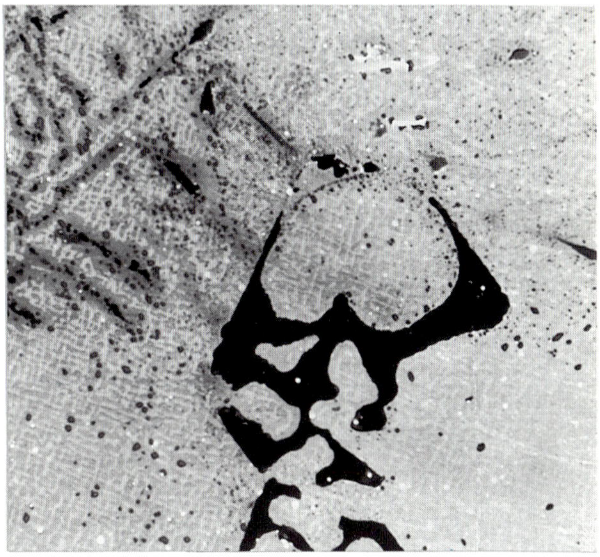

176 500 : 1

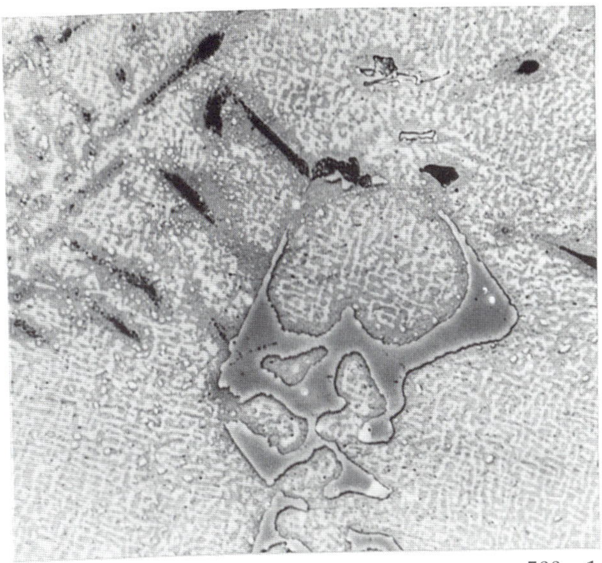

177 500 : 1

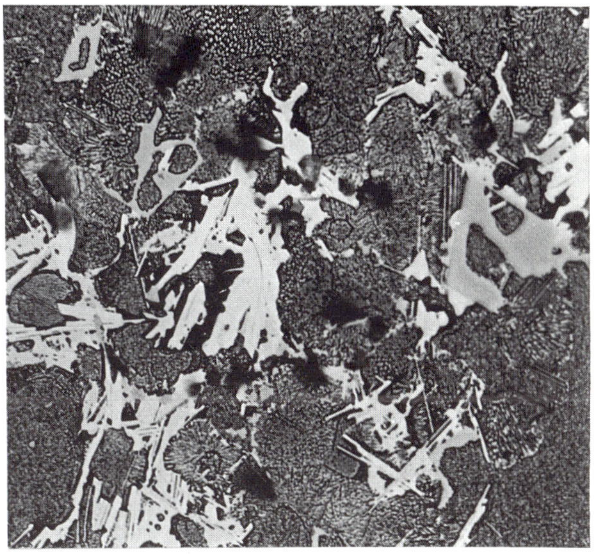

178 650 : 1

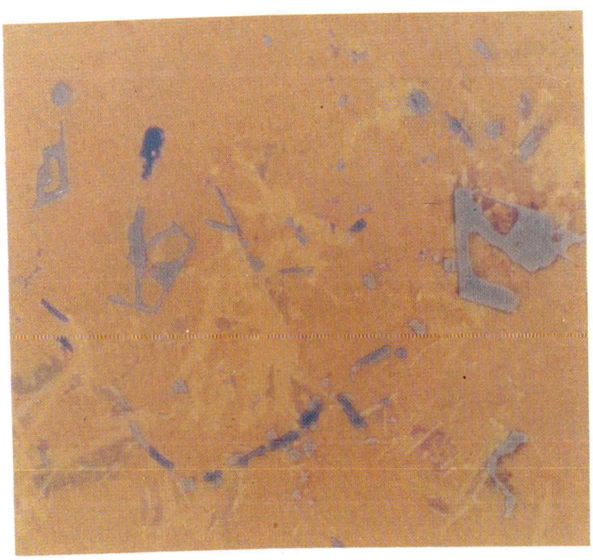

179 650 : 1

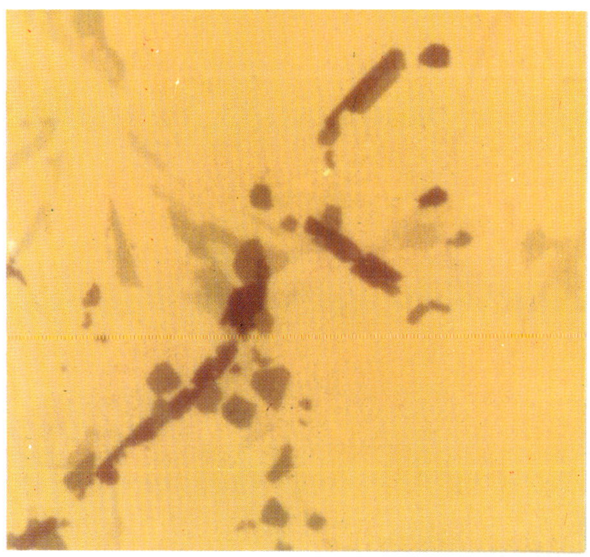

180 1700 : 1

2.1.4 Hard metals

2.1.4.1 General introduction

Hard metal alloys are important examples of composite materials produced powder metallurgically by liquid phase sintering. Carbides, principally of the elements tungsten, titanium, tantalum, niobium and hafnium are employed for the hard metal. The most common alloys usually contain tungsten monocarbide (WC) with the other elements mainly in the form of carbide solid solutions. Cobalt or nickel are the most common bonding agents. A wide range of properties can be obtained by altering the proportions of the various phases – commercial alloys contain between 3 and 30% of the bonding metal – and by suitable choice of sintering conditions.

Coated hard metals must also be included in a metallographical study of the hard metals. Single or multiple layers only a few µm thick consisting of carbides and/or nitrides, mainly of titanium are deposited on the hard metal part to increase the wear resistance. Following the latest developments, Al_2O_3 is also used as the outermost layer.

Hard metals can thus be seen as model materials for the large class of composites (see also section 2.1.8) in which the desired properties can be achieved in the finished product by suitable choice of the individual components. In essence, the metallography of sintered composite hard metals in both the coated and uncoated condition involves investigations into microstructure, pore volume and defects liable to cause failure.

As regards the production of optimum contrasts for microstructural investigations using the light microscope, one fundamental limitation must be mentioned, namely, the often very small size of the carbide particles (< 0.5 µm) and in particular the thinness of the intermediate layers of the bonding phase (< 0.01 µm) which in many cases necessitate the use of electron microscopy. An investigation of the microstructure by optical microscopy is therefore impossible or at best very unsatisfactory for a large proportion of the commercially manufactured hard metal alloys. In the case of the intermediate layers, however, the production of oblique sections can be helpful (c.f. section 2.1.4.5).

2.1.4.2 Microstructure

A number of etching solutions have proved valuable for the metallographic investigation of the simplest WC-Co alloys and these are also used in routine investigations (quality control). No advantage is gained by using interference layer techniques.

In the case of hard metal alloys containing mixed carbides in varying proportions, multiple stage etching is often successful but interference layer methods offer decided advantages: in most cases all the phases in multicomponent alloys – possibly even inhomogeneities in the mixed carbides (e.g. (Ti,W)C) – can be shown up simultaneously in good contrast. This provides not only a quick general idea of the microstructure; the specimens are also in a suitable condition for an automatic quantitative microstructural analysis (c.f. Figs. 189–192).

The microstructure of hard metal alloys containing mixed carbides can also be shown up by a temper etch (c.f. Figs. 181 to 184) – this method has been used successfully for years – but in the majority of cases the contrasts between the different phases are inferior to those obtained with interference layers: in particular slight variations in the composition of the mixed carbides are not revealed. An example demonstrating the possibilities and limitations of temper etching is given in Figs 183 and 184. In particular, attention should be drawn to one decisive disadvantage of temper etching, namely that in this case it is not possible to distinguish with sufficient reliability between Co-regions (black) and pores.

The use of monochromatic light of various wavelengths, best obtained by means of a graduated interference line filter, not only brings about a distinct improvement in the contrast between the individual phases, it also permits any one of the phases present to stand out from all the other phases and to appear alone on the corresponding black and white photograph. In this way pictures can be obtained giving an overall view of the distribution of the individual phases and thus of the homogeneity of the specimen. This type of picture is particularly suitable for quantitative evaluations including the use of image analysers. An example of this is a hard metal containing mixed carbides coated with ZnSe shown in Figs. 189–192.

2.1.4.3 Porosity

In determinations of the porosity, doubts often arise, even when the polished sections have been carefully prepared, as to whether the pores are genuine or in fact graphite particles or non-metallic inclusions of similar size. This is particularly true where matrix porosity of type A of the ASTM classification is involved. In such cases interference layers can often bring clarification. Whereas true pores still appear dark or black after deposition of an interference layer, impurity phases appear bright (c.f. examples in section 2.2.3).

2.1.4.4 Material failure

Hard metal alloys sometimes contain embrittling, usually carbon-deficient carbide phases which cannot always be identified definitively by etching. The use of interference layer techniques yields unequivocal results in such cases. Fig. 195 shows a particularly well marked defect: the presence of η-phase in the vicinity of WC, TiC and cobalt.

2.1.4.5 Coated hard metals

The enormous diversity of phases occurring at the interface between the hard metal and its wear resistant protective layer makes a metallographic assessment using etching very difficult. Once again in this special field, which has practical importance for adhesion of layers, the advantages of the interference layer technique must be exploited.

Investigations of coated hard metal specimens are carried out preferably on $5^0 50'$ oblique sections (10 : 1 distortion).

Fig. 193 illustrates the microstructure and the layer structure of a hard metal specimen with multiple coatings by means of an oblique section.

Recently attempts have been made to produce wear resistant layers based on TiN by high pressure treatment with nitrogen at high temperatures. The only way that the resulting diffusion zones and TiN-rich layers can be rendered visible is by means of interference layers. Wear resistant layers of this type can be shown up in good contrast in oblique sections (Fig. 194).

Literature to 2.1.4

1) Pepperhoff, W.: Gefügeentwicklung durch Interferenz-Aufdampfschichten.
Arch. Eisenhüttenwes. **32** (1961) 269

2) Peter, W., Kohlhaas E. and Jung, O.: Sichtbarmachung der Gefüge von Hartmetallen durch Interferenz-Aufdampfschichten.
Prakt. Metallographie **4** (1967) 284

3) Arndt, R. and Hillnhagen, E.: Spezielle Verfahren zur metallografischen Untersuchung von Hartmetallen.
Prakt. Metallographie **9** (1972) 309

4) Kohlhaas, E. and Fischer, A.: Kontraststeigerung beim Interferenzschichten-Verfahren durch die Verwendung eines Verlauf-Linienfilters.
Prakt. Metallographie **6** (1969) 339

5) Kohlhaas, E. and Jung, O.: Metallographische Untersuchungen an Karbidpulvern für Hartmetalle.
Prakt. Metallographie **5** (1968) 552

Figs. 181 to 188: Use of temper etching and coating for revealing microstructures of hard metals

Figs. 181 to 188 compare the development of the microstructure on a temper etched hard metal specimen and a sputtered specimen of the same material.

181: Temper etched (3% nitric acid, 2% sodium carbonate and 10 min 500^{0}C/air), observed in bright field.

tungsten carbide: white

cobalt: very finely dispersed, black

(Ti,Ta)C: medium grey, various grey levels due to changing TiC-content.

182: Same specimen, same area of the microstructure and same etch as in Fig. 181, bright field.

tungsten carbide: light yellow (2 : 6 : 2)

cobalt: light yellow (2 : 2 : 1)

(Ti,Ta)C: dark brown to dark yellow, (2 : 5 : 2) to (5 : 4 : 4)

183: Same specimen and same treatment as in Fig. 181, but with a thicker temper etched layer.

tungsten carbide: white

cobalt: finely dispersed, dark grey

(Ti,Ta)C: medium grey

The cobalt and the (Ti,Ta)C are difficult to distinguish.

184: Same specimen, same treatment as in Fig. 181, and same coating layer as in Fig. 183.

tungsten carbide: reddish brown (5 : 2 : 2)

cobalt: red (8 : 5 : 3)

(Ti,Ta)C: violet (11 : 3 : 4)

Figs. 185–188: See page 118

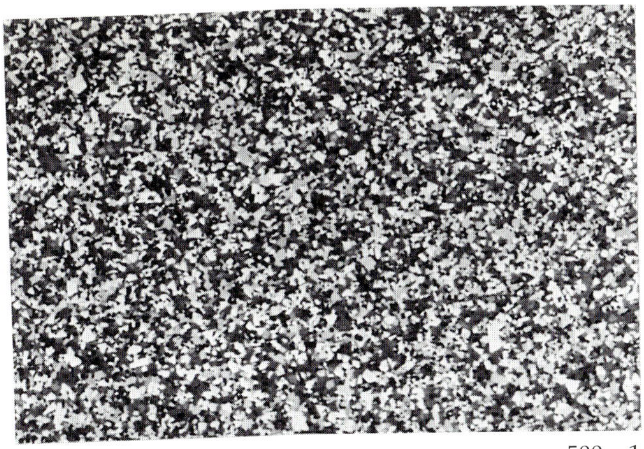

181 500 : 1

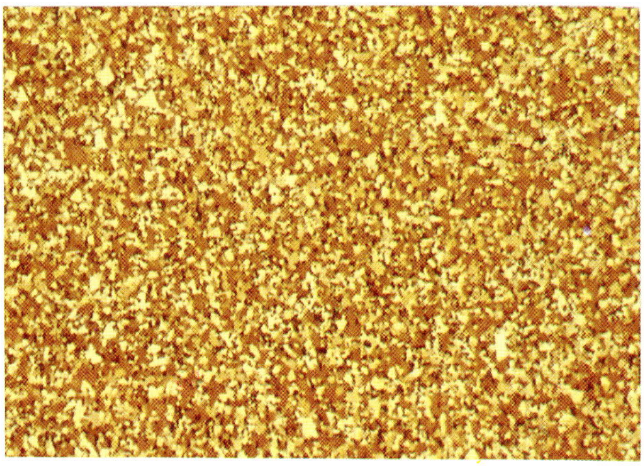

182 500 : 1

183 500 : 1

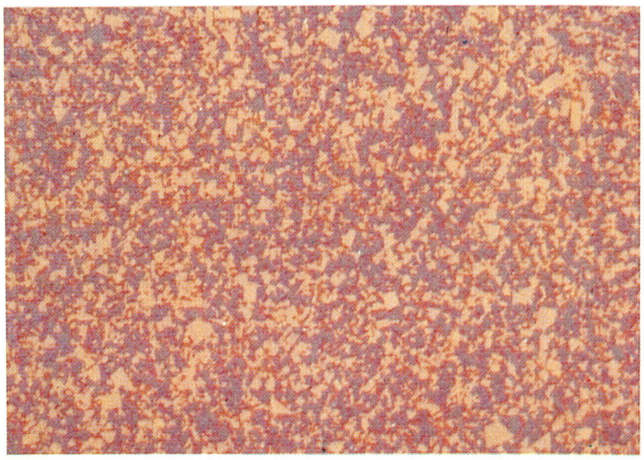

184 500 : 1

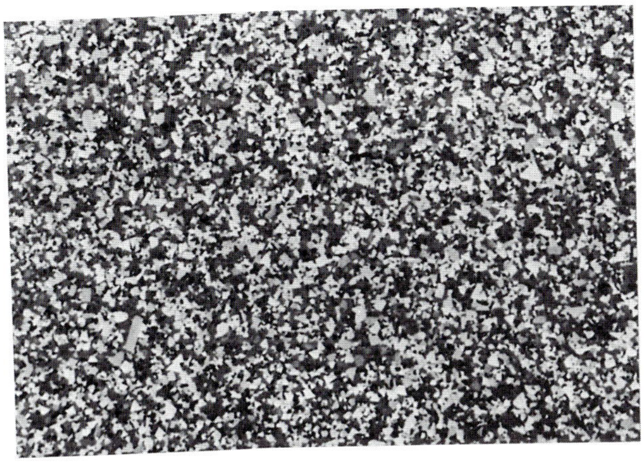

185 500 : 1

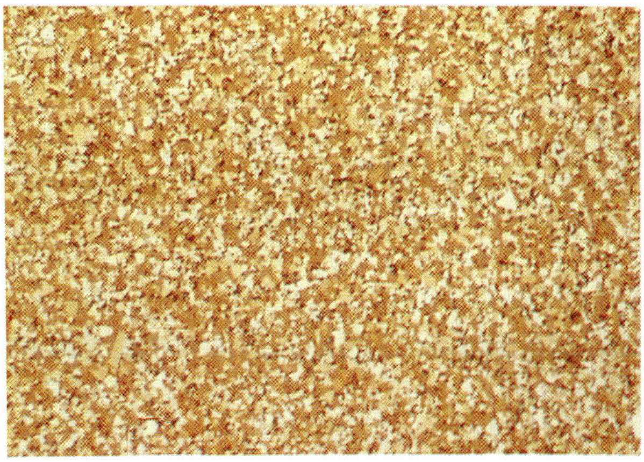

186 500 : 1

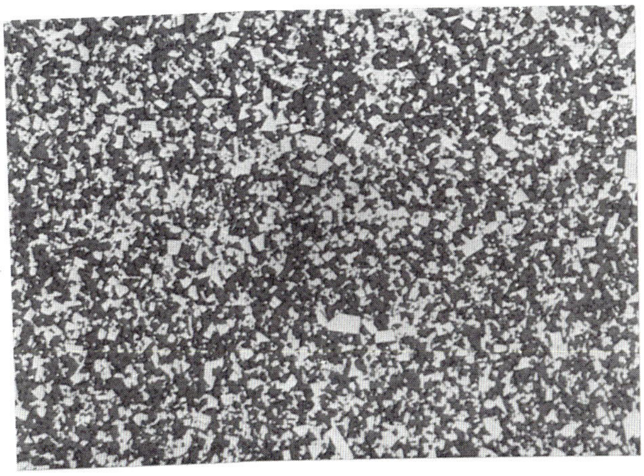

187 500 : 1

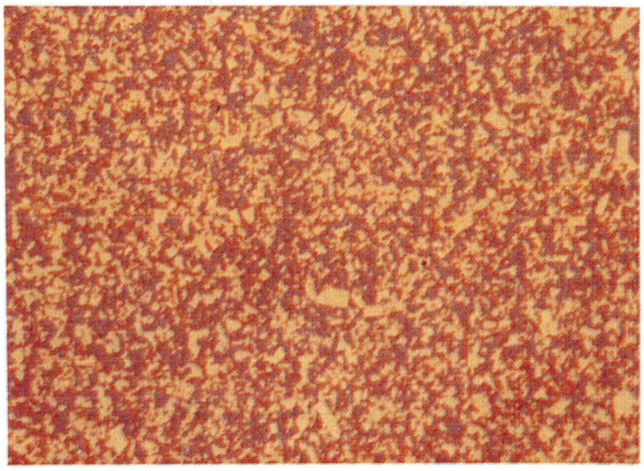

188 500 : 1

185: Same specimen and same treatment as in Fig. 181, but coated with a sputtered layer using an iron cathode and oxygen as reaction gas, bright field.

tungsten carbide: medium grey to white

cobalt: black

(Ti,Ta)C: dark to medium grey, cf. Fig. 181.

186: Same specimen, same treatment as in Fig. 181 and same specimen preparation as in Fig. 185.

tungsten carbide: light yellow (2 : 2..3 : 1)

cobalt: dark red (8 : 3 : 5)

(Ti,Ta)C: dark brown to dark yellow (2 : 4 : 2) to (4 : 4 : 3)

187: Same specimen and same treatment as in Fig. 181, but with a thicker layer than in Fig. 185, bright field.

tungsten carbide: white

cobalt: dark grey

(Ti,Ta)C: dark grey

In parts the tungsten carbide and the cobalt cannot be differentiated.

188: Same specimen, same treatment, same area of the microstructure and same specimen preparation as in Fig. 187.

tungsten carbide: yellow (4 : 5 : 2)

cobalt: red (8 : 5 : 3)

(Ti,Ta)C: violet (11 : 3 : 4)

An unequivocal assignment of colour code was very difficult. The results of this comparative study of temper etched and sputtered specimens demonstrates that as the layer thickness increases (Figs. 181 and 183) the micrographs of the temper etched specimens are less well contrasted because of the increased self absorption. The sputtered specimens (Figs. 185 to 188) show similar contrasts to those produced by temper etching. A comparison of Figs. 181 and 183 and of 185 and 187 shows that for black and white photographs taken in bright field, the layer thickness must be set at an optimum value. The contrasts can be accentuated further by the use of monochromatic light. More definite results may be obtained by deposition of absorption-free ZnSe, cf. Figs. 189 to 192.

Figs. 189 to 192: **Hard metal WC-(Ti,Ta)C-Co,**
coated with a vacuum deposited layer of ZnSe, zero order. After coating the section appears purplish.

189: Photographed in bright field, coated as described. The individual phases appear in the following grey tones:

tungsten carbide: grey

cobalt: white

(Ti,Ta)C: black

Figs. 190–195: See page.120

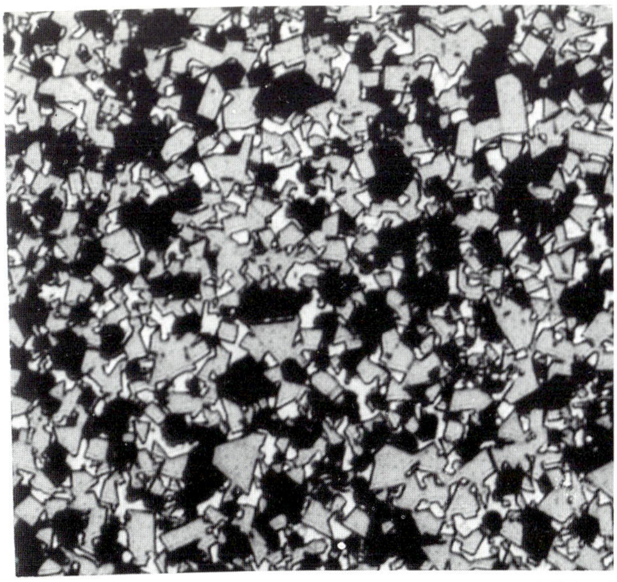

189 1500 : 1

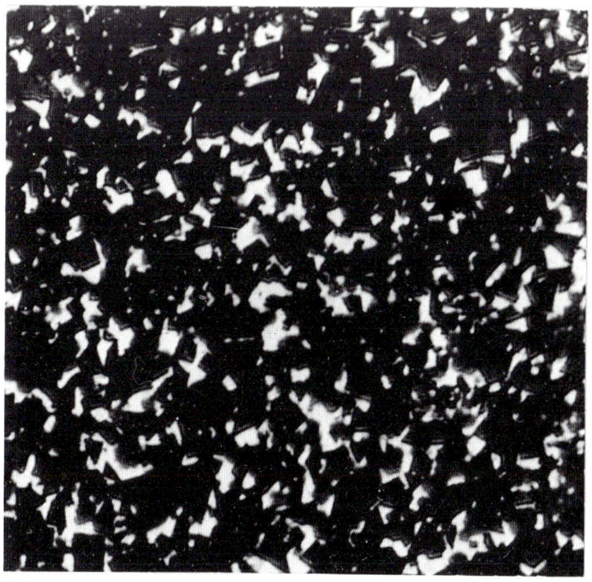

190 1500 : 1

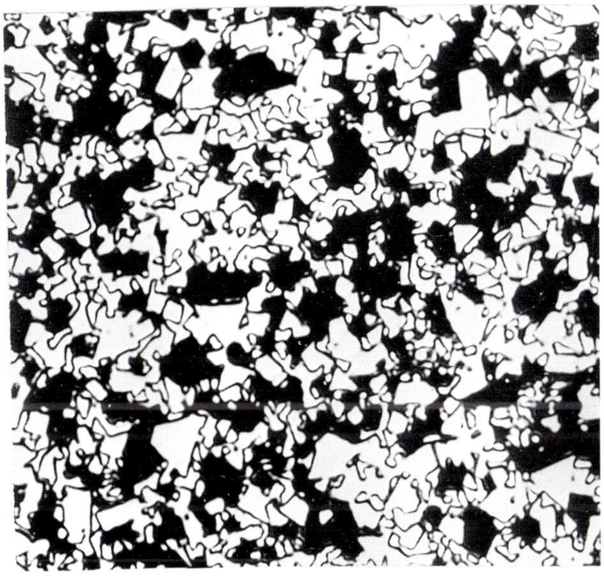

191 1500 : 1

192 1500 : 1

193 1500 : 1

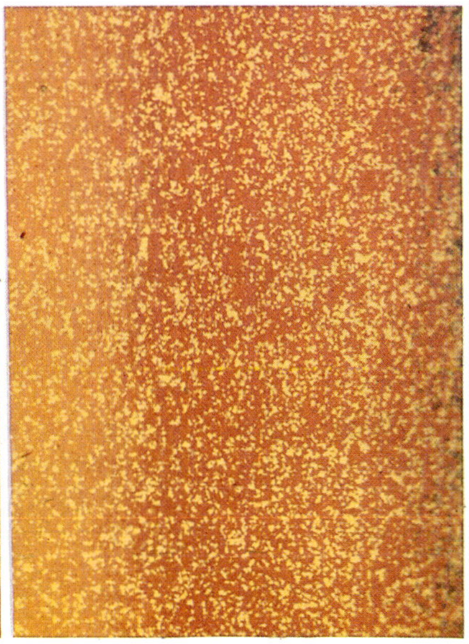

194 1500 : 1

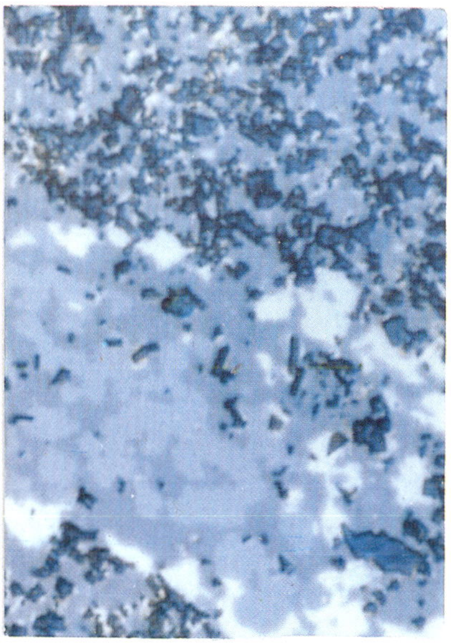

195 1500 : 1

190: Same specimen, same area of the microstructure and same coating as in Fig. 189.

Photographed at 405 nm.

The cobalt bonding phase stands out white against the other microstructural components, which cannot be differentiated.

191: Same specimen, same area of the microstructure and same coating as in Fig. 198.

Photographed at 436 nm.

In this micrograph the tungsten carbide and cobalt bonding phase are white. (Ti,Ta)C is black.

192: Same specimen, same area of the microstructure and same coating as in Fig. 189.

Photographed at 520 nm.

Tungsten carbide is the only black phase and can be quantitatively evaluated in relation to the other microstructural components which do not differ in reflectivity at this wavelength.

Fig. 193: **Hard metal specimen with wear resistant layer.**

Oblique section, vacuum deposited ZnSe layer, zero order, bright field.

The specimen contains the following phases, from left to right:

hard metal (multiphase)

titanium carbide: yellowish brown (3 : 6 : 2)

titanium nitride: red (7 : 5 : 3)

Fig. 194: **Wear resistant layer based on titanium nitride produced by high pressure treatment with nitrogen at high temperatures.**

Oblique section, vacuum deposited ZnSe layer, zero order, bright field.

On the left hand side of the picture, hard metal with a predominance of (Ti,Ta)C: yellowish brown (3 : 6 : 2)

In the centre of the picture an intermediate layer rich in titanium nitride: red (7 : 5 : 3)

Fig. 195: **P25-Hard metal with a brittle phase.**

Section coated with vacuum deposited ZnSe, zero order, bright field.

η-phase: light blue (predominantly in the centre of the figure: 17 : 1 : 2)

tungsten carbide: dark blue (17 : 4 : 4)

titanium carbide: medium blue (17 : 2 : 2)

cobalt: white

2.1.5 Copper alloys

2.1.5.1 General introduction

Unlike most metals, copper exhibits strong dispersion of the reflectivity and of the optical constants, namely refractive index and absorption coefficient, in the region of the visible spectrum. This is expressed in the characteristic colour of the metal and to some extent also of its alloys, e.g. bronzes and brasses.

The optical constants of copper are summarised in Figures 196–198. These show that the reflectivity increases from 50 to about 85% in the region of the visible spectrum between 450 and 650 nm. The refractive index, on the other hand, decreases from about 1.2 to values of 0.1, whilst the absorption coefficient shows a slight increase from 2.1 to 3.5. The rather large scatter in some of the experimental values can safely be attributed to differences in the specimen preparation or chemical purity of the metals employed.

In general, the microstructure of copper alloys can be revealed perfectly well using conventional metallographic techniques. The microstructure can be described quite adequately thanks to the characteristic colours of the different phases and the ease with which they can be etched. Accentuation of the contrast of the microstructure by surface layers, produced either by evaporation or by gas-ion-reaction, is of no particular advantage except in special cases where chemical attack of the specimen surface is to be avoided.

There are however a series of examples in which interference layer microscopy can be employed successfully in the identification of phases and in the production of contrast. The following examples are taken from the copper-indium-tin system and from the aluminium bronzes.

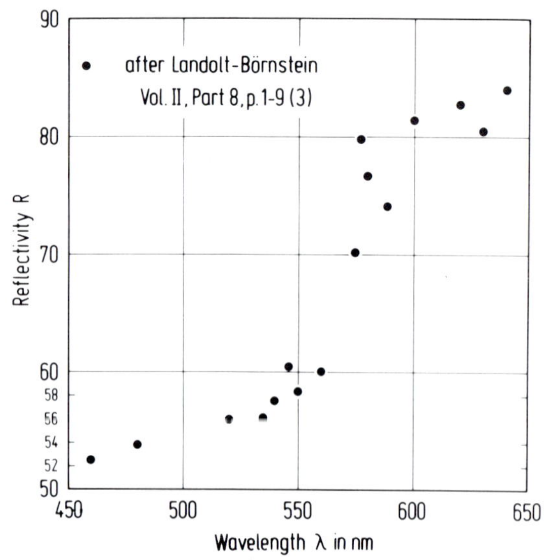

Fig. 196: Reflectivity of copper in the visible region of the spectrum.

Fig. 197

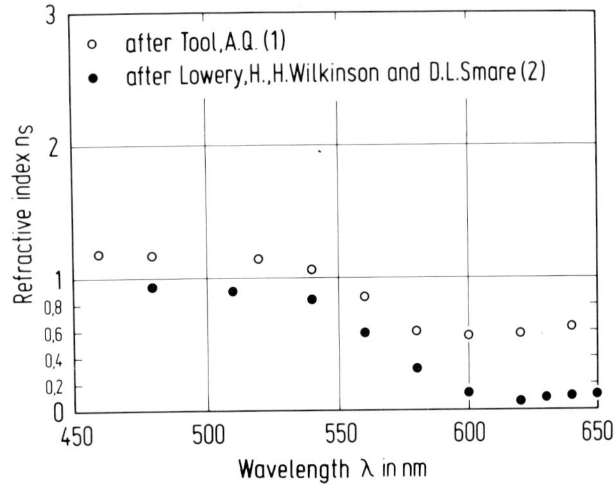

Fig. 198

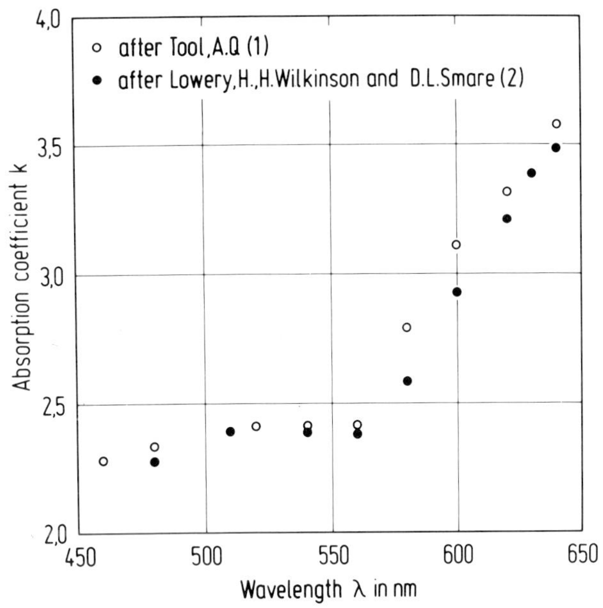

Figs. 197 and 198: Refractive index (Fig. 197) and absorption coefficient (Fig. 198) of copper in the visible region of the spectrum.

Literature to 2.1.5

1) Tool, A. Q.: Phys. Rev. **31** (1910) 1

2) Lowery, H., Wilkinson, H. and Smare, D. L.: Reference see under 3)

3) Landolt-Börnstein, New Series, Vol. II, Part 8, pp. 1–9 (on optical constants, see here for further references)

4) Schlüter, P., Aldinger, S. and Elssner, G.: Vergleich der Kontrastierung von Phasen in Nichteisenmetallen und Verbundsystemen durch Gasionenbeschichten und durch chemisches Ätzen.
Prakt. Metallographie, Special Volume **9,** ed. by W.-U. Kopp and H.-E. Bühler, Dr. Riederer-Verlag Stuttgart (1978) 57

5) Roberts, E. F. J. and Clarke, K. M.: The colour characteristics of gold alloys, Gold Bulletin **12** (1969) 9

2.1.5.2 Copper alloys, copper-indium-tin system

Figs. 199–205: **Alloy with 77% Cu, 21% In and 2% Sn.**

199: As cast, polished, photographed at 640 nm.

The as cast microstructure, produced by slow cooling from the melt, consists of a $Cu_{11}In_2Sn$ phase (light grey), α-Cu (dark grey), [1], and a eutectic consisting of α-Cu (dark grey) and Cu_7Zn_3 (black) [2]. In this photograph the eutectic is very difficult to see (cf. Figs. 201 and 203). In addition, black pores are visible, as also in Figs. 200–205.

200: Same alloy and area of microstructure as in Fig. 199, polished, bright field. In bright field, the colours are pale and the contrast is poorer than in Fig. 199.

201: Same alloy and area of microstructure as in Fig. 199. Etched with a solution of 100 ml methanol, 5 ml hydrochloric acid and 1 g picric acid.

Compared with Fig. 199, chemical etching has improved the contrast, particularly between the Cu_7In_3 phase and α-Cu, with the result that the eutectic shows up more clearly [2].

202: Same alloy, same area of microstructure and same preparation as in Fig. 201, bright field. This shows an improvement over Fig. 200. The eutectic stands out from the α-Cu.

203: Same alloy and area of microstructure as in Fig. 199. The specimen has been sputtered with an iron cathode using oxygen as reaction gas. Photographed at 540 nm. There is a considerable improvement in the contrast as compared with Fig. 199. The eutectic stands out clearly from the α-Cu.

204: Same alloy, same area of microstructure and same surface layer as in Fig. 203, photographed at 520 nm. The light-dark contrast is very strong at a wavelength of 520 nm. The intermetallic compound $Cu_{11}In_2Sn$ [3] appears black against the other microstructural constituents. The Cu_7In_3 [2] in the eutectic stands out white against the light grey α-Cu [1].

205: Same alloy, same area of microstructure and same surface layer as in Fig. 203, bright field. The different phases appear in the colours:

α-Cu: pale red (9 : 2 : 3)

Cu_7In_3: yellow (2 : 3 : 1)

$Cu_{11}In_2Sn$: dark red (10 : 3 : 3)

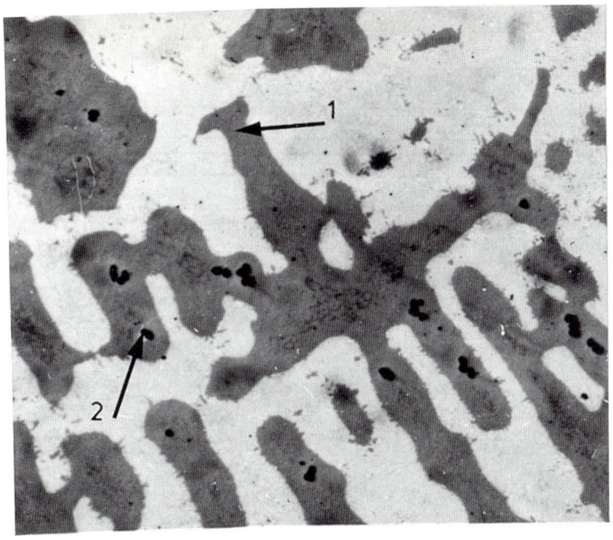

199 200 : 1

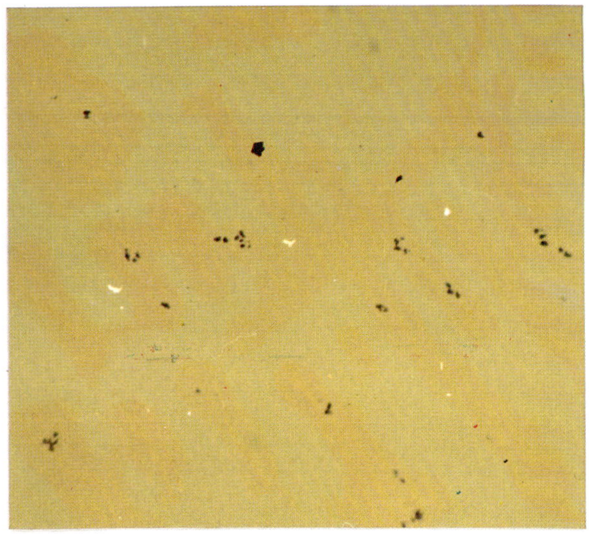

200 200 : 1

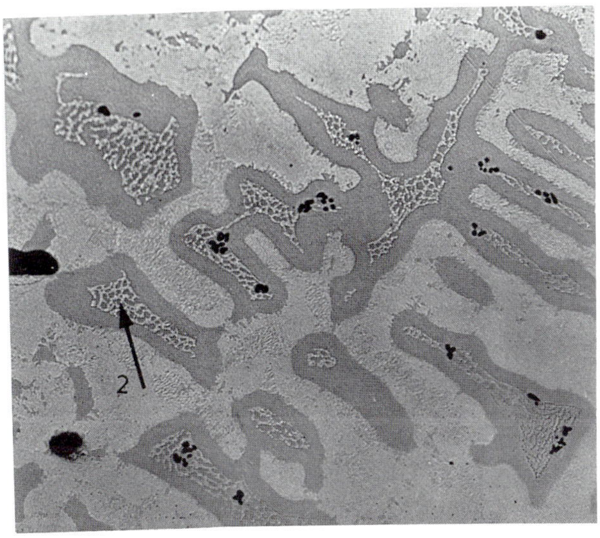

201 200 : 1

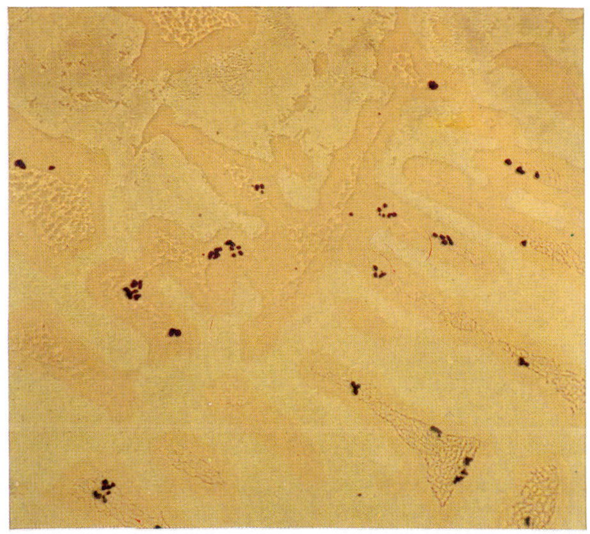

202 200 : 1

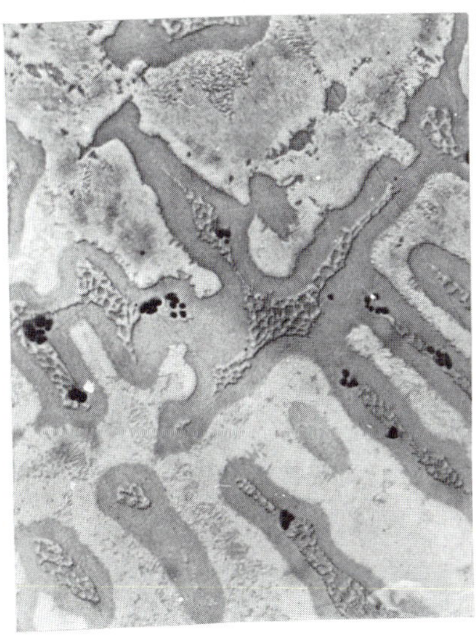

203 200 : 1

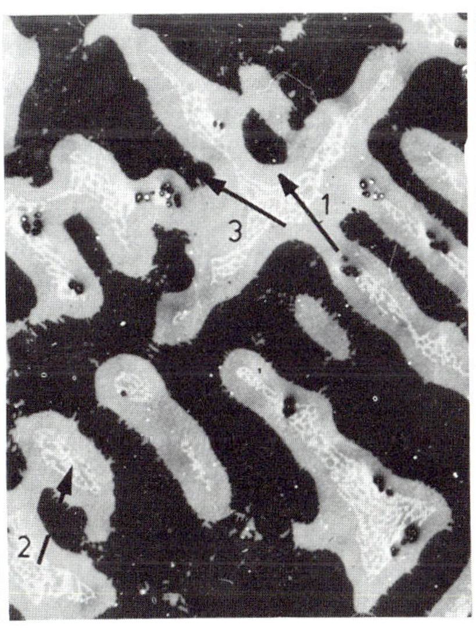

204 200 : 1

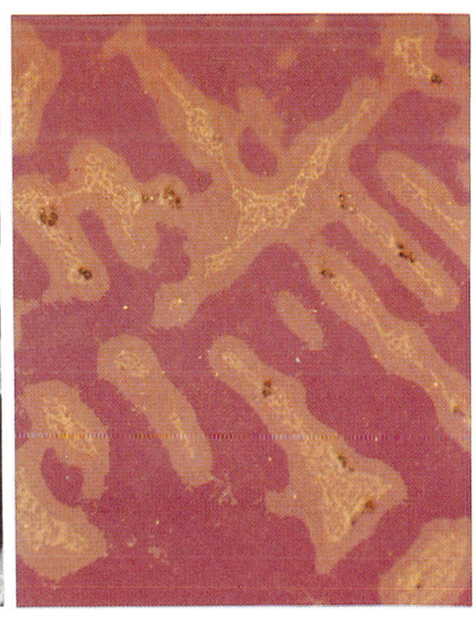

205 200 : 1

2.1.5.3 Copper alloys, bronzes

Figs. 206–207: **Alloy with 77.2% Cu, 9.0% Al, 8.8% Mn, 3.0% Fe and 2.0% Ni,** as cast.

206: Etched in 10% ammonium persulphate solution, bright field.

The white grains consist of α-Cu [1] containing a finely dispersed precipitate of γ-iron. On etching, the region of the residual melt takes on a greyish colour (β-phase with included α-Cu precipitates, [2]). The black primary precipitates in Fig. 206 consists of a Mn-rich compound.

207: Same alloy and treatment as in Fig. 206. Surface layer produced by gas-ion reaction with an iron cathode and oxygen as reactions gas, bright field.

manganese-rich phase: red (7 : 2 : 4)

α-Cu: light beige (3 : 1 : 2)

β-phase: dark yellow-grey (3 : 1 : 3)

In contrast to the chemically etched section (Fig. 206) the finely divided dispersed γ-phase is not so clearly visible in the α-Cu grains after production of the surface coating. This is possibly due to self-absorption by the surface layer.

Figs. 208–212: **Bronze consisting of 64.9% Cu, 19.0% Mn, 5.1% Ni, 3.8% Fe, 3.8% Al and 3.4% Zn,** as cast.

208: Polished, bright field.

The specimen consists of two different phases, an α-Cu matrix (light grey) and the β-phase which is visible in the polished section as a dark grey fine-grained structure [1]. A further manganese-rich phase is not visible. Fine pores show up black.

209: The same alloy, treatment and area of the microstructure as Fig. 208, bright field, etched with 10% ammonium persulphate solution. The manganese-rich phase stands out black against the other microstructural constituents. As in the polished section in Fig. 208, the β-phase [1] is dark grey. The α-Cu matrix is light grey.

210: The same alloy, treatment and area of microstructure as Fig. 209, surface layer produced by gas-ion-reaction with an iron cathode and oxygen as reaction gas, bright field.

α-Cu matrix: light red (9 : 1 : 3)

manganese-rich phase: dark red (10 : 4 : 4)

β-phase: greyish blue

The pores appear yellowish white.

211: Same material, same treatment, same area of the microstructure and same preparation as Fig. 210. Photographed at 540 nm.

The manganese-rich phase is black, the β-phase is light grey [1]. The tone of the α-Cu matrix is between those of the other two microstructural constituents.

212: Same material, same treatment, same area of the microstructure and same preparation as Fig. 210.

Photographed at 630 nm.

The β-phase is dark compared with the other microstructural constituents; there is very little contrast between the α-Cu and manganese-rich phases.

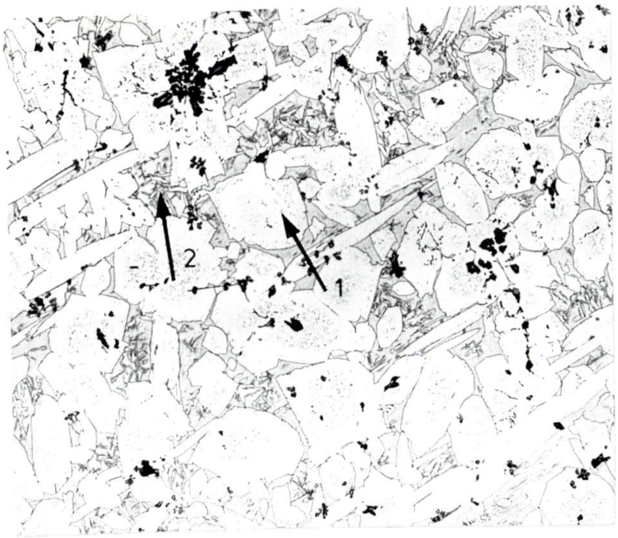

206 200 : 1

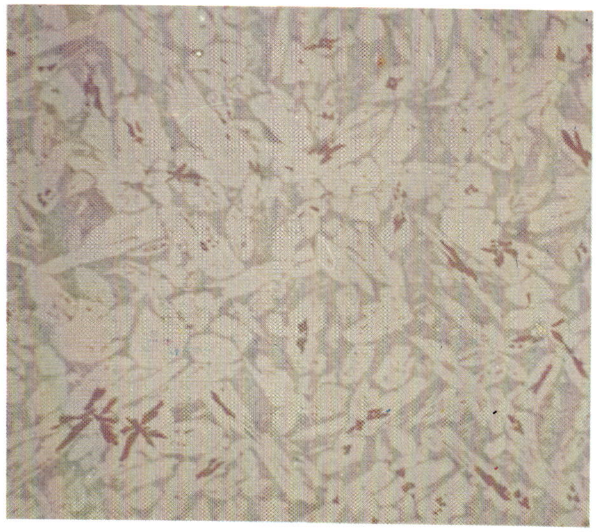

207 200 : 1

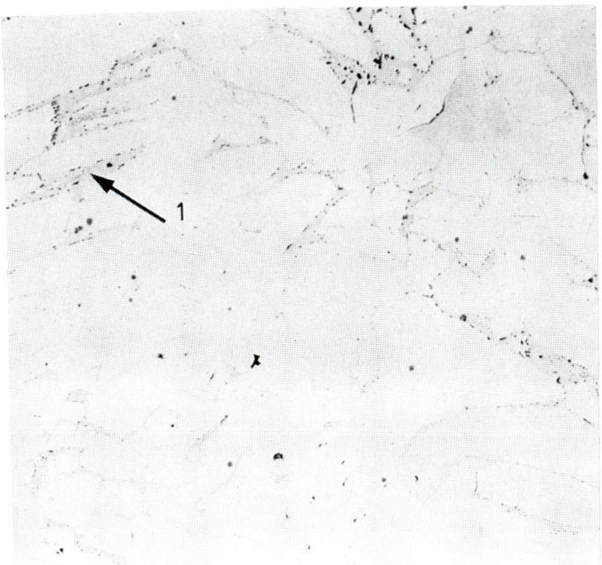

208 200 : 1

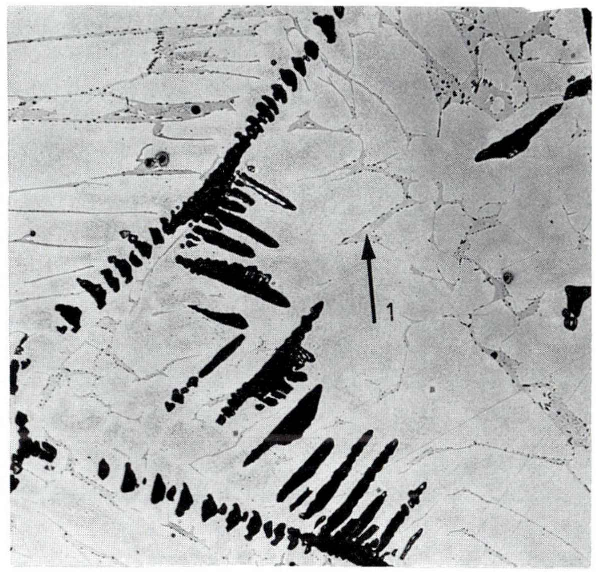

209 200 : 1

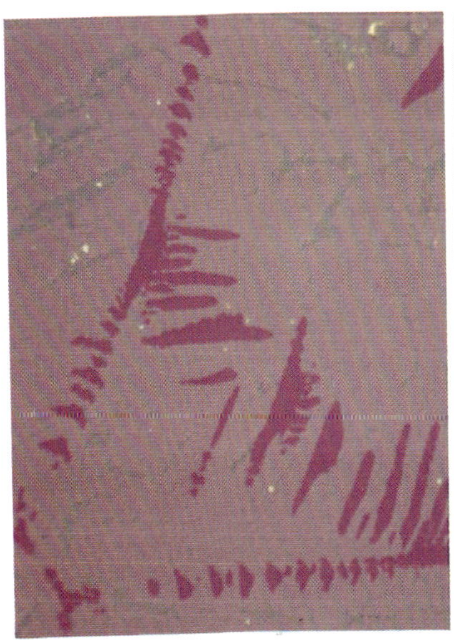

210 200 : 1

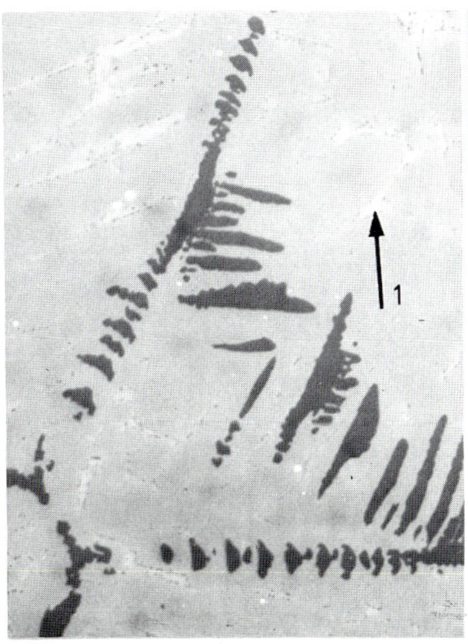

211 200 : 1

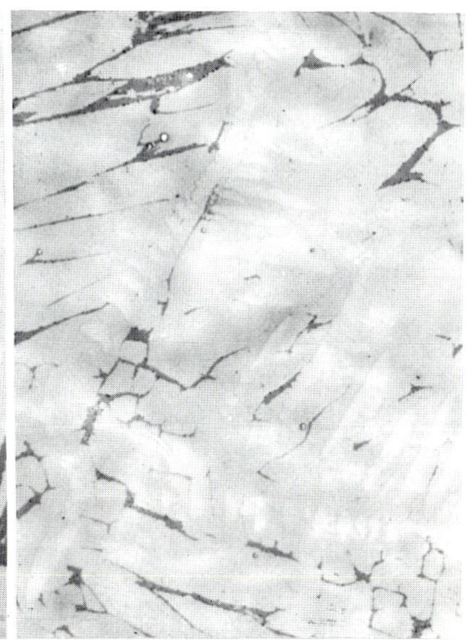

212 200 : 1

2.1.6 Aluminium alloys

2.1.6.1 General introduction

As a consequence of the low solubility of most elements in aluminium, pure technical aluminium and aluminium alloys, even those with only small additions of alloying elements, generally contain a large number of intermetallic compounds. Apart from a few exceptions, these intermetallic compounds have a much lower reflectivity than the highly reflecting aluminium matrix and are therefore visible even on an unetched section. Similarly, good contrast can normally be achieved between the matrix and intermetallic phases using simple etching techniques. Complete metallographic identification of the phases using prescribed etching techniques is, however, seldom possible even by a worker experienced in aluminium metallography, in spite of extensive tabulated data 1, 2).

Because of their very high reflectivity (up to 90%), contrast enhancement on aluminium alloys by means of interference layers neccessitates highly refracting, non-absorbing or absorbing layer materials (c.f. section 1.2.3).

Non-absorbing vacuum deposited layers such as ZnSe or ZnS or layers produced by sputtering with an iron cathode yield a good contrast in the case of the less reflecting intermetallic phases, whereas the aluminium matrix, because of its high reflectivity, becomes only faintly coloured. In the case of low alloy specimens, the microscopic tinting due to ZnSe or ZnS layers is so pale that it is essential for a suitable reference specimen, e.g. α-Fe, to be coated simultaneously so that the growth of the layer in the coating unit can be followed visually.

If, however, weakly absorbing but highly refracting ZnTe layers (zero order) are deposited on the specimen (cf. Figs. 62 and 63 and Tables 7 and 8), highly saturated colours with good contrast are produced. This is because the amplitude condition can be satisfied approximately both for the aluminium matrix and for the less reflecting phases.

Both vacuum deposition and sputtering yield a broad range of possibilities for contrast enhancement; many phases can be identified directly by their characteristic interference colours. Here again, however, it is not always possible to differentiate **all the phases** unequivocally, partly because the differences in optical constants are too small, partly due to inadequate reproducibility of the specimen preparation, layer composition and layer thickness.

Furthermore the extinctions shown by doubly refracting phases when a section is rotated between crossed polarizing prisms are reinforced by the interference layer. (Reinforcement of anisotropy effects c.f. section 2.2.2).

For quantitative purposes observations can be made in monochromatic light, in which case phases can be made to stand out against the matrix or to disappear. Similarly microsegregations in the aluminium matrix can be rendered visible by suitable highly refracting layers.

Some properties of the phases investigated are given in Table 10b (Chapter 5). The usual phase description, their structure and composition range, the interference colour when coated with a 24 nm layer of ZnTe (zero order) and the magnitude of the birefringence of optically anisotropic phases are given.

2.1.6.2 Optical constants

The known optical constants of aluminium alloys are summarised in Table 10a (Chapter 5). The published values for high purity aluminium fluctuate considerably; the most recent values obtained on specimens measured in ultra-high vacuum 6) lie considerably higher than those measured in air on aluminium surfaces covered with an oxide layer. Using the polishing techniques employed in the present work, the reflectivity of the aluminium matrix should lie between 85 and 90% at k-values between 5 and 6. Dissolved Cu and Mn reduce the reflectivity of the solid solution 7), dissolved Mg causes it to increase slightly 8).

Detailed data 9, 10) are available for the strongly dispersive semiconductors silicon and Mg_2Si, whereas, unfortunately, very few optical constants are known for the intermetallic phases occurring in technical alloys.

Fig. 213 shows the data contained in Table 10a plotted in the n-k diagram. Most of the other phases with unknown optical constants occurring in aluminium alloys can most probably be placed in the shaded region in Fig. 213. This was estimated qualitatively by comparisons of colour and intensity before and after deposition of a coating layer (cf. section 2.1.6.3).

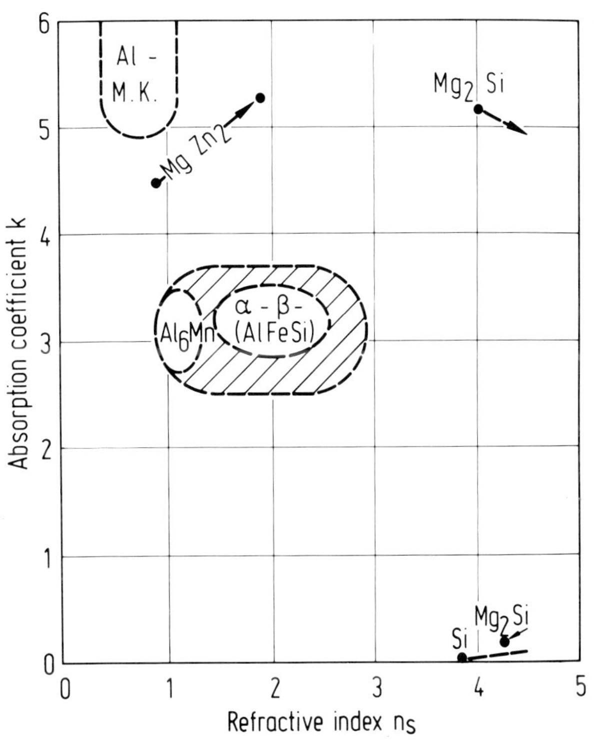

Fig. 213: Optical constants of aluminium and the phases occurring in aluminium alloys.

2.1.6.3 Production of contrast by vacuum deposited ZnTe layers

2.1.6.3.1 Calculated contrasts

Fig. 214 shows the calculated interference colours and intensities at the reflection minimum for ZnTe layers 20 and 25 nm thick 12). The optical constants of ZnTe were taken from the literature 14). The obtainable colours and intensities can easily be estimated or explained by comparison with Fig. 213.

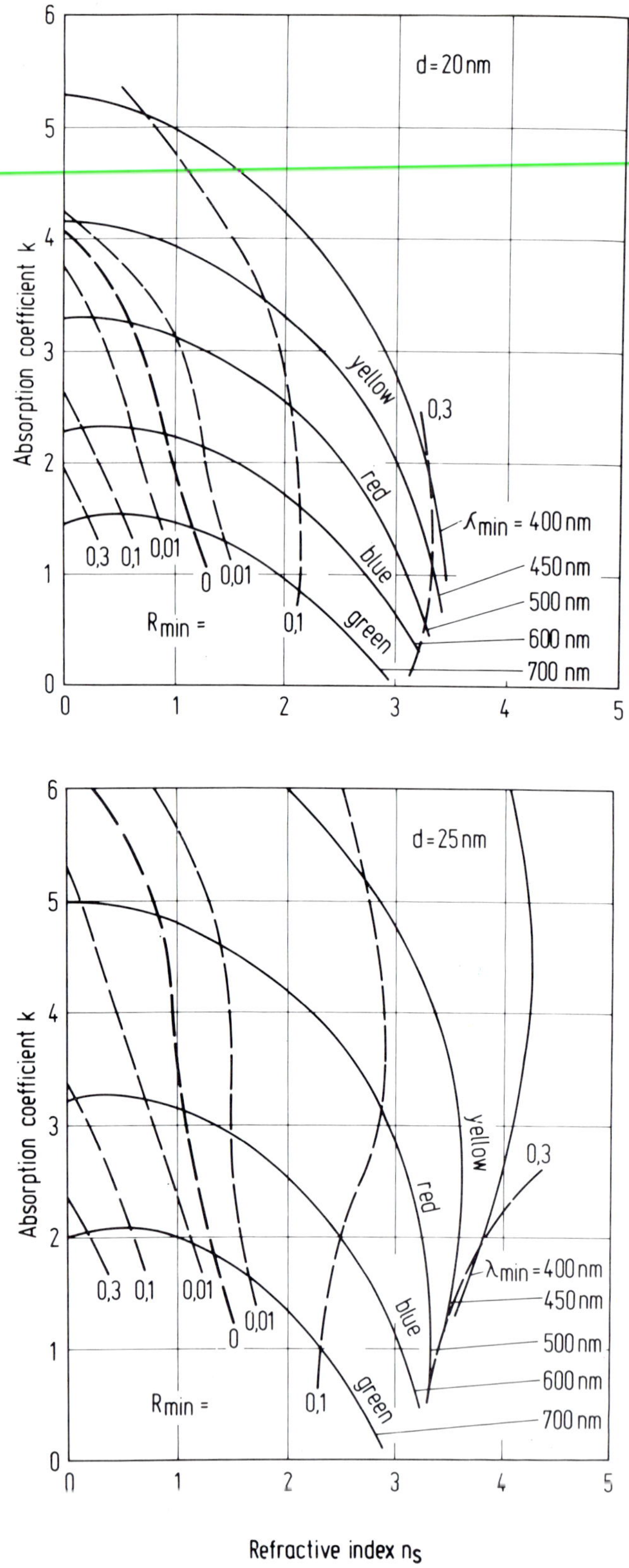

Fig. 214: Calculated contrasts between aluminium and the phases given in Fig. 213 for ZnTe interference layers 20 nm and 25 nm thick. Curves of constant reflectivity at the interference minimum R_{min} (dashed) and curves of constant wavelength at the interference minimum λ_{min} (full lines, values in nm). The colours correspond to the complementary colours of λ_{min}. The totally different shape of the R_{min} = const. curves as compared with those for non-absorbing layers is a consequence of the dispersion of the absorption constant of ZnTe 9).

At the experimental layer thickness of 24 nm the aluminium matrix appears yellowish to reddish whilst most of the less reflecting intermetallic phases have taken on a darker, brownish red to violet colour. If the layer thickness is reduced the family of curves λ_{min} = const. in Fig. 214 becomes shifted downwards and all the interference colours move in the direction of "yellow", whereas if the layer thickness is increased, there is a shift towards the "blue".

The optical constants of silicon lie in an area where the λ_{min} = const. curves converge. Independent of the layer thickness, therefore, the interference colour of silicon is grey with just a touch of colour.

In the case of Mg_2Si, the strong dispersion of k (λ) at high values of n results in high intensity and a slightly bluish tinge.

2.1.6.3.2 Experimental results for aluminium alloys

The compositions of the alloys were chosen to cover the maximum number of phases occurring in technical alloys using only a small number of test alloys. Coarse dispersions of the phases were produced by sand casting, casting into thick walled iron moulds preheated to 500^0C or in some cases directly by controlled cooling at ca. 0.5^0C/min in the melting furnace.

Figures 215 to 239 show examples of the contrast-enhanced test alloys. The phases were determined from published phase diagrams or by experiment. A good contrast between the phases and the Al-matrix was achieved in all cases. Some phases, such as Mg_2Si, Si, Al_3Mg_2, Al_2Cu, $MgZn_2$ or $CaSi_2$ can easily be distinguished from one another and hence identified by means of their characteristic interference colours. In contrast, α-(AlFeSi), β-(AlFeSi), α-(AlMnSi), β-(AlMnSi), Al_7Cr, Al_3Ti and Al_3Zr cannot be distinguished by means of their interference colours. In these cases, however, further differentiation is often possible by observation of the birefringence of non-cubic phases, which is increased by the layer (cf. section 2.2.2) (12).

In the case of very strongly birefringent phases such as Al_2Cu or Al_4Ca, the interference colours are somewhat different even in unpolarized light, according to the orientation (bright field anisotropy).

On account of the marked tendency to chemical reaction building up a coating layer on Mg_2Si, this transparent phase appears in different colours according to the mode of preparation (cf. Figs. 227 and 228).

Similarly the apparent optical properties of the highly reflecting phases Al_3Mg_2, Al_2Cu and $MgZn_2$ are more strongly dependent on the polishing conditions than those of less reflecting phases. Al_2Cu, for example, can take on any shade from light to dark blue.

Minimal fluctuations in the layer thickness or composition of the layer material can result in colour shifts because the R_{min} = const. and λ_{min} = const. curves in Fig. 214 depend very sensitively on these parameters. A change in the colour of the aluminium matrix can also be caused by the formation of solid solutions.

Literature to 2.1.6

1) Hanemann, H. and Schrader, A.: Atlas Metallographicus
Vol. III.1: Binäre Legierungen des Aluminiums, Verlag Gebr. Bornträger, Berlin (1941)
Vol. III.2: Ternäre Legierungen des Aluminiums,
Verlag Stahleisen, Düsseldorf (1952)

2) Lymann, T.: Metals Handbook, Vol. 8, American Society for Metals, Metals Park Ohio (1973)

3) Jeglitsch, F. and Novak, H.: in: 3. Int. Leichtmetall-Tagung Leoben/ Wien (1975), Aluminium-Verlag Düsseldorf,
Aluminium **51** (1975) 768

4) Bartz, G.: Prakt. Metallographie **10** (1973) 311

5) Back, H. and Pabst, R.: Leitz-Mitt. Wiss. u. Techn. **6** (1976) 230

6) Allen, T. H.: J. Vac. Sci. Technol. **13** (1976) 112

7) Beaglehole, D. and Wihl, M.: J. Phys. F: Metal Phys. **2,** (1972) 43

8) Wulff, J.: J. Opt. Soc. Am. **24,** (1934) 223

9) Landolt-Börnstein, Zahlenwerte und Funktionen Vol. II, Part 8, Springer-Verlag, (1962)

10) Scouler, W. J.: Phys. Rev. **178,** (1969) 1353

11) Paufler, P., Eichler, K., Siegel, S. and Schulze, G. E. R., in: Elektronenstruktur und phys. Eigenschaften metallischer Werkstoffe, Autorenkollektiv, Deutscher Verlag für Grundstoffindustrie, Leipzig, (1972), 73

12) Zogg, H.: in: Prakt. Metallographie Special Volume 6,
Dr. Riederer-Verlag, Stuttgart (1976) 163

13) Zogg, H., Weber, S. and Warlimont, H.: Prakt. Metallographie **14,** (1977) 553

14) Bühler, H.-E.: Radex-Rundschau **3/4,** (1967) 672

15) Pearson, W. B.: A Handbook of Lattice Spacings and Structures of Metals and Alloys, Vol. 2, Pergamon Press (1967)

16) Landolt-Börnstein, New Series III.6, Springer-Verlag (1971)

Fig. 215: **Aluminium alloy with 7.4% Ca, 0.8% Si and 0.2% Ti,** chill casting, vacuum deposited ZnTe layer, zero-order, bright field.

Al solid solution: beige (4 : 3 : 1)

Al_4Ca: blue (16 : 4 : 3) (A 16 : 4 : 3..4)

$CaSi_2$: reddish violet (10 : 4 : 4) (A 10 : 4 : 3..4)

The proportions of the aluminium solid solution, coarsely formed $CaSi_2$ and finely dispersed Al_4Ca correspond approximately to the composition of the ternary eutectic.

Fig. 216: **Aluminium alloy with 4.2% Cr,** sandcasting, vacuum deposited ZnTe layer, zero-order, bright field.

Al_7Cr: violet (partly peritectic, partly acicular) (11 : 3 : 3) (A 11 : 3 : 3)

$Al_{11}Cr_2$: violet (enclosed in Al_7Cr) (12 : 2 : 4)

Primary $Al_{11}Cr_2$ has been precipitated in the aluminium solid solution and surrounded by peritectic Al_7Cr. In the course of solidification the residual melt becomes depleted in chromium so that there is the possibility of primary precipitation of Al_7Cr as needles.

Fig. 217: **Aluminium alloy with 10% Cu and 2% Mg,** chill casting, coated with ZnTe, zero-order, bright field.

Al solid solution: beige (5 : 2 : 2)

Θ-Al_2Cu: blue (18 : 5 : 3) (A 18 : 5 : 3)

Primary aluminium solid solution with a binary eutectic of aluminium and Θ-Al_2Cu and a further ternary eutectic of aluminium, Θ-Al_2Cu and S-Al_2CuMg at the cell boundaries. The ternary eutectic cannot be resolved clearly into colours.

The white spots in the eutectics are polishing effects.

Fig. 218: **Aluminium alloy with 2.3% Cu and 14% Mg,** chill casting, coated with ZnTe, zero-order, bright field.

Al solid solution: beige (4 : 3 : 2) (A 4 : 2..3 : 2)

T-Al_6CuMg_4: blue (16..17 : 4 : 2) the variation in brightness and saturation is the result of segregations

β-Al_3Mg_2: whitish blue (16 : 1 : 2)

The section comprises a primary aluminium solid solution, a binary eutectic of aluminium and T-Al_6CuMg_4 and a second eutectic of aluminium and Θ-Al_3Mg_2 at the cell boundaries. On account of its extensive range of stability, the T-phase appears in different saturations.

Fig. 219: **Aluminium alloy with 6% Cu and 3.5% Mn,** chill casting, coated with ZnTe, zero-order, bright field.

Al solid solution: beige (4 : 3 : 2) (A 4 : 2..3 : 2)

T-$Al_{20}Cu_2Mn$: reddish violet (10 : 3 : 4)

Al_6Mn: yellow (3 : 5 : 2) (A 3 : 5 : 2)

Al_2Cu: light blue (16 : 1 : 2)

Polishing effects white as in Fig. 217.

As well as the aluminium matrix there are primary precipitates of Al_6Mn which has undergone peritectic transformation to T-$Al_{20}Cu_2Mn$ in places at the edges. A binary eutectic of Θ-Al_2Cu and aluminium has formed at the cell boundaries. The various shades of colour in the aluminium solid solution indicate segregations.

Fig. 220: **Aluminium alloy with 3.7% Cu and 3% Fe,** sand casting, coated with ZnTe, zero-order, bright field.

Al solid solution: light violet (10 : 0..1 : 3) (A 10 : 1 : 3)

$(Al,Cu)_6(Fe,Cu)$: brown-yellow (4 : 2 : 3)

Al_7Cu_2Fe: reddish violet (10 : 4 : 4) (A 10 : 4 : 3..4)

Θ-Al_2Cu: light blue (14 : 1 : 2)

As well as the aluminium solid solution, two eutectic phases can be seen in this section: $(Al,Cu)_6(Fe,Cu)$ and Al_7Cu_2Fe (the original primary precipitate of Al_3Fe has undergone complete transformation). Al_7Cu_2Fe and Θ-Al_2Cu can be seen in the remaining eutectic.

Fig. 221: **Aluminium alloy with 9.2% Mg and 1.5% Mn,** chill casting, annealed 55 h at 400⁰C, coated with ZnTe, bright field.

Al solid solution: orange (6 : 2 : 3)

T-$Al_{18}Mg_3Mn_2$: dark blue (17 : 4 : 3) (A 17 : 4 : 3), difficult to resolve

β-Al_3Mg_2: light blue (18 : 3 : 3) (A 18 : 3 : 2..3)

Al_6Mn: brown (6 : 4 : 4) (A 6 : 4 : 3..4)

The alloy lies in the 3-phase region aluminium solid solution-Al_6Mn-T-$Al_{18}Mg_3Mn_2$ but is not yet in equilibrium because the β-Al_3Mg_2 has not yet completely dissolved.

Fig. 222: **Aluminium alloy with 5.2% Mg and 22% Zn,** sand casting, annealed 120 h at 400⁰C, coated with ZnTe, bright field.

Al solid solution: light brown (4 : 2 : 2)

$(Al,Zn)_{49}Mg_{32}$: violet (12 : 3 : 3)

$MgZn_2$: light blue (17 : 2 : 1)

In this section, a eutectic of aluminium and $(Al,Zn)_{49}Mg_{32}$ has formed together with the aluminium solid solution. As a result of the subsequent anneal, $(Al,Zn)_{49}Mg_{32}$ and $MgZn_2$ have been precipitated from the supersaturated solid solution.

Fig. 223: **Aluminium alloy with 6% Fe and 2% Mn,** sand casting, coated with ZnTe, bright field.

Al solid solution: grey, greyish violet

Al_3Fe: blue (15 : 2 :4)

$Al_6(Fe,Mn)$: violet (12 : 4 : 4) (A 12 : 4 :4)

As well as the aluminium matrix, this section contains the intermetallic compound Al_3Fe with $Al_6(Fe,Mn)$ crystallized on to it. Furthermore, colour shifts are visible in the aluminium solid solution occuring as a result of a depletion in manganese in the vicinity of $Al_6(Fe,Mn)$ precipitates.

Fig. 224: **Aluminium alloy with 6% Ni and 0.3% Fe,** chill casting, coated with ZnTe, bright field.

Al solid solution: yellowish brown (4 : 3 : 2) (A 4 : 2..3 : 2)

T-Al_9FeNi: yellowish red (6 : 4 : 3) (A 6 : 4 : 3..4)

Al_3Ni: yellowish red (6 : 4 : 3) (A 6 : 4 : 3..4)

The alloy has the approximate composition of the ternary eutectic: aluminium solid solution, T-Al_9FeNi and Al_3Ni (degenerate eutectic).

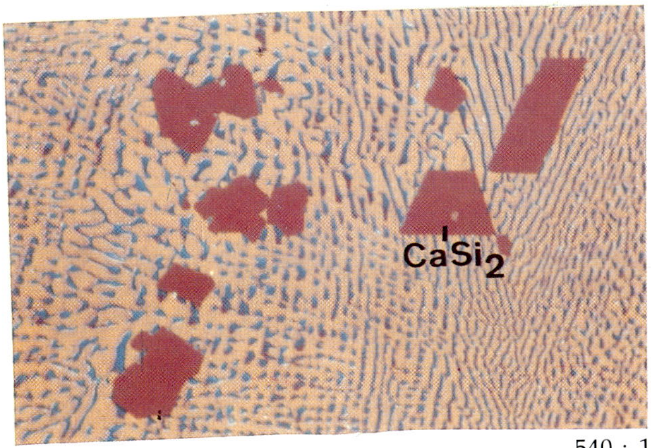

215 540 : 1

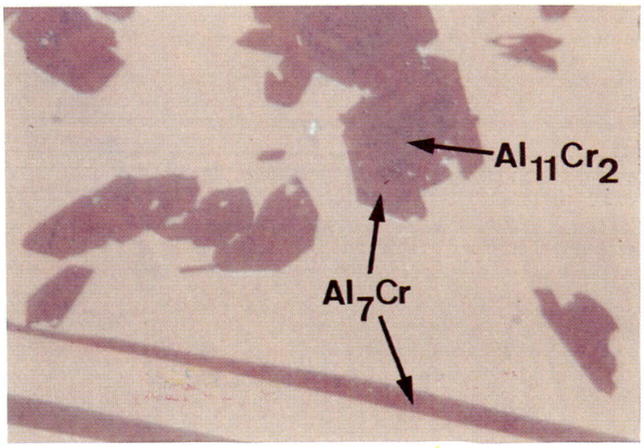

216 540 : 1

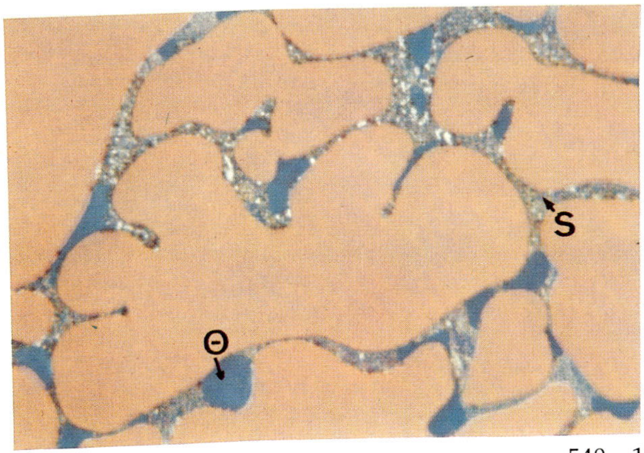

217 540 : 1

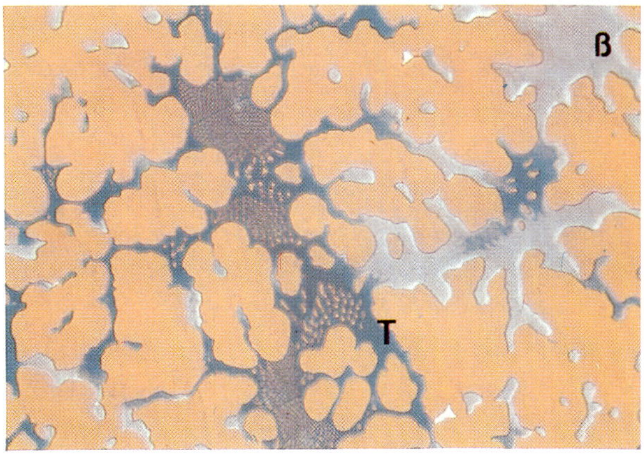

218 215 : 1

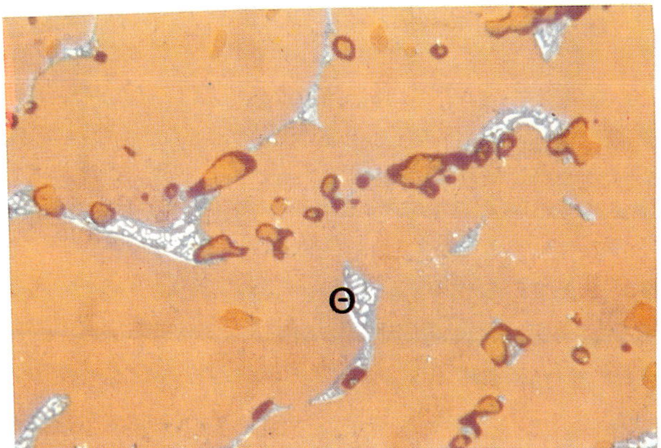

219 540 : 1

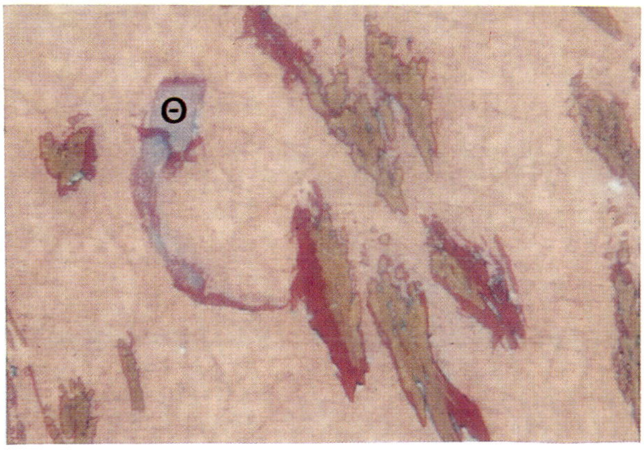

220 540 : 1

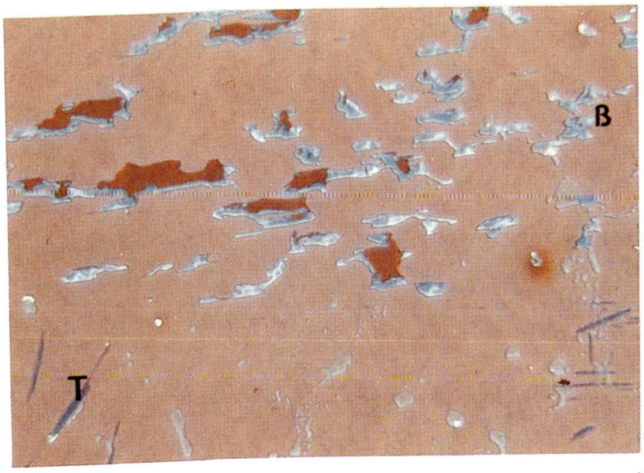

221 215 : 1

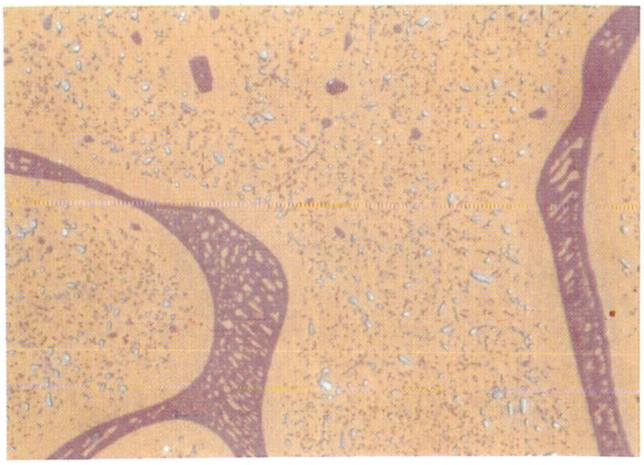

222 540 : 1

Figs. 223 and 224: See page 134

Fig. 225: **Aluminium alloy with 4.35% Fe and 5.2% Si,** sand casting, coated with ZnTe, bright field.

Al solid solution: pink (9 : 2 : 3)

Al_3Fe (enclosed in α-AlFeSi): bluish grey (14 : 1 : 3)

α-$Al_{12}Fe_3Si$ (= α-AlFeSi): violet (12 : 2 : 4)

β-$Al_9Fe_2Si_2$ (= β-AlFeSi): bluish grey (needles), (14 : 1 : 3)

silicon: pale green, colour cannot be resolved definitely

The primary precipitate Al_3Fe is surrounded by peritectic α-AlFeSi. As a result of depletion of the residual melt in iron during the course of solidification, primary polyhedral β-AlFeSi is also present in this section. Solidification was complete at the ternary eutectic point aluminium solid solution-α-AlFeSi-silicon.

Fig. 226: **Aluminium alloy with 0.96% Mn and 11% Si,** chill casting, coated with ZnTe, bright field.

Al solid solution: orange (6 : 2 : 3)

α-$Al_{12}Mn_3Si$: brown (5 : 4 : 4)

β-Al_9Mn_3Si: brown (needles), (5 : 4 : 4)

silicon: pale gren (24 : 1 : 2..3)

α-AlMnSi, looking rather like Chinese writing, needle-like β-AlMnSi and silicon are present in an aluminium solid solution matrix; α and β cannot be distinguished by colour.

Fig. 227: **Alumnium alloy with 0.2% Fe, 0.35% Mg and 10% Si,** sand casting, diamond polished, coated with ZnTe, bright field.

Al solid solution: greyish violet

$Al_8FeMg_3Si_6$: violet (12 : 3 : 5) colour not easily resolved

β-$Al_9Fe_2Si_2$: violet-blue (enveloped in $Al_8FeMg_3Si_6$), (16 : 4 : 5)

Mg_2Si: white

silicon: pale green (~ 24 : 1 : 3)

The main component is once again the aluminium solid solution, into which the β-AlFeSi phase, surrounded by $Al_8FeMg_3Si_6$ has precipitated. Mg_2Si and silicon are also visible as discrete phases. (β and Al_8FeMg_3Si can easily be distinguished by the difference in birefringence 9).

Fig. 228: **Aluminium alloy with 4.6% Mg and 13.2% Si,** chill casting, polished with alumina, coated with ZnTe, bright field.

Al solid solution: pink

Mg_2Si: blue (18 : 4 : 1)

silicon: pale green (24 : 1 : 3)

The composition of this alloy is approximately that of the ternary eutectic aluminium solid solution-Mg_2Si-silicon (degenerate eutectic). This specimen had only a preliminary polish so that the Mg_2Si appears light blue (18 : 4 : 2) (cf. Fig. 227).

Fig. 229: **Aluminium alloy with 2.3% Ti and 1.4% Si,** sand casting, coated with ZnTe, bright field.

Al solid solution: pale violet (10 : 1 : 2)

Al_3Ti: violet (12 : 2 : 3)

silicon: pale green (24 : 1 : 2)

A primary precipitate of Al_3Ti has formed in the aluminium solid solution matrix. Silicon has crystallised on to it in various places.

Figs. 230 and 231: See page 138

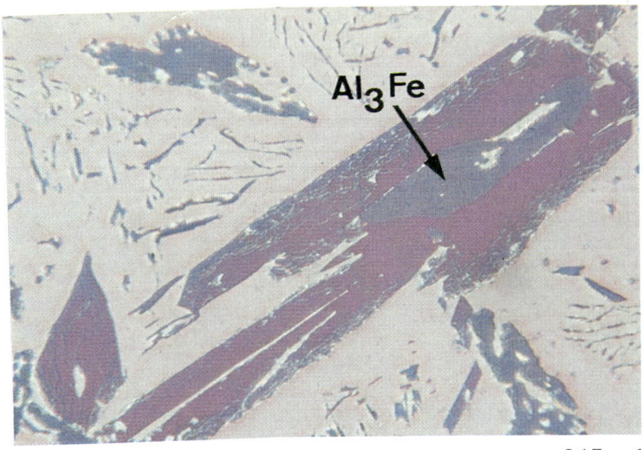

223 215 : 1

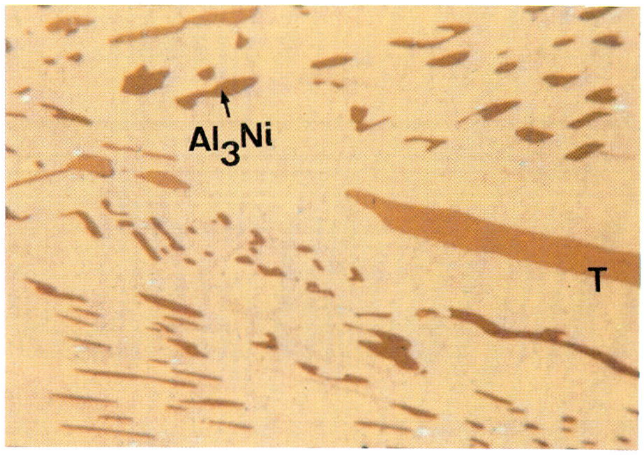

224 540 : 1

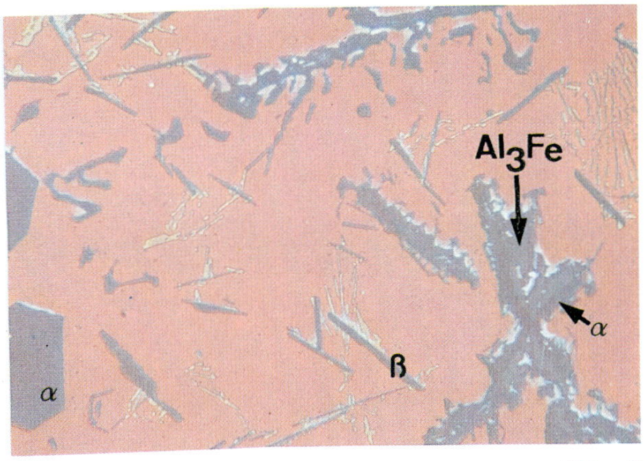

225 215 : 1

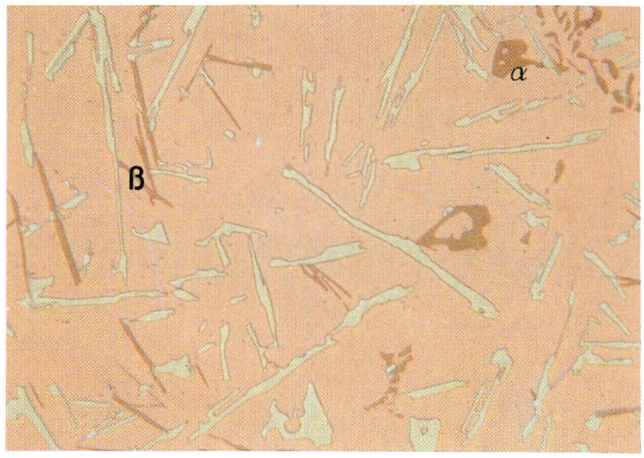

226 540 : 1

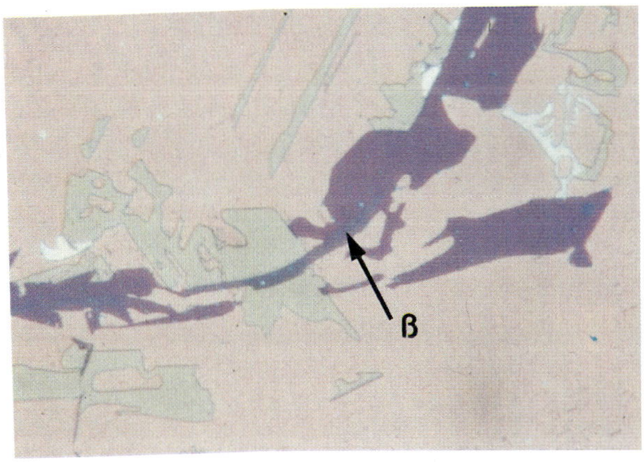

227 540 : 1

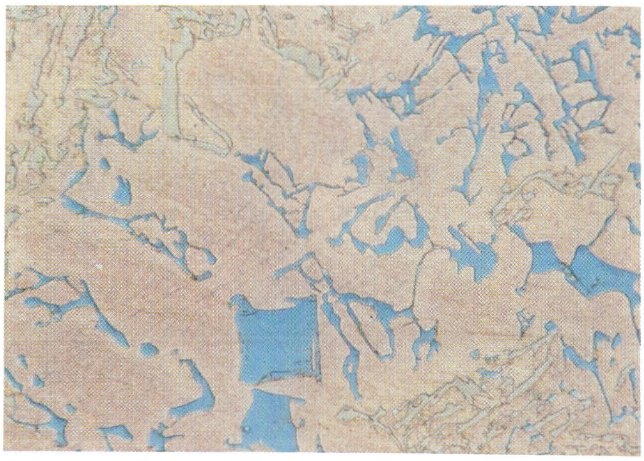

228 215 : 1

229 540 : 1

230 215 : 1

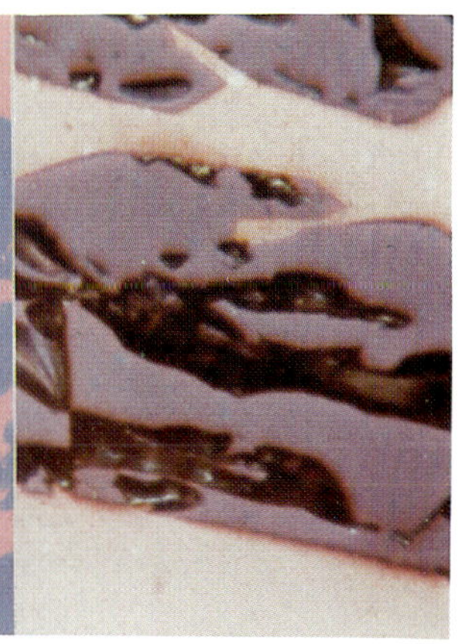

231 215 : 1

Fig. 230: **Aluminium alloy with 0.6% V,** cooled at 0.5°C/min, coated with ZnTe, bright field.

Al solid solution: pink (10 : 2 : 3) (A 10 : 2 : 2..3)

$Al_{10}V$: blue (16 : 2..3 : 2..3) (A 16 : 2..3 : 3)

The intermetallic compound $Al_{10}V$ has precipitated in the aluminium solid solution matrix.

Fig. 231: **Aluminium alloy with 2.8% Zr,** cooled at 0.5°C/min, coated with ZnTe, bright field.

Al solid solution: light grey

Al_3Zr: violet (12 : 1 : 3)

cracks: black

Primary Al_3Zr has precipitated in the aluminium solid solution matrix. Dark pores and cracks are also visible.

Figs. 232–236: **Aluminium alloy with 2.3% Cu and 14% Mg,** same specimen as in Fig. 218.

The section contains a primary aluminium solid solution, a binary eutectic of aluminium and T-Al_6CuMg_4 and a second eutectic of aluminium and Θ-Al_3Mg_2 at the cell boundaries. On account of its wide range of stability, the T-phase appears in different shades of colour.

232: Same specimen as in Fig. 218, unetched, no surface coating, polished, bright field.

Al solid solution: light grey

eutectic: dark grey, cannot be resolved

The contrast between the different phases is very poor.

233: Same specimen, same area of microstructure as in Fig. 232, etched in 0.5% HF for 15 s, bright field.

Al solid solution: light grey

T-phase: black

eutectic: medium grey, cannot be resolved in some areas.

234: Same specimen, same area of microstructure as in Fig. 232, coated with ZnTe, layer thickness 24 nm, photographed at 500 nm.

Al solid solution: dark blue (18 : 5 : 3) (A 18 : 5 : 3)

eutectic and T-phase: almost uniformly light blue (18 : 4 : 2)

At this wavelength the reflectivity of the aluminium solid solution is a minimum, the brightness of this phase is therefore less than those of the two intermetallic compounds at the same wavelength. At 500 nm the reflectivities of the T-phase and Al_3Mg_2 are virtually equal.

Figs. 235–239: See page 140

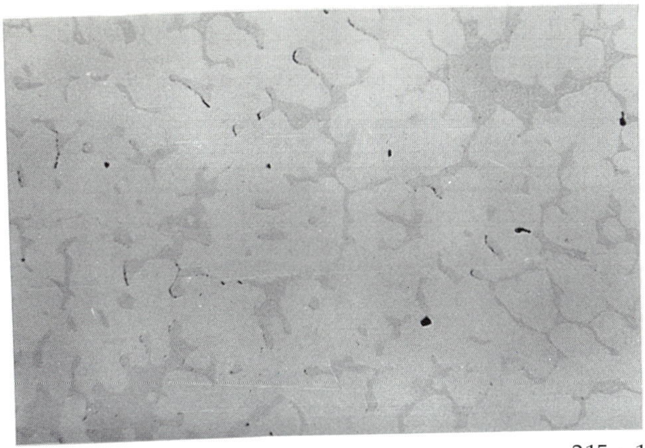

232 215 : 1

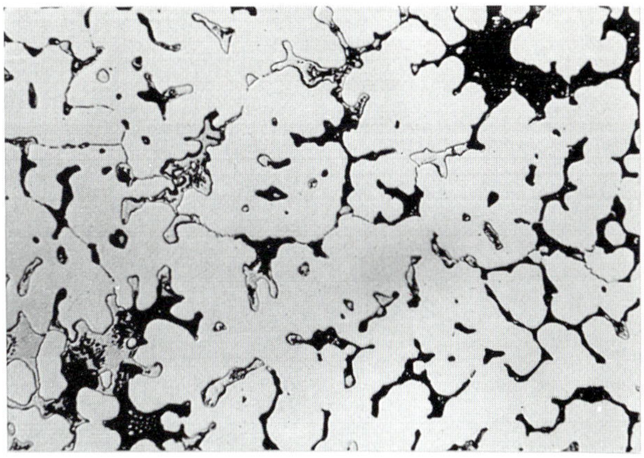

233 215 : 1

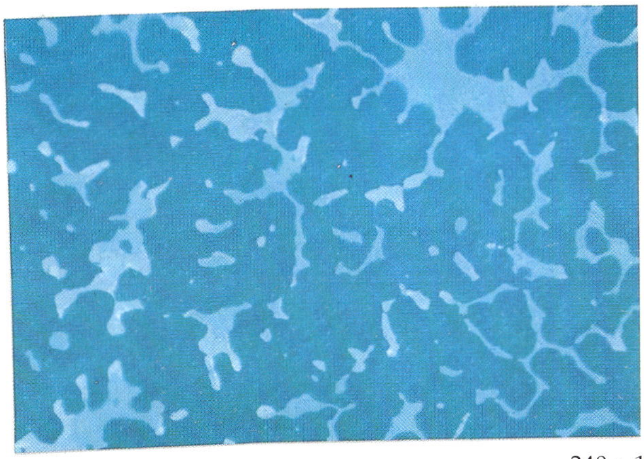

234 240 : 1

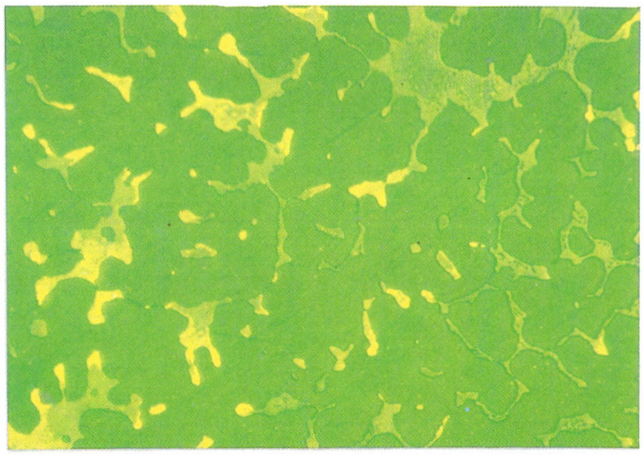

235 240 : 1

236 240 : 1

237 550 : 1

238 550 : 1

239 550 : 1

235: Same specimen, same area of the microstructure and same surface coating as in Fig. 234, but photographed at a wavelength of 550 nm.

Al solid solution: dark green (23 : 6 : 3)

Θ-Al_3Mg_2 (eutectic): light yellow (2 : 6 : 1) (A 2 : 6 : 1)

T-phase: yellowish green (1 : 6 : 2) (A 1 : 6 : 2)

The reflectivity of the aluminium is still a minimum. The T-phase has a lower reflectivity than the Al_3Mg_2.

236: Same specimen, same area of microstructure and same surface coating as in Fig. 234, but photographed at a wavelength of 600 nm.

Al solid solution: red (7 : 6 : 2)

T-phase: dark red (8 : 5 : 3)

Θ-Al_3Mg_2 (eutectic): light red/yellow (4 : 6 : 2)

In the long wave range the T-phase has the lowest reflectivity. The regions in which this phase occurs stand out dark even against the aluminium solid solution matrix. The brightest and hence the most highly reflecting is the Al_3Mg_2 phase.

Figs. 237–239: **Aluminium alloy with 0.3% Fe and 6% Ni,** same specimen as in Fig. 224.

237: Same specimen as in Fig. 224, no surface coating, unetched, bright field.

Al solid solution: light grey

T-Al_9FeNi and Al_3Ni: medium grey

The contrasts between the phases present in this specimen: aluminium solid solution, T-Al_9FeNi and Al_3Ni, are very poor, especially that between the two intermetallic compounds.

238: Same specimen and same area of the microstructure as in Fig. 237, etched with 0.5% HF for 15 s, bright field.

Al solid solution: light grey

T-Al_9FeNi: black

Al_3Ni: dark grey

After this etch, the T-phase is a darker shade than the Al_3Ni.

239: Same specimen and same area of the microstructure as in Fig. 237, coated with ZnTe, layer thickness 24 nm, photographed at 625 nm.

Al solid solution: white

T-Al_9FeNi: dark grey

Al_3Ni: black

At this wavelength the Al_3Ni is darker than the T-phase. Furthermore both intermetallic compounds stand out clearly from the almost white aluminium matrix.

2.1.7 Other non-ferrous alloy systems

Literature to 2.1.7

1) Bartz, G.: Zur Kontrastierung mikroskopischer Objekte in einer Gasionenreaktionskammer.
Prakt. Metallographie **10** (1973) 311

2) Ondracek, G. and Spieler, K.: Gasätzen, ein neues Kontrastierungsverfahren in der Gefügepräparation.
Prakt. Metallographie **10** (1973) 324

3) Back, H. and Pabst, H.: Das Kontrastieren mit der Leitz-Kontrastiereinrichtung in der Metallographie.
Leitz-Mitteilungen für Wissenschaft und Technik Vol. VI, (1976), 230

4) Schlüter, P., Aldinger, S. and Elssner, G.: Vergleich der Kontrastierung von Phasen in Nichteisenmetallen und Verbundsystemen durch Gasionenbeschichten und durch chemisches Ätzen.
Prakt. Metallogr., Special Vol. 9, ed. by W.-U. Kopp and H.-E. Bühler, Dr. Riederer-Verlag, Stuttgart 1978, 57

5) Back, H. and Dietrich, R.: Zur Kontrastierung von Dentalamalgamgefügen.
Prakt. Metallographie **11** (1974) 117

6) Bungardt, W. and Liewald, G.: Beitrag zur Metallographie technischer Cer-Mischmetall-Eisen-Legierungen.
Prakt. Metallographie **5** (1968) 419

7) Semlitsch, M.: Interferenz-Aufdampfschichten bei Mikrogefügeanalysen von Werkstoffen mit der Elektronenstrahl-Mikrosonde.
Prakt. Metallographie **8** (1971) 259

2.1.7.1 Electron beam welded aluminium-nickel joint

Fig. 241: Specimen coated using an iron cathode and oxygen as reaction gas, reflection minimum of nickel lies at 490 nm, bright field.

A broad spectrum of colours can be seen in the weld zone between the aluminium and nickel due to concentration differences in the welded zone. The yellow coloured aluminium is visible on the left hand side of the picture, whilst the nickel appears orange. To gain an impression of the distribution of elements in the welded zone, the concentration profiles for aluminium and nickel were measured by a microprobe along the trace drawn in Fig. 241. The results of these measurements are shown in Fig. 240.

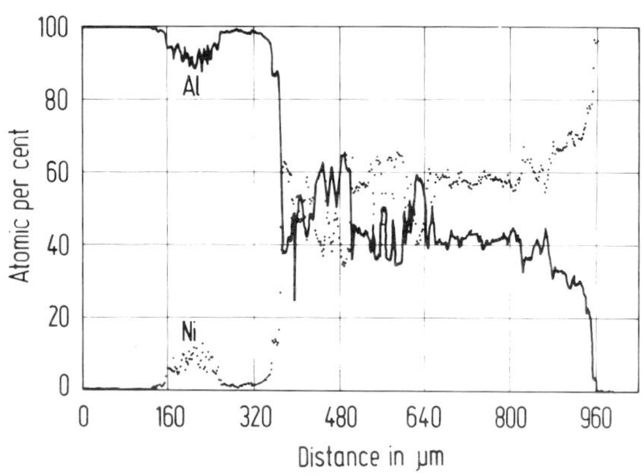

Fig. 240: Element concentration profile for aluminium and nickel along the line A-A in Fig. 241, 4).

Fig. 241: aluminium: yellow (2 : 5 : 1) (A 2..3 : 5 : 1..2)

nickel: orange (5 : 4 : 3)

welded zone: transition colours orange to reddish brown

oxide inclusions in the welded zone: white, blue (17 : 1 : 2), cannot be resolved in some cases.

Figs. 242 and 243: **Wood's metal (50% Bi, 25% Pb, 12.5% Cd, 12.5% Sn)**

242: Specimen etched in 2% alcoholic nitric acid, bright field.

Primary spear-like cadmium precipitates: light blue (18 : 5 : 2)

Cd-phase: light blue (18 : 5 : 2)

Sn-phase: yellowish brown (3 : 4 : 3)

Bi-phase: purple (11 : 4 : 5) (A 11 : 4 : 4..5)

The microstructure consists of a quaternary eutectic in which three phases can be identified:

The Cd-phase in the form of small black rods of the same tone as the primary cadmium precipitates; the light grey Sn-phase and finally bismuth which forms the grey background.

The fourth phase of the quaternary eutectic, the ε-phase of the lead-bismuth binary system is not visible.

243: Same alloy, same area of the microstructure as in Fig. 242, coated using an iron cathode and oxygen as reaction gas, bright field.

primary spear-like cadmium precipitates: light blue (A 18 : 5 : 2)

Cd-phase: light blue (18 : 5 : 2)

Sn-phase: yellowish brown (3 : 4 : 3)

Bi-phase: purple (11 : 4 : 5) (A 11 : 4 : 4..5)

ε-phase: blue (18 : 5 : 3) (A 18 : 5 : 3)

Figs. 244–247: **Aluminium-cobalt alloy with 70% aluminium and 30% cobalt**

244: Polished specimen, bright field.

$Al_{13}Co_4$: primary precipitate; dark grey, surrounded by a seam of peritetic Al_9Co_2

Al_9Co_2: light grey

α-aluminium solid solution: white

245: Same alloy, same area of the microstructure as in Fig. 244, etched in aqueous 25% nitric acid, bright field.

$Al_{13}Co_4$: grey

Al_9Co_2: light grey-white grains (very fine dispersion)

α-aluminium solid solution: greyish-black (etched out in places)

246: Same alloy, same area of the microstructure as in Fig. 244, coated using an iron cathode and oxygen as reaction gas, reflection minimum of $Al_{13}Co_4$ lies at 540 nm, bright field.

$Al_{13}Co_4$: dark violet (11 : 4 : 5) (A 11 : 4 : 4..5)

Al_9Co_2: pale violet (11 : 3 : 3)

α-aluminium solid solution: white (the light blue areas in the aluminium solid solution are preparation effects).

247: Same alloy, vacuum deposited ZnSe layer, reflection minimum of $Al_{13}Co_4$ lies at 540 nm, photographed at 540 nm.

$Al_{13}Co_4$: black

Al_9Co_2: grey (possibly optical anisotropy in bright field)

α-aluminium solid solution: white

241 160 : 1

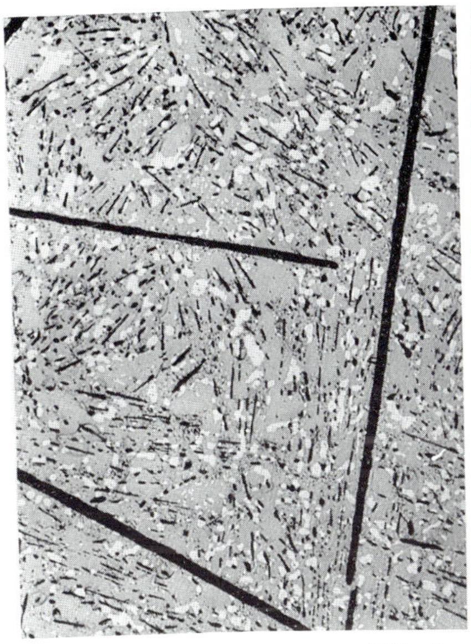

242 200 : 1

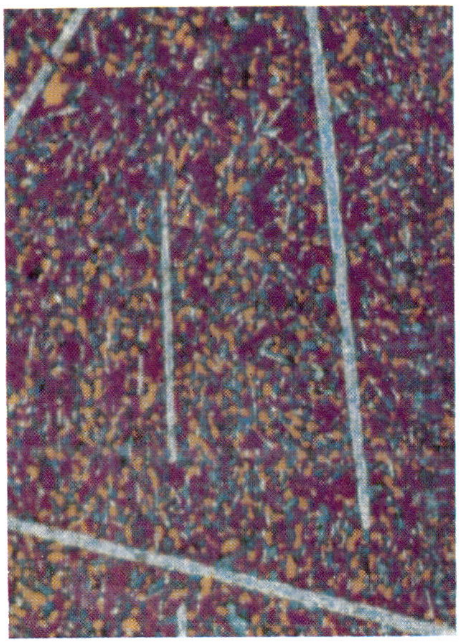

243 200 : 1

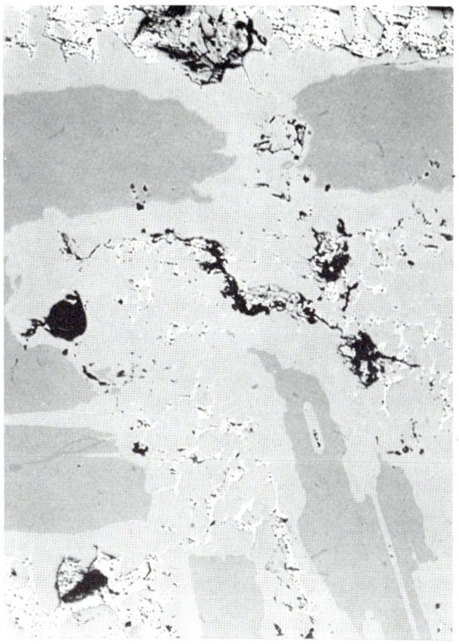

244 100 : 1

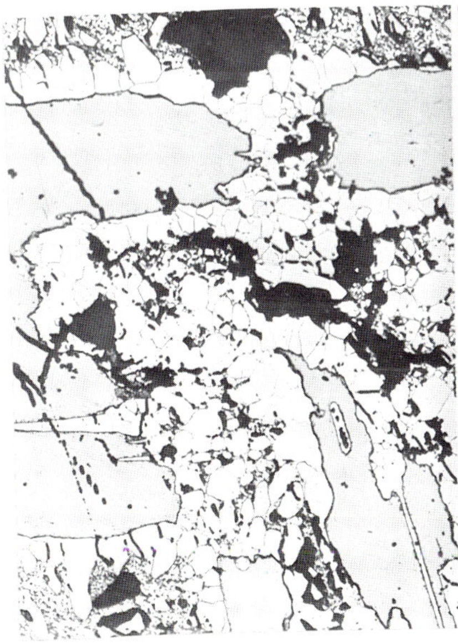

245 100 : 1

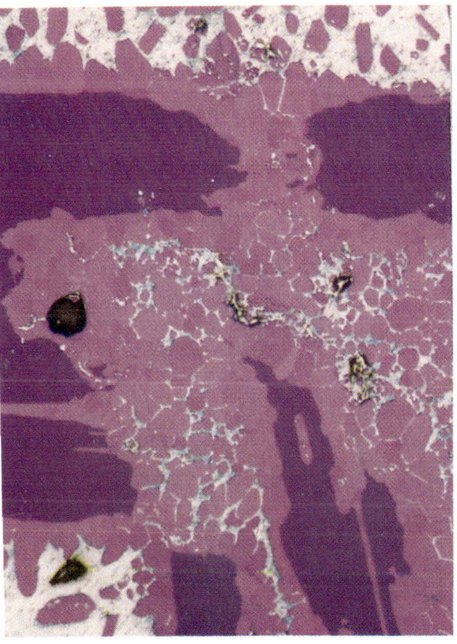

246 100 : 1

247 100 : 1

2.1.7.2 Cerium alloy flints

Figs. 248–250: Primary $CeFe_5$ has precipitated in the cerium alloy flints. Further phases which can be observed in this section are $CeFe_2$, α-cerium and a eutectic of γ-cerium and a cerium-magnesium solid solution.

248: Polished, bright field.

$CeFe_5$: pale yellow (2 : 3 : 1) [1], surrounded by $CeFe_2$: light brown (3 : 4 : 2) [2]

The background consists of α-cerium (2..3 : 3 : 3) (A 2 : 4 : 3) [0] and the γ-cerium/cerium-magnesium solid solution eutectic (almost uniformly dark brown, predominantly (2 : 5 : 3) (A 2..3 : 5 : 3)).

249: Same alloy, same area of the microstructure, vacuum deposited ZnSe layer, zero-order, reflection minimum of $CeFe_2$ lies at 545 nm, bright field.

$CeFe_5$: white [1]

$CeFe_2$: purple (11 : 3 : 4) (A 11 : 3 : 4) [2]

α-cerium: blue (16 : 2 : 3) (A 16 : 2..3 : 3) [0]

eutectic: predominantly dark violet (12 : 3 : 4) (A 12 : 3 : 4) [E]

250: Same alloy, same area of the microstructure, same treatment as in Fig. 249, photographed at 600 nm.

$CeFe_5$: black

$CeFe_2$: grey

Background consisting of α-cerium and eutectic: light grey (the individual phases cannot be distinguished).

Figs. 251–254: Same alloy, but a different area of the microstructure from Figs. 248 to 250.

251: Polished, bright field.

$CeFe_2$: brown plates or needles (2 : 4 : 2) [2]

α-cerium: dark brown (2 : 4 : 4) (A 2..3 : 4 : 4) [0]

γ-cerium/cerium-magnesium solid solution eutectic: (2 : 3 : 4) (A 2..3 : 3 : 4) [E]

The phase marked [X] in the top left of the picture cannot be identified. In terms of the colour code it is (2 : 3 : 2).

252: Same alloy, same area of the microstructure, polished, bright field.

$CeFe_2$: light grey

α-cerium: medium grey

cerium-magnesium solid solution in the eutectic: white

The unidentified phase [X] in Fig. 251 also appears light grey.

253: Same alloy, same area of the microstructure as in Fig. 251, vacuum deposited ZnSe layer, zero-order, reflection minimum of $CeFe_2$ lies at 545 nm, bright field.

$CeFe_2$: purple (11 : 3 : 4) (A 11 : 3 : 4)

α-cerium: blue (16 : 2 : 3) (A 16 : 2..3 : 3)

Eutectic of γ-cerium: violet-blue (11 : 2 : 2) and cerium-magnesium solid solution: pale red

Unknown phase [X] as in Fig. 251: reddish violet (11 : 2 : 3)

254: Same alloy, same area of the microstructure and preparation as in Fig. 253.

$CeFe_2$: light grey [2]

α-cerium: dark grey [0]

Eutectic of γ-cerium (medium grey) [E] and cerium-magnesium solid solution [e] (almost white).

Unidentified phase [X]: white.

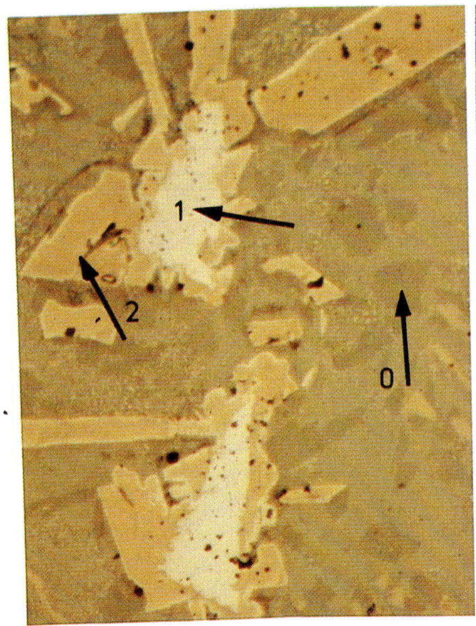

248 200 : 1

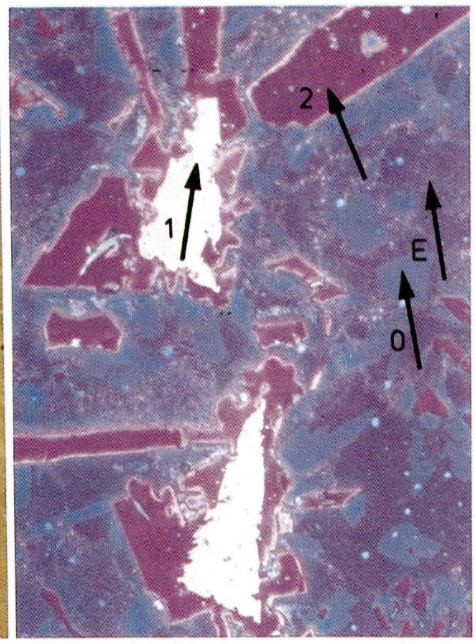

249 200 : 1

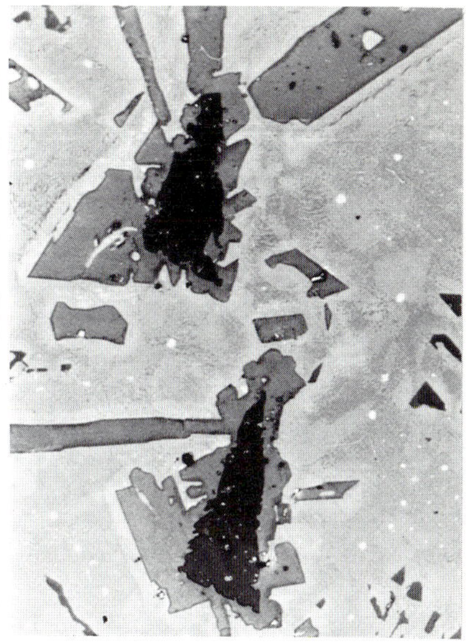

250 200 : 1

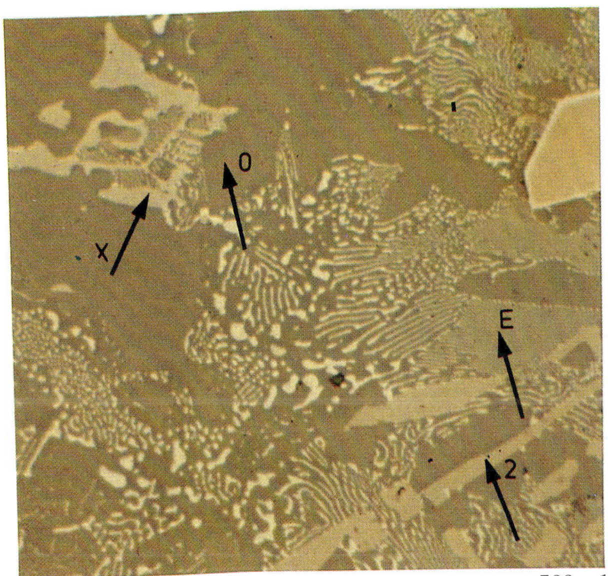

251 500 : 1

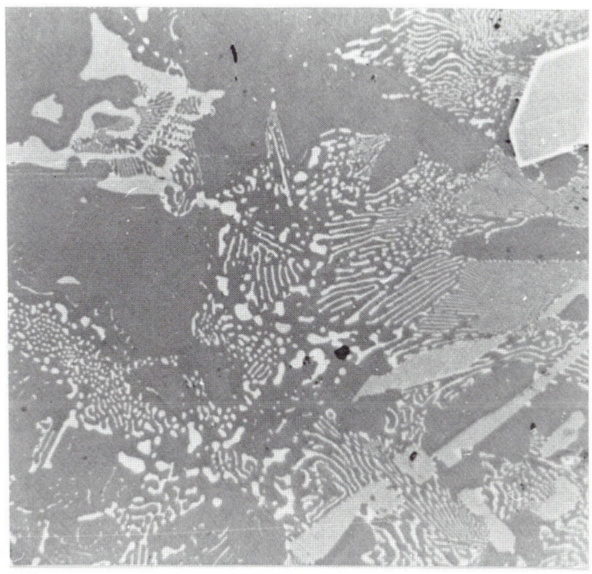

252 500 : 1

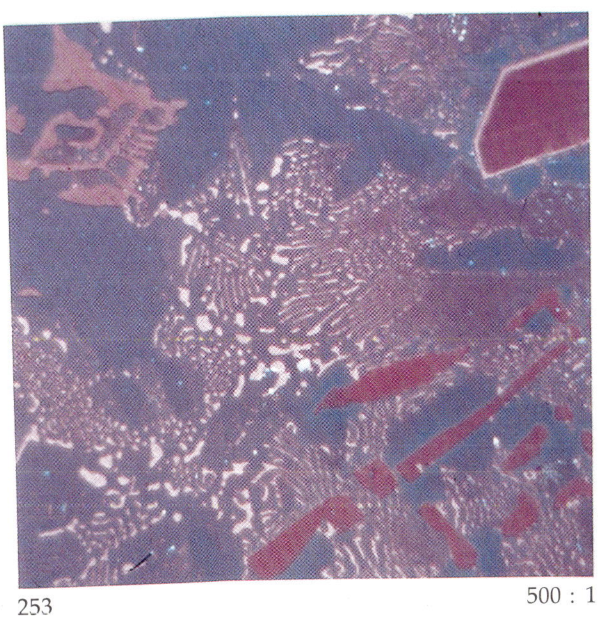

253 500 : 1

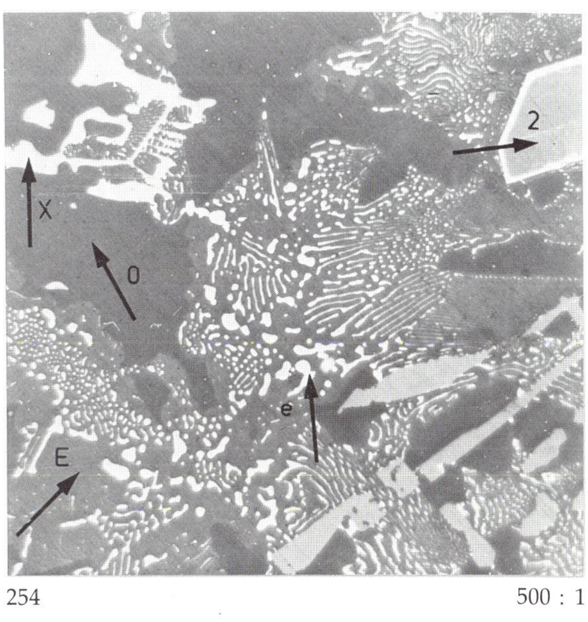

254 500 : 1

2.1.8 Composites (metallic and non-metallic coatings, plasma layers, metal-ceramic composites)

2.1.8.1 General introduction

Metal surfaces can be protected from corrosion by a variety of coating techniques. The basic principle is the production of layers capable of withstanding high mechanical and/or thermal stresses on a less resistant and therefore cheaper substrate material. Construction parts which have suffered wear or corrosion can often be renovated quickly and economically by this method. In the following, three systems have been selected from the wide range of coating techniques and layer combinations. These examples demonstrate in characteristic fashion the possibilities of alloying and processing techniques, but they also give an insight into the application of interference metallography in those cases in which the phases to be examined have widely diverging optical properties.

2.1.8.2 Surface-treated steels (steel-aluminium, steel-zinc system etc.)

In its practical applications unalloyed steel is subject to a wide variety of corrosive attack. In order to prevent or at least limit corrosion, unalloyed steel is coated with materials which are more resistent to corrosive media, principally the metals zinc, aluminium, tin, chromium and lead. The numerous techniques for the production of protective layers such as hot dipping, electroplating, vacuum deposition or cladding, indicate their importance. Clearly the properties of surface coated materials will be determined by the properties of the different components of the system. Thus the forming properties, weldability, colouring and colour reproducibility are important characteristics of a coated steel as well as its corrosion resistance. Most of these properties are closely related to the structure of the deposited layer and the properties of the steel/metal interface 1).

From among the materials used for protecting steel sheet, aluminium has been selected for closer consideration below. Hot aluminized sheet is being used increasingly because of its good corrosion resistance in certain climatic conditions and at high temperatures and its good forming and welding properties.

Aluminium-clad steel is a metallographically less interesting combination because it involves a cold weld. In the case of hot aluminized sheet, however, compound layers form which are influenced by solidification and reaction processes. The nature of the structure of the alloy layers in hot aluminized strip is determined by the operating conditions at the plant and the chemical composition of the aluminium melt. A distinction is made between a coating of pure aluminium, and one for which 7–10% Si was added to the aluminium bath. The structure of the alloy layer is different for these two surface treatments and is discussed in Figs. 255–261.

2.1.8.3 Plasma layers

In plasma spraying, a metallic or non-metallic material in wire or powder form is melted, atomized by means of a carrier gas, and sprayed on the surface to be coated. On impact with the cleaned and roughened substrate material a layer is formed which is normally characterized by a heterogeneous microstructure.

For wear resistance, an important application of thermally sprayed layers, materials such as CoCrW and NiCrBSi alloys, tungsten carbide-cobalt alloys and metallic oxides are available 1). The common metallic and non-metallic materials used in plasma-spraying can be found in the specification sheet "Pulver zum thermischen Spritzen" (Powders for thermal spraying) DIN 32529, sheet 1, February 1976, listed according to their chemical composition and grain size range.

Plasma-sprayed layers have a lamellar structure in which lamellae of various phases lie parallel to the substrate surface. Due to the numerous pores and many different metallic and non-metallic phases in the plasma layer, contrast production by conventional chemical etching is very difficult. Interference layer metallography is virtually the only method of producing a reliable picture of all the different microstructural constituents. On account of the increasing importance of thermally sprayed layers for corrosion and wear resistance, particular attention has been paid to the metallography of these layer systems in this book.

2.1.8.4 Metal-ceramic composites

In metal ceramic composites, combinations of materials are often used in which the different components have widely differing optical properties. Highly reflecting metals and practically absorption-free oxides must therefore be differentiated by means of an adequate light-dark or colour contrast. The composite systems Al_2O_3-$TaAl_3$-Ta and Al_2O_3-Nb have been selected as representative of this group of materials.

Literature to 2.1.8

1) Elssner, G. and Petzow, G.: Metallographie von keramischen Werkstoffen und Metall/Keramik-Verbundsystemen
in: Metallographie und Keramographie – Fortschritte in der Präparationstechnik, Prakt. Metallographie, Special Vol. **9,** ed. by W.-U. Kopp and H.-E. Bühler, Dr. Riederer-Verlag, Stuttgart (1978), 207

2) Steffens, H.-D. and Sepold, G.: Beitrag zur Beschichtung von Metallen. Z. f. Werkstofftechnik **8,** (1977), 22

3) Steffens, H.-D.: Metallographie oberflächenveredelter Stähle – heterogene Schichtsysteme. In: Metallographie – Oberflächenveredelte Stähle, Schweiß- und Lotverbindungen.
Prakt. Metallographie, Special Vol. **7,** Dr. Riederer-Verlag GmbH, Stuttgart (1977), 28

4) Bühler, H.-E., Baumgartl, S., Meyer, L., Morck, G., Nikoleizig, A. and Robusch, G.: Metallographie oberflächenveredelter Stähle – metallische Schichten, reference as 3), but p. 5

5) Jackel, G., Bühler, H.-E. and Meyer, L.: Über Bildung und Eigenschaften von Phasen im Dreistoffsystem Eisen-Zink-Kohlenstoff im Temperaturbereich von 500^0-1100^0C.
Proc. 9th Int. Conference on Galvanization Düsseldorf 1970

6) Schlüter, P., Aldinger, S. and Elssner, G.: Vergleich der Kontrastierung von Phasen in Nichteisenmetallen und Verbundsystemen durch Gasionenbeschichten und durch chemisches Ätzen.
In: Metallographie und Keramographie – Fortschritte in der Präparationstechnik, Prakt. Metallographie, Special Vol. **9,** Dr. Riederer-Verlag GmbH, Stuttgart (1978), 57

2.1.8.2 **Figs. 255–261: Surface treated (corrosion resistant) steels (steel-aluminium system)**

Figs. 255–257: Hot aluminized sheet (no addition of silicon to aluminium bath)

255: Polished, bright field.

steel: medium grey (lower half of picture)

Fe_2Al_5: dark grey (palisade-like configuration, centre of picture)

aluminium surface layer: light grey (with embedded Fe_2Al_5 crystals)

256: Same material, same treatment as in Fig. 255, etched with dilute alcoholic nitric acid, bright field.

The etching leads to the same structural assessment as Fig. 255.

257: Same material as in Fig. 255, vacuum deposited ZnSe layer, zero order, reflection minimum of steel lies at 540 nm, photographed at 540 nm.

steel: dark grey

Fe_2Al_5: black

aluminium surface layer: white (embedded Fe_2Al_5 crystallites: black)

Figs. 258–261: Hot aluminized sheet with an aluminium-silicon surface layer

258: Polished, etched with dilute alcoholic nitric acid, bright field.

steel: medium grey with black pearlite islands (lower half of picture)

aluminium-silicon solid solution: white (upper half of picture)

α-AlFeSi: medium grey (about the same grey level as the steel matrix, compact layer between aluminium and steel)

The grey level of the steel is such that it cannot be differentiated from the α-AlFeSi layer or from the peritectic and eutectic α-AlFeSi and silicon crystals which have precipitated in the aluminium-silicon solid solution.

259: Same specimen as in Fig. 258, vacuum deposited ZnSe layer.

steel matrix: greenish blue (19..20 : 3 : 3)

α-AlFeSi: reddish violet (10 : 3 : 5)

aluminium: beige (2 : 3 : 1)

Peritectic or eutectic comprising aluminium-silicon solid solution, α-AlFeSi (cf. Fig. 225) and silicon: thread-like configuration in which the α-AlFeSi cannot be distinguished. The silicon is yellowish green.

260: Same specimen, same coating as in Fig. 259.

steel (edge, lower half of picture): light blue (18 : 4 : 3)

α-AlFeSi: both as a layer in the lower half of the picture: dark violet (15 : 2 : 6) and as a eutectic or peritectic precipitate in the aluminium-silicon solid solution (17 : 3 : 4).

silicon: brown, thread-like, angular crystals: (4 : 5 : 3). The aluminium-silicon solid solution appears in a blueish white shade with this vacuum deposited coating.

261: Same specimen, same coating as in Fig. 260, photographed at 540 nm.

steel and α-AlFeSi: black

aluminium-silicon solid solution: white

silicon: medium grey

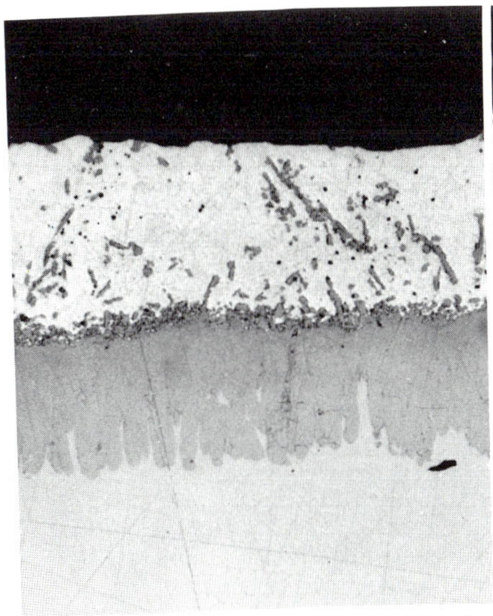

255 500 : 1

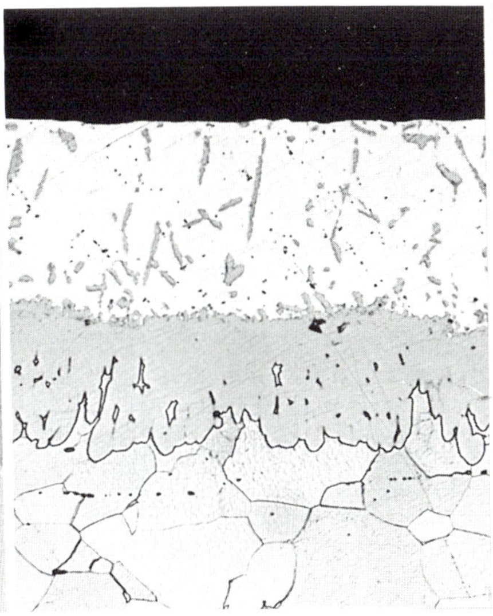

256 500 : 1

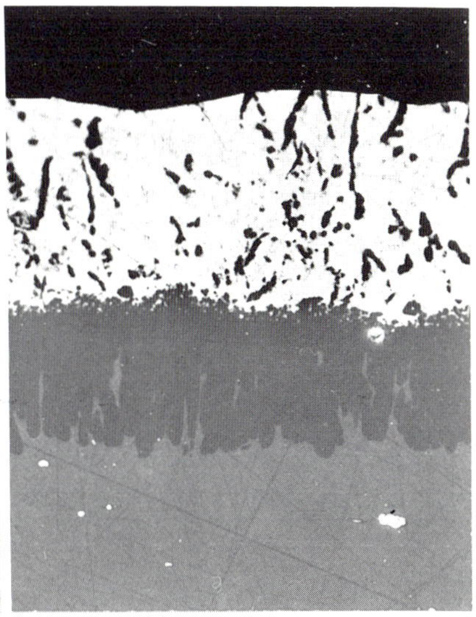

257 500 : 1

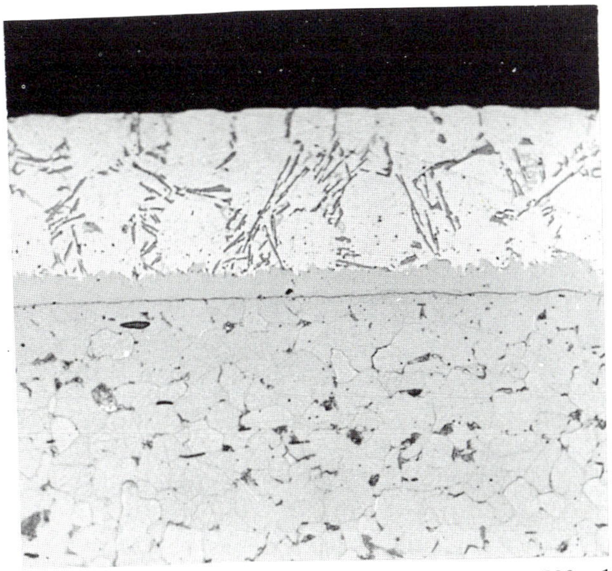

258 500 : 1

259 200 : 1

260 1000 : 1

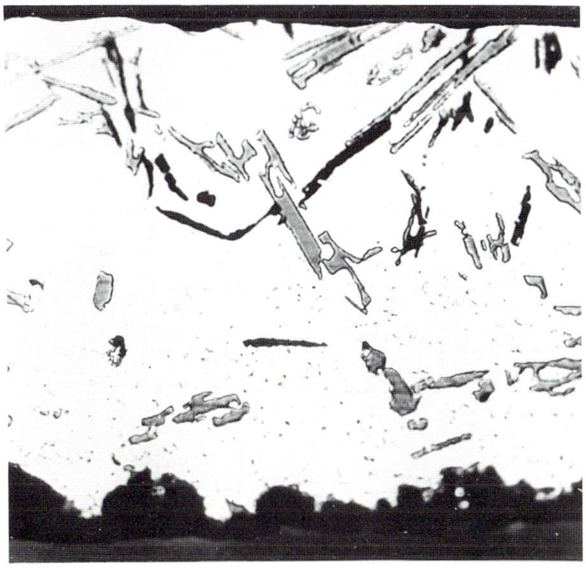

261 1000 : 1

2.1.8.3 Plasma sprayed layers

Figs. 262 to 264: **Steel specimen with plasma sprayed of Al_2O_3 ceramic (with additions of TiO_2) and of a Ni-Cr-Al alloy (20% Cr, 6% Al, remainder Ni).**

The specimen was exposed to combustion gas over long periods and at high temperatures.

262: Polished, bright field.

In the polished, uncoated plasma sprayed layer, one can distinguish only the metallic and non-metallic microstructural constituents on account of their different reflectivities. The metallic components appear bright because of their higher reflectivity.

263: Same specimen and same area of the microstructure as in Fig. 262, sputtered with an iron cathode and oxygen as reaction gas, zero order, reflection minimum of the steel at 530 nm, bright field.

steel matrix (lower half of picture): dark red (9 : 3 : 5)

nickel-chromium-aluminium alloy (predominantly in the upper half of the picture): light red (7 : 4 : 4)

Al_2O_3 ceramic with TiO_2: greenish grey (24 : 0 : 1)

Al_2O_3 without TiO_2: light green (24 : 3 : 5)

oxides enriched in Cr, Al, Si: violet threads (12 : 4 : 5)

264: Same specimen, same area of microstructure and same preliminary treatment as in Fig. 263, enlarged section.

nickel-chromium-aluminium alloy: red (7 : 4 : 4)

Al_2O_3 with TiO_2: green (23 : 1 : 2) [1]

Al_2O_3 without TiO_2: green (23 : 1 : 4) [2]

oxides enriched in Cr, Al, Si: violet (11 : 4 : 5)

After coating, more phases may be distinguished than in the polished condition (Fig. 262).

Figs. 265 and 266: **Steel St 37 with a plasma sprayed layer of 50% Fe and 50% Fe_3O_4.**

265: Specimen polished, bright field.

steel matrix: white

iron in the plasma layer: white

Fe_3O_4: grey

pores: black

266: Same specimen and area of the microstructure as in Fig. 265, vacuum deposited PbF_2 layer, reflection minimum of steel at 520 nm, bright field.

steel matrix: light blue (15 : 0..1 : 2)

iron in the plasma layer: light blue (15 : 0..1 : 2)

Fe_3O_4: dark violet (11 : 2 : 6)

oxide inclusions (Al_2O_3 with calcium and sodium): green (24 : 4 : 2)

pores: black

After coating, two oxide phases may be seen which are indistinguishable in the polished condition (Fig. 265).

Figs. 267 and 268: See page 152

262 200 : 1

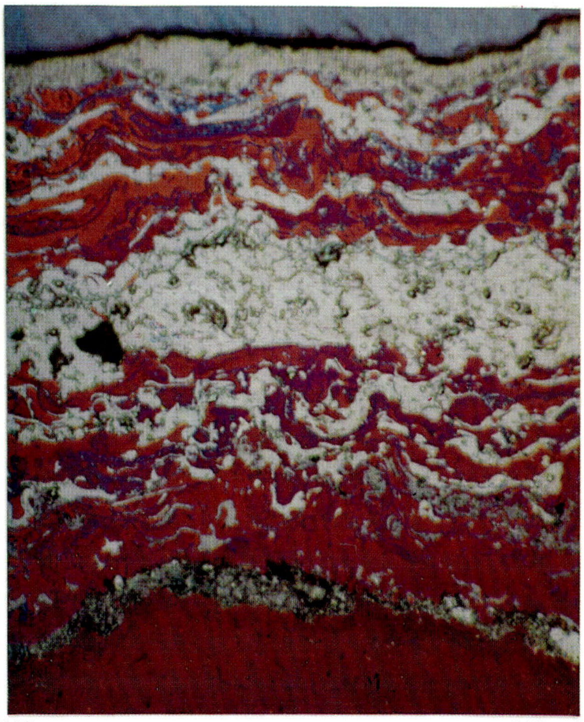

263 200 : 1

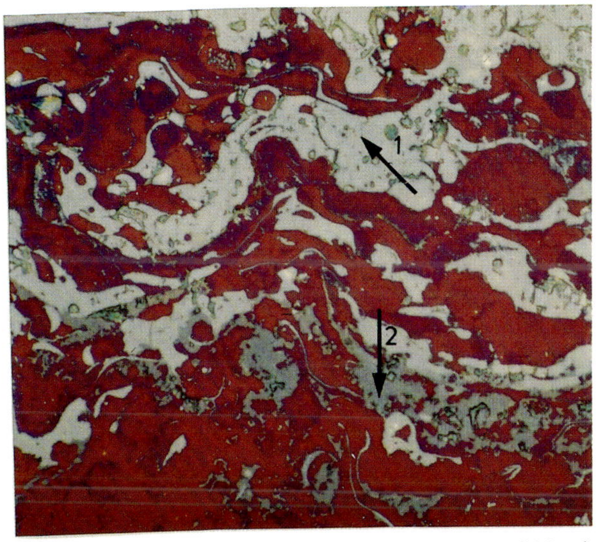

264 500 : 1

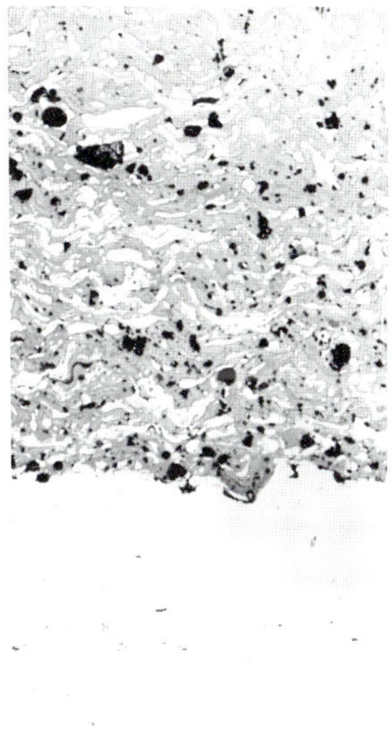

265 500 : 1

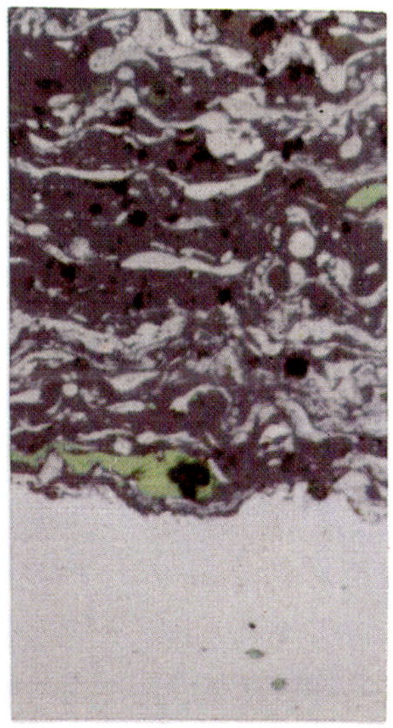

266 500 : 1

267 200 : 1

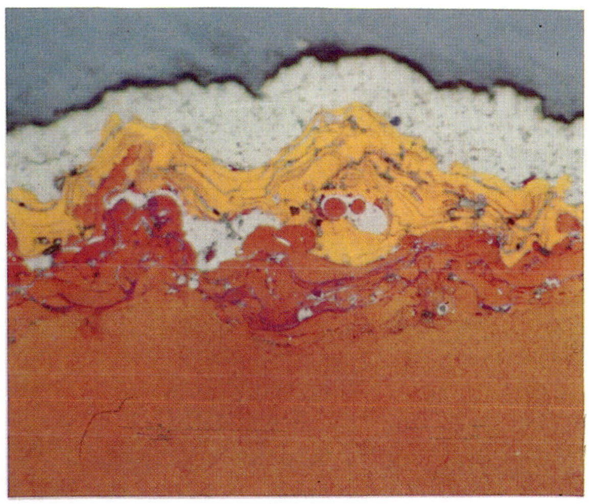

268 200 : 1

Figs. 267 and 268: **Steel specimen with plasma sprayed layers in the order: Nickel-chromium-aluminium alloy** (20% Cr, 6% Al, remainder Ni), **Chromium and ($ZrO_2 + Y_2O_3$)-mixture.**

267: Specimen polished, bright field.

In the polished condition the metallic constituents steel, nickel-chromium alloy, and chromium cannot be differentiated on the basis of reflectivity.

ZrO_2 and Y_2O_3 appear grey

268: Same specimen, same area of the microstructure as in Fig. 267, sputtered with an iron cathode and oxygen as reaction gas, zero order, reflection minimum of steel at 490 nm, bright field.

steel matrix: red (7 : 4 : 3)

nickel-chromium-aluminium alloy: brown to red (7 : 4 : 3 to 10 : 4 : 4)

chromium: yellow (3 : 6 : 2)

$ZrO_2 + Y_2O_3$: pale green

The mounting medium in the upper half of the picture is a bluish colour. The different constituents of the layer can be distinguished clearly.

Figs. 269–276: **Plasma sprayed nickel layer on St 37**

269: Polished, unetched, bright field.

The different phases cannot be differentiated. All that can be seen on the section are various shades of grey.

270: Same material as in Fig. 269, sputtered with an iron cathode using oxygen as reaction gas, zero order, reflection minimum of the steel at 485 nm, bright field.

steel matrix: light grey (lower half of picture)

nickel layer: medium to dark grey in the region of the sprayed zone

oxides: white (at the steel/plasma layer interface and in the sprayed layer)

The different regions of the layer can be distinguished much better than in Fig. 269.

271: Same material, same area of the specimen and same coating layer as in Fig. 270.

steel matrix: yellow (2 : 4 : 2)

nickel layer: yellowish brown (3 : 4 : 2)

oxides: red (10 : 4 : 4) (streaky, embedded in the nickel layer) and light blue (17 : 3 : 2) (at the steel matrix/plasma layer interface)

272: Same material, same specimen, same area of the microstructure and same coating as in Fig. 271, photographed at 540 nm.

steel matrix: light grey

nickel layer: dark grey to black

oxides: white, especially in the region of the steel/plasma layer interface

At the 540 nm filter setting, a noticeably better contrast is possible than in bright field as in Fig. 270.

Figs. 273–276: See page 154

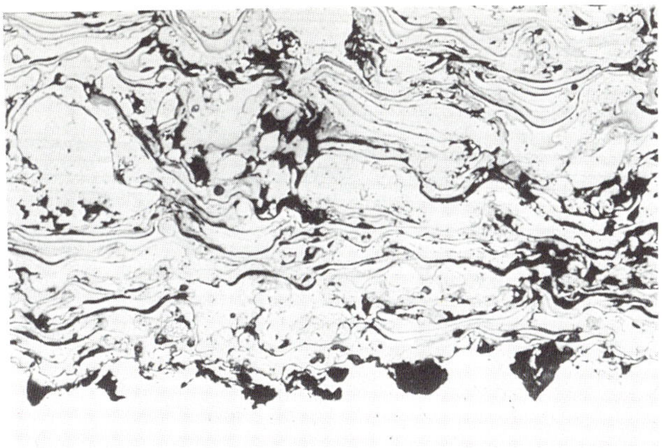
269 200 : 1

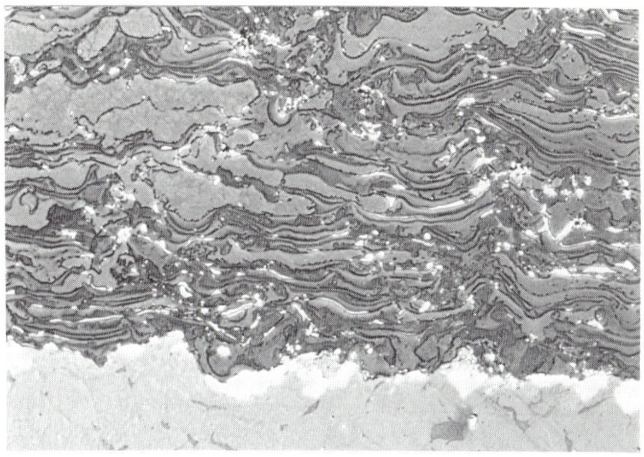
270 200 : 1

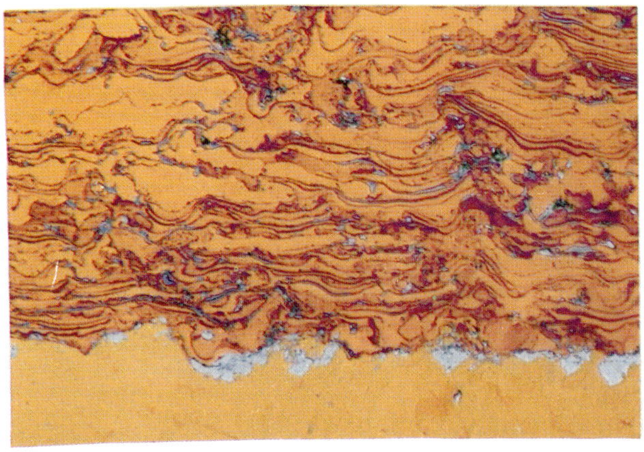
271 200 : 1

272 200 : 1

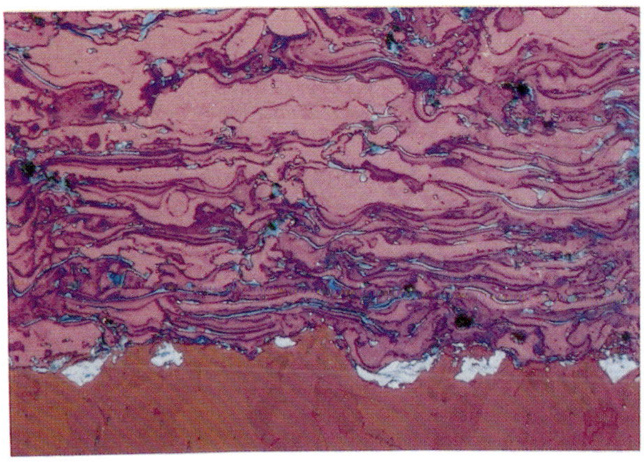
273 200 : 1

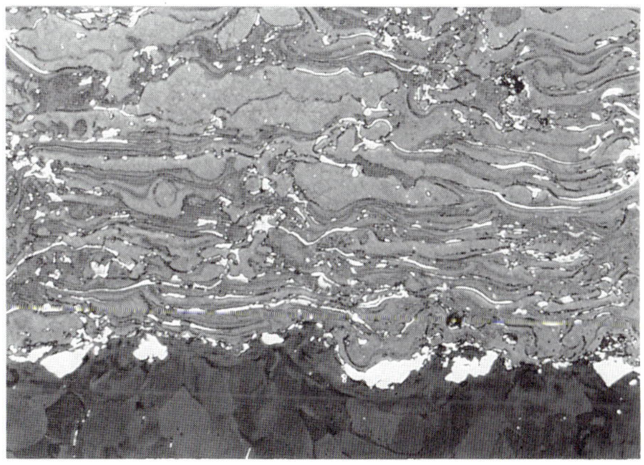
274 200 : 1

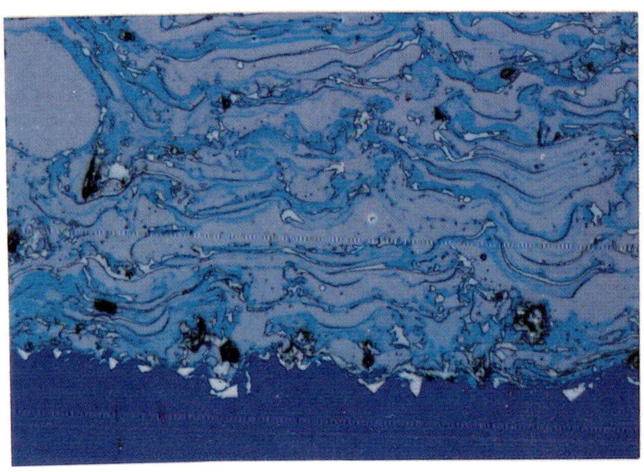
275 200 : 1

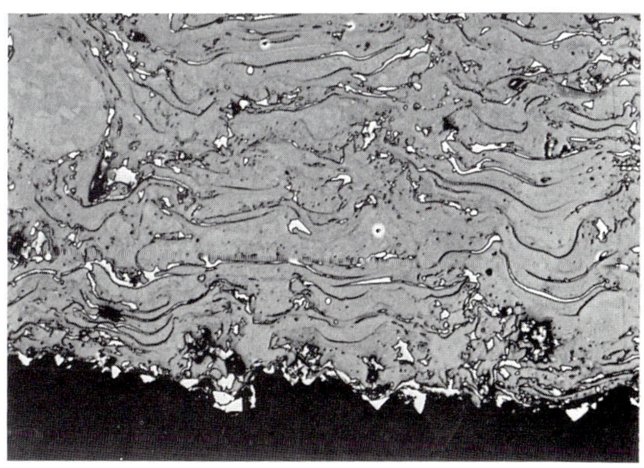
276 200 : 1

273: Same specimen and same coating as in Fig. 270, reflection minimum of the steel at 485 nm, bright field.

steel matrix: red (7 : 2 : 3)

nickel layer: light violet (10..11 : 2 : 3)

oxides: turquoise (18 : 3 : 2) and dark violet (11 : 3 : 4)

Due to the somewhat greater layer thickness the difference in hue between the phases has been increased.

274: Same specimen and same coating as in Fig. 273, bright field.

steel matrix: dark grey

nickel layer: light to medium grey

oxides: white and dark grey

275: Same specimen and same coating as in Fig. 273, reflection minimum of the steel matrix lies at 570 nm, bright field.

matrix: dark blue (16 : 5 : 4)

nickel layer: light blue (17..18 : 2 : 2)

oxides: bluish white to medium blue (18 : 4 : 2)

276: Same specimen, same coating and same area of the microstructure as in Fig. 275, bright field.

steel matrix: black

nickel layer: uniformly grey

oxides: white

By means of this coating the black and white contrast in bright field has been further improved.

2.1.8.4 Metal ceramic composites

Figs. 277 to 279: **Al_2O_3-$TaAl_3$-Ta composite**

277: Polished, bright field.

tantalum: light grey, white

$TaAl_3$: white

Al_2O_3: dark grey

pores: black

Because of its transparency (low absorption), the Al_2O_3 phase on the left hand side of the picture appears dark compared with the highly reflecting tantalum. The intermetallic compound $TaAl_3$, located between the two phases Al_2O_3 and tantalum, is difficult to see on a polished specimen.

278: Same material, same area of the microstructure, sputtered with an iron cathode and oxygen as reaction gas, reflection minimum of tantalum at 530 nm, photographed at 560 nm.

Al_2O_3: light grey

$TaAl_3$: white

tantalum: dark grey

pores: grey to black

The coating layer reduces the reflectivity of the highly reflecting metal (tantalum) and increases that of the transparent Al_2O_3. The intermetallic compound $TaAl_3$ now stands out clearly against the two other components in good light-dark contrast.

279: Same material, same area of the microstructure as in Fig. 277, same coating as in Fig. 278, bright field.

Al_2O_3: grey (– : 0 : 1.5)

$TaAl_3$: red (7 : 4 : 3)

tantalum: yellow (3 : 4 : 2)

Figs. 280 and 281: **Al_2O_3-copper/silver/copper-Al_2O_3 composite**

280:

Polished, bright field.

Al_2O_3: black with pores

copper: dark grey

silver: light grey

The aluminium oxide looks dark on the polished section because of its low absorption whereas the metallic components appear bright on account of their high reflectivity. Silver is more highly reflecting than copper. The copper inclusions in the silver layer appear dark grey.

281:

Same material, same area of the microstructure as in Fig. 280, sputtered with an iron cathode and oxygen as reaction gas, zero order, bright field.

Al_2O_3: brownish red (6 : 2 : 4)

copper: light red (8 : 5 : 4)

silver: dark yellow (3 : 5 : 2)

The contrast between the phases is considerably greater than in Fig. 280. The pores may be identified easily.

Fig. 282: **Al_2O_3 (97.5%) + SiO_2 (2.5%) -niobium composite**

Polished, sputtered with an iron cathode and oxygen as reaction gas, zero order, bright field.

Al_2O_3: yellow (2 : 4 : 2)

niobium: red (7 : 5 : 3 or 9 : 3 : 5) colour variable

pores: black

The diffusion of metallic niobium into the ceramic mass along the grain boundaries of the Al_2O_3 is worthy of note.

Figs. 283 and 284: **Al_2O_3 (99.7%)-niobium-molybdenum-niobium-Al_2O_3 (99.7%) composite**

283:

Polished, bright field.

Al_2O_3: dark grey with black pores

molybdenum: almost white

niobium: white

The comment as regards contrast and the difficulty in identifying the phases in this photograph is the same as for Fig. 277 or Fig. 280.

284:

Same area of the microstructure and same material as in Fig. 283, vacuum deposited ZnS layer, bright field.

Al_2O_3: light green (24..1 : 1 : 2)

niobium: violet (10 : 4 : 4)

molybdenum: brownish red (7 : 4 : 3)

A phase has formed within the niobium similar in colour to the molybdenum. This phase was not visible before coating. Its chemical composition is unknown.

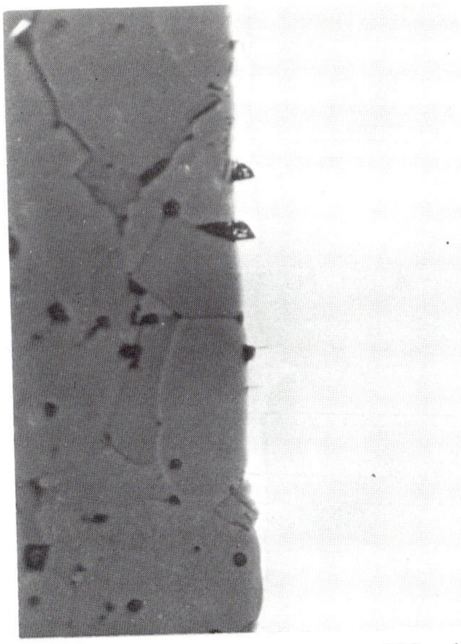

277 500 : 1

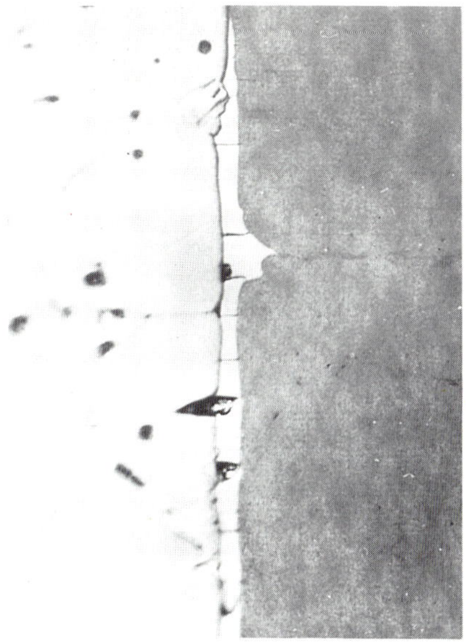

278 500 : 1

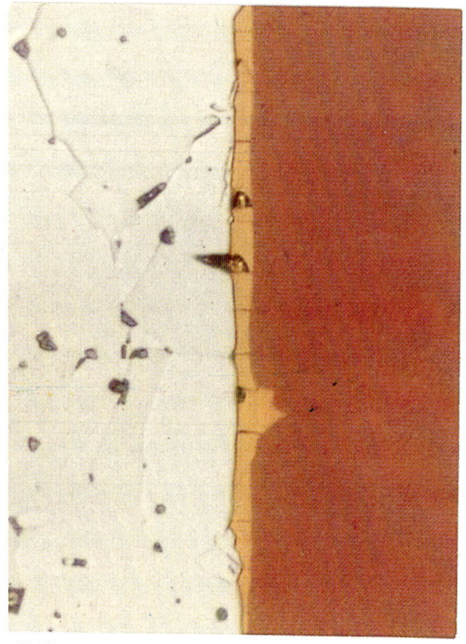

279 500 : 1

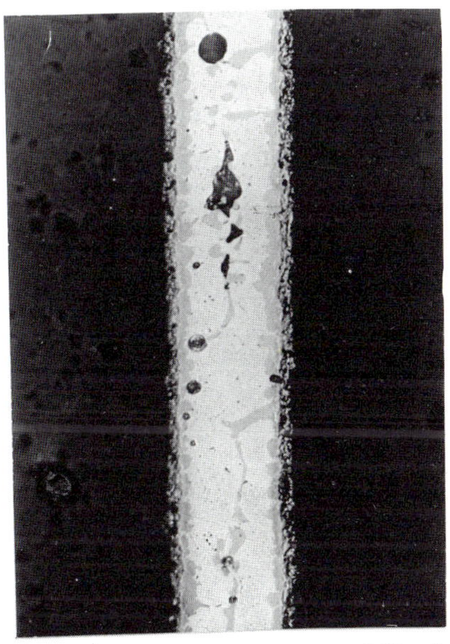

280 200 : 1

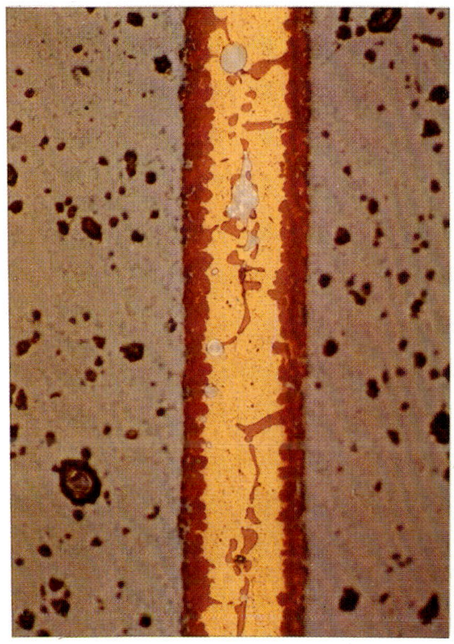

281 200 : 1

282 100 : 1

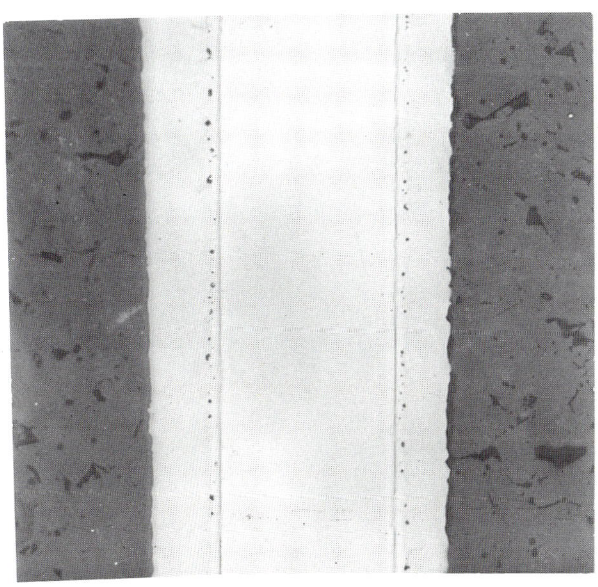

283 200 : 1

284 200 : 1

2.1.9 Brazed joints

2.1.9.1 General Introduction

The metals and alloys involved in brazed joints very often possess widely differing optical properties. This is a good prerequisite for the successful use of interference layer metallography (c.f. sections 2.1.7 and 2.1.8).

High alloy steels and nickel alloys brazed with high melting nickel base brazing alloys provide suitable examples of brazed joints. The joining of steels to tantalum by means of Ag-Pd-Cu brazing metals and the metallographic preparation of a brazed "sandwich" will also be discussed.

Literature to 2.1.9

1) Draugelates, U., Wielage, B. and Hartmann, K.-H.: Cracking susceptibility and microstructure of NiCr 20 TiAl brazed joints. Welding Journal **54** (1975) 344

2) Hartmann, K.-H., Draugelates, U. and Steffens, H.-D.: Beeinflussung der Lötnahteigenschaften bei großen Spaltbreiten durch Wärmebehandlung. DVS-Berichte **38,** 72. Deutscher Verlag für Schweißtechnik, Düsseldorf (1975)

3) de Paoli, A. and Colbus, J.: Gefügeaufbau von Lötverbindungen aus Hochtemperaturwerkstoffen mit Loten auf Ni-Cr-Basis und dessen Einfluß auf die Zähigkeit der Verbindung. DVS-Berichte **32,** 115. Deutscher Verlag für Schweißtechnik, Düsseldorf (1974)

4) Guard, R. W. and Smith, E. A.: Constitution of Nickel-Base-Ternary Alloys. V. Nickel-Chromium-Silicon-System. Inst. of Metals, Journal, **88** (1960) 373

5) Lange, H.: Anwendung des Hochtemperaturlötens im Vakuum zum stoffschlüssigen Verbinden warmfester Legierungen. Metall **26** (1972) 814

6) Gräf, I. and Draugelates, U.: Phasenidentifizierung in Nickelloten mit modernen metallographischen Techniken. Special Volume **7** of Prakt. Metallographie. Metallographie oberflächenveredelter Stähle, Schweiß- und Lötverbindungen, ed. by H.-E. Bühler, Hillnhagen, E. and E. Horn. Riederer-Verlag (1977) 311

7) Lang, W.: Differential-Interferenzkontrastmikroskopie nach Nomarski. Zeiss Informationen **16** (1968) 114

8) Gräf, I., Hartmann, K.-H. and Draugelates, U.: Metallographische Untersuchung von Lötnähten aus Nickelbasislegierungen mit Kontrastierverfahren. Prakt. Metallographie **14** (1977) 87

9) Steffens, H.-D. and Hartmann, K.-H.: Untersuchungen an Hochtemperaturlötverbindungen aus NiCr 20 TiAl. Schweißen und Schneiden **29** (1977) 431

10) Petzow, G.: Metallographisches Ätzen. 5th edition, Verlag Gebr. Bornträger, Berlin-Stuttgart (1976)

2.1.9.2 Brazed joints between X2 NiCoMo 18 8 5 and NiCr20TiAl using BNi-1 or BNi-5

High alloy steels and nickel alloys which are difficult to weld can in many cases be brazed with high melting nickel base alloys. The brazed joint must be highly resistant to thermal stresses. As the brazing metal and parent metal cool, numerous hard phases can precipitate, depending on the composition of the alloys and rate of cooling. The nature and configuration of these phases have a decisive influence on the properties of the brazed joint 1–5).

The metallographic assessment of these precipitates is very tedious because it is not possible to reveal all the phases at once by conventional etching methods 6).

The phases occurring in the brazing metal are often of different hardness with the result that a relief is formed on polishing. This relief can be shown up by differential interference contrast 6–8). If polishing is prolonged the relief may be even more pronounced and under certain circumstances this may give a false picture of the microstructure. Identification of the phases is not possible by this method.

Deposition of a surface layer, however, allows all the phases to be differentiated quickly and clearly 6,8,9). In addition, a preliminary etch with oxalic acid shows up all the grain boundaries in the parent metal. Before coating, the specimens must be prepared carefully and under constant conditions so that any relief resulting from the different hardness of the individual phases is avoided and the true microstructure can be observed.

Figs. 299 to 318 show microstructures of NiCr20TiAl or X2NiCoMo 18 8 5 vacuum brazed with BNi-5 or BNi-1. The best black and white contrasts are obtained with these brazing metals if after deposition of a coating layer, either evaporated or sputtered with an iron cathode in oxygen, the background is light yellow and the phases appear dark yellow to red.

Figs. 291 to 296 illustrate the effect of an evaporated layer on the appearance of the phases in brazing metal BNi-5. This alloy is well suited to such a comparison on account of the five phases present. The influence of increasing the thickness of a sputtered layer has already been illustrated by Figs. 30 to 35. Figs. 291 to 296 show the microstructure after evaporation of the slightly absorbing ZnSe to give zero and first order layer. Compared with the sputtered layer produced using an iron cathode in oxygen, c.f. Figs. 30 to 35, the saturation of the colours is greater because the absorption coefficient of ZnSe is only small. There is scarcely any change in the colour saturation between zero order (Figs. 291 to 293) and first order (Figs. 294 to 296). In contrast to the specimens shown in Figs. 24 to 35 the specimen shown in Fig. 291 received a second coating until the colour shown in Fig. 292 was reached. Because of this, slight shifts are possible as compared with coatings applied continuously.

The chemical compositions of the materials described in the following are given in the table below.

Material	Chemical composition wt. %									
	Ni	Cr	Si	B	Fe	C	Ti	Al	Co	Mo
NiCr20TiAl	72	20	—	—	4.0	0.06	2.4	1.4	—	—
X2NiCoMo 18 8 5	18	—	—	—	69	0.02	0.45	0.1	8	4.7
Braze metal BNi-5	71	19	10	—	—	0.1	—	—	—	—
Braze metal BNi-1	73	14	4.5	3.5	4.5	0.7	—	—	—	—

Microstructure of the braze metal

Figs. 285–290: **Brazing metal BNi-5**

A brazed joint was made in a vacuum furnace at 10^{-5} mb and 1190°C. On account of the long period allowed for cooling, ca. 5 h, the phases grew to large plates. Only the microstructure of the brazing metal is shown in the photographs. See Figs. 299–307 for the brazing metal – parent metal transition zone.

285: Brazing metal BNi-5

Sputtered with an iron cathode in oxygen, bright field.

Four different silicides can be seen in the matrix 9):

a cube-like silicide with 42% Ni, 45% Cr and 11% Si with a microhardness of 1350 HV 0.05, colour (3 : 6 : 2) [1];

a silicide with 68% Ni, 15% Cr and 15% Si, hardness 1000 HV 0.05, colour (4 : 5 : 2) [2];

a silicide with 75% Ni, 7% Cr and 16% Si, hardness 870 HV 0.05, colour (7 : 5 : 3) [3];

a phase which has yet to be identified, in the form of needles, colour (2 : 5 : 1) [4].

The green patches are impurities which are also found in the parent metal.

286: Same material, same brazing conditions, same area of the microstructure, same preparation as in Fig. 285.

Photographed at 540 nm.

The phases can be differentiated just as easily in the black and white photograph by the different grey tones as in the colour photograph in Fig. 285.

287: Same material, same brazing conditions, same area of the microstructure as in Fig. 285.

Etched electrolytically with 5% oxalic acid, sputtered with an iron cathode in oxygen, differential interference contrast.

The preliminary etch and also the use of differential interference contrast accentuate the relief more strongly so that the outlines of the phases stand out better against the matrix and the contrast is even better than after simple coating. The needle-like phase shows up particularly clearly in this photograph.

288: Same material, same brazing conditions, same area of the microstructure and same preparation as in Fig. 287.

Photographed at 540 nm.

The contrast is better than in Fig. 286 but the large cube-shaped precipitates are shadowed from the edge to the middle which could lead to errors in an automatic structure evaluation. These shadows do not appear in Fig. 286. In a quantitative structure assessment the relief can lead to errors in the case of small particles e.g. of the areal proportions.

289: Same material, same brazing conditions and same area of the microstructure as in Fig. 285.

Etched electrolytically with 5% oxalic acid, differential interference contrast.

Compared with Figs. 285 to 288, the contrast is so slight that clear differentiation of the 4 phases from the matrix is not possible although the outlines of the phases stand out well against the matrix as a result of the preliminary etch.

290: Same material, same brazing conditions and same area of the microstructure as in Fig. 285.

Polished only, bright field.

The light coloured, needle-like precipitate is clearly visible. Differentiation of the other 3 phases is still possible from the shape but not from the contrast or grey tone.

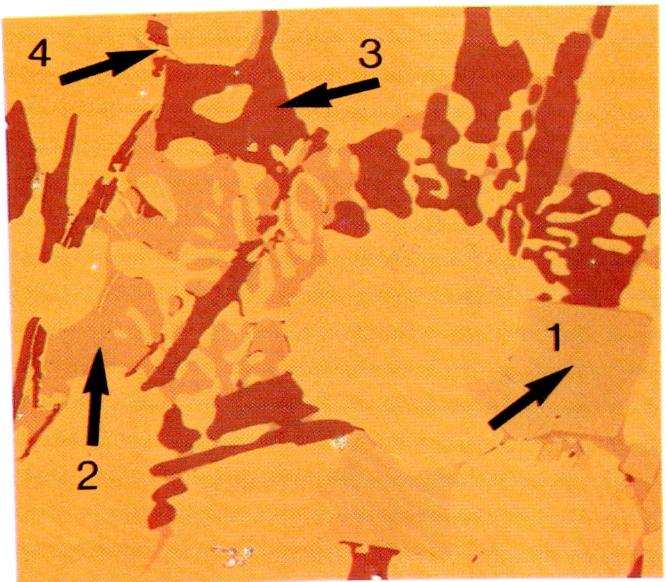

285 500 : 1

286 500 : 1

287 500 : 1

288 500 : 1

289 500 : 1

290 500 : 1

Figs. 291–296 **Brazing metal BNi 5**

291: Brazed in vacuum furnace at $5 \cdot 10^{-5}$ mb and 1190^0C. Light electrolytic etch with 5% oxalic acid, coated with ZnSe, zero order, photographed in differential interference contrast.

As in Figs. 285 to 290, four phases are visible within the matrix but the colour differences are not so great. The colours of the phases are given in the table below.

292: Same material, same brazing conditions, same area of the microstructure same preparation and same photographic conditions as in Fig. 291.

Given a second coating of ZnSe, zero order, layer thickness greater than in Fig. 291. The colours of the phases are given in the table below.

293: Same material, same brazing conditions, same area of the microstructure, same preparation and same photographic conditions as in Fig. 292.

Given a third coating of ZnSe, zero order, layer thickness greater than in Fig. 292. The colours of the phases are given in the table below. Colours were taken from the central area of the photograph.

Fig.	Colour Matrix	Phase 1	Phase 2	Phase 3	Phase 4
291	3...4:6:1	4:5:1..2	not determ.	6:5:2	not determ.
292	6:5:3	8:5:4		11:3..4:5	
293	22:5:4	22:6:4		23:5:3	

294: Same material, same brazing conditions but a different area of the microstructure, same preparation and photographic conditions as in Fig. 291.

Coated with ZnSe, first order, i.e. the layer is thicker than in Fig. 293. The colours of the phases are given in the table below.

In this area of the specimen and for the given conditions only one other phase can be differentiated from the matrix on the basis of colour.

295: Same material, same brazing conditions, same area of the microstructure, same preparation and same photographic conditions as in Fig. 294.

Given a second coating of ZnSe, first order, layer thicker than in Fig. 294. The colours of the phases are given in the table below.

Two other phases are visible besides the matrix.

296: Same material, same brazing conditions, same area of the microstructure, same preparation and same photographic conditions as in Fig. 295.

Given a third coating of ZnSe, first order, layer thicker than in Fig. 295. The colours of the phases are given in the table below.

As in Fig. 294 only one other phase is visible besides the matrix.

Fig.	Colour Matrix	Phase 1	Phase 2
294	3:5:2	3:6:1..2	3:6:1..2
295	6:4:3	6:3:2..3	4..5:4:2
296	22...23:5:3	22:6:3	22:6:3

This sequence shows how the ease of differentiation of phases depends on layer thickness for a given layer material and illustrates the differences between this and other types of coating, e.g. Figs. 285 to 290.

Figs. 297 and 298: See page 164

291 500 : 1

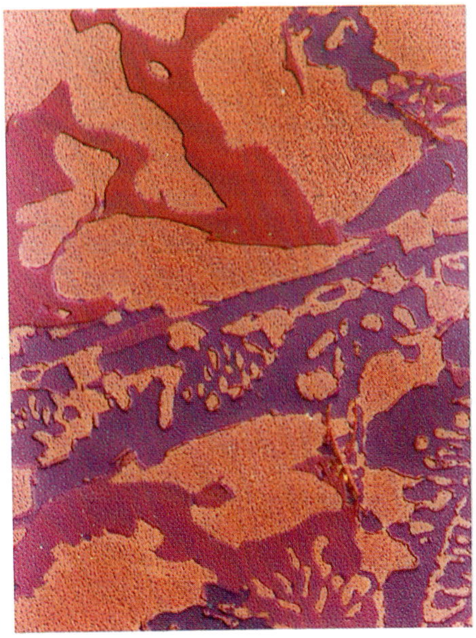

292 500 : 1

293 500 : 1

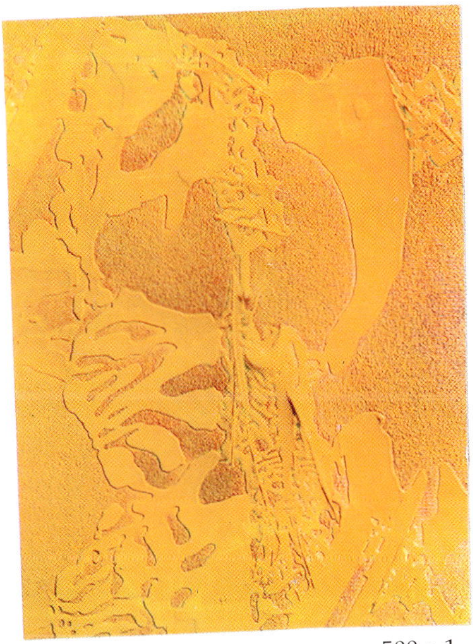

294 500 : 1

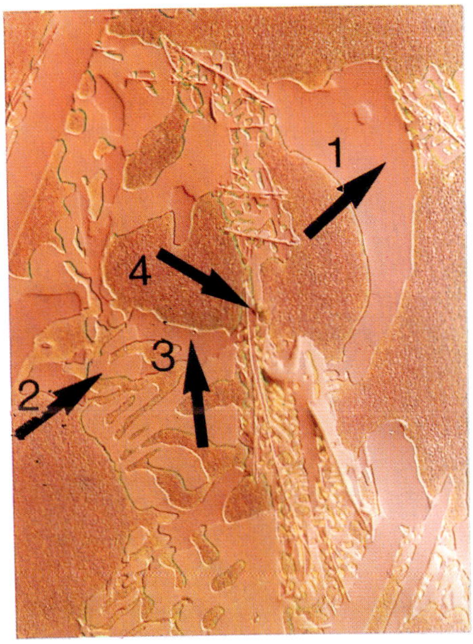

295 500 : 1

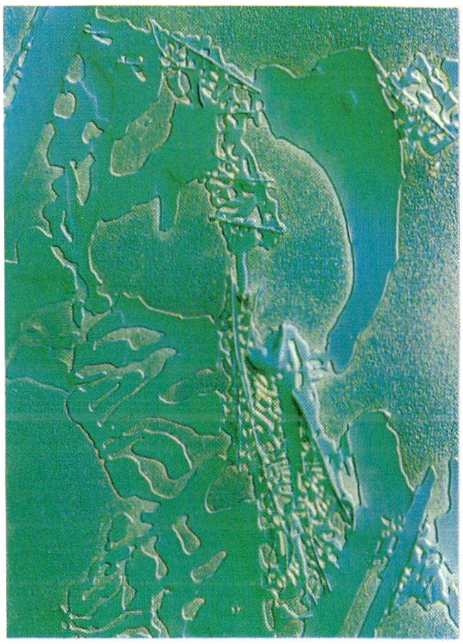

296 500 : 1

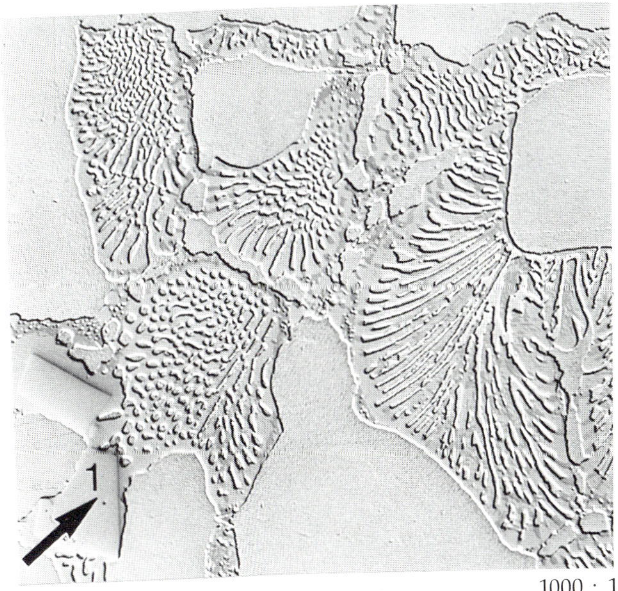

297 1000 : 1

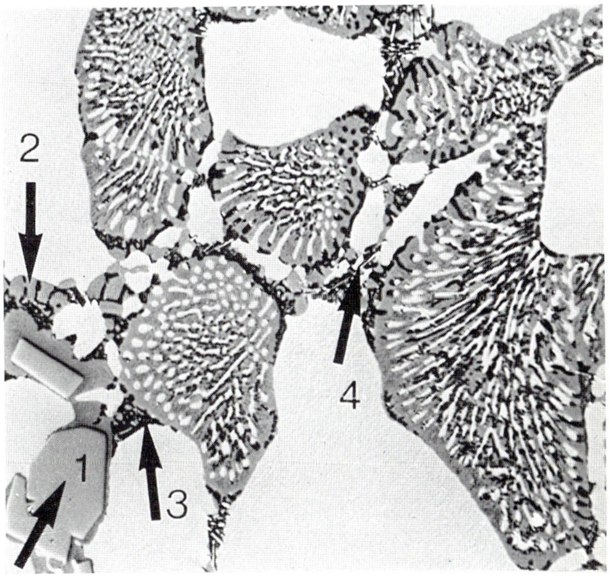

298 1000 : 1

Induction brazing

Brazed joints can also be produced by inductive heating in which pressure and temperature are comparable with those in vacuum brazing. After induction brazing, however, the rate of cooling is more rapid than after brazing in the vacuum furnace. The phases that precipitate out on cooling are therefore very finely dispersed and can be differentiated only by coating. If the specimen is first etched with oxalic acid, it may mean that after coating the microstructure can no longer be resolved.

Figs. 297 and 298: **Brazing metal BNi-5**

297: Induction brazed at $5 \cdot 10^{-5}$ mb and 1190^0C. Etched electrolytically with 5% oxalic acid, differential interference contrast.

After etching a microstructure is visible within the brazing metal but the number of phases cannot be determined.

298: Same material, same brazing conditions and same area of the microstructure as in Fig. 297.

Sputtered with an iron cathode in oxygen, differential-interference contrast.

After coating, four silicides designated [1] to [4] can once again be detected. Apart from the cube-shaped phase they are very finely dispersed. The white needles described in Figs. 285 to 288 can just be seen. Compared with Fig. 297 the resolution is much improved.

2.1.9.3 **Evaluation of the microstructure of the brazing metal/parent metal transition zone**

In the assessment of a brazed joint the bonding to the parent metal is as important as the microstructure of the brazing metal. An oxalic acid etch is necessary to reveal the brazing metal/parent metal transition zone. This simultaneously shows up the grain boundaries in the parent metal. Since an assessment of the brazing metal/parent metal transition is not concerned with fine structure, such an etch is eminently reasonable. In the notes to the following photographs, therefore, the phases within the brazing metal will not be discussed. They have already been described in Figs. 285 to 298.

Figs. 299–307: **NiCr20TiAl brazed with BNi-5**

299: Parent metal: steel NiCr 20 TiAl (right hand side).

Brazing metal: BNi-5 (left hand side).

Vacuum brazed at $5 \cdot 10^{-5}$ mb and 1190^0C.

Unetched, differential interference contrast.

The boundary between brazing metal and parent metal is only faintly visible.

300: Same materials, same brazing conditions and same area of the microstructure as in Fig. 299.

Sputtered with an iron cathode in oxygen, differential interference contrast.

Colour of the parent metal matrix: (2 : 6 : 2), colour of the brazing metal matrix: (2 : 5 : 2). The difference in colour between brazing metal and parent metal is still only slight even after sputtering.

301: Same materials, same brazing conditions, same preparation and same area of the microstructure as in Fig. 300.

Photographed at 540 nm.

In monochromatic light the contrast between brazing metal and parent metal is so much enhanced that the boundary between them is easily visible.

Figs. 302–307: See page 166

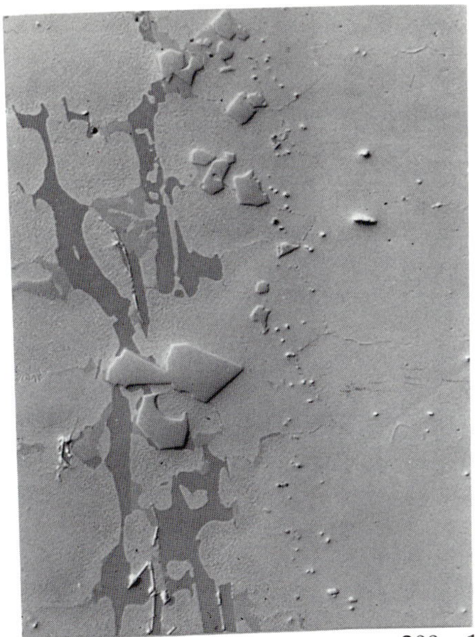

299 200 : 1

300 200 : 1

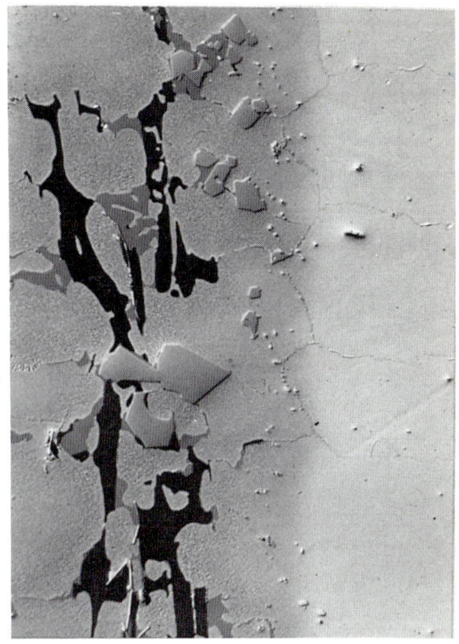

301 200 : 1

302 200 : 1

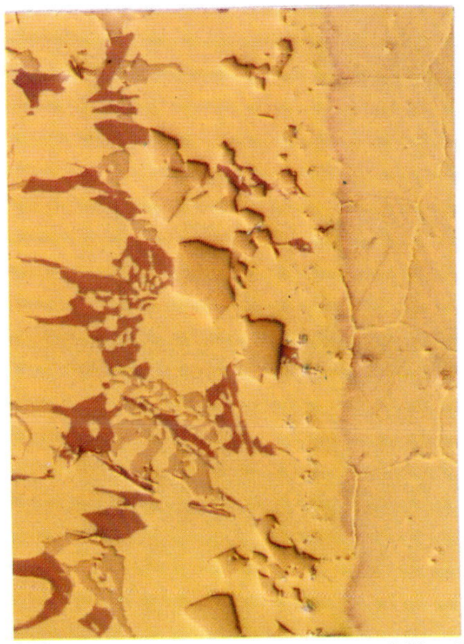

303 200 : 1

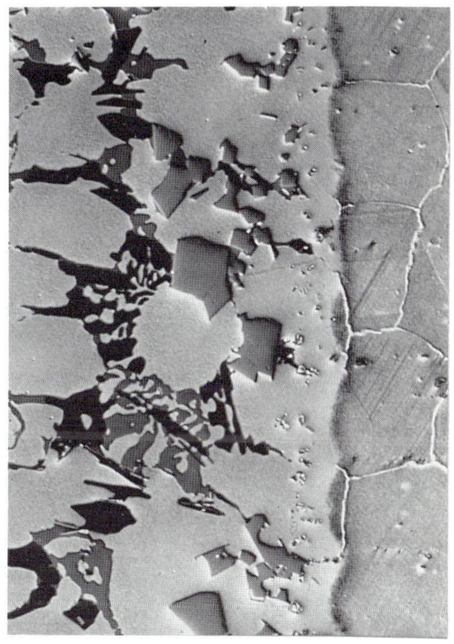

304 200 : 1

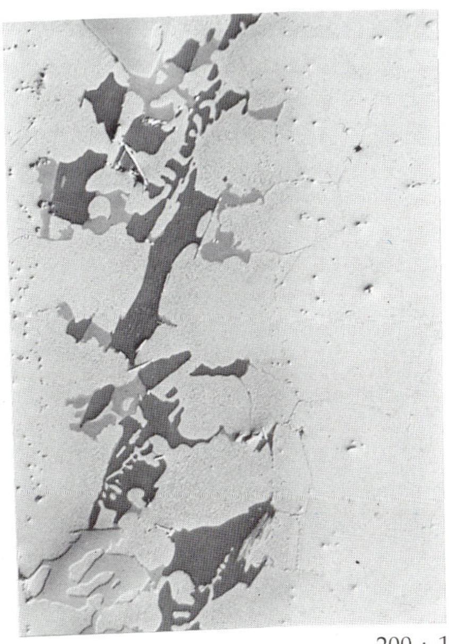

305 200 : 1

306 200 : 1

307 200 : 1

302: Same materials, same brazing conditions and same relative positions of parent metal and brazing metal as in Fig. 299.

Etched electrolytically with 5% oxalic acid, differential interference contrast, photographed at 540 nm.

Two precipitated phases can be distinguished in the brazing metal on account of their different brightness (c.f. Fig. 298). The brazing metal-parent metal boundary is easily visible. The grain boundaries show up clearly in the parent metal.

303: Same materials, same brazing conditions and same area of the microstructure as in Fig. 302.

Etched electrolytically with 5% oxalic acid, sputtered with an iron cathode in oxygen. Differential interference contrast.

Colour of the parent metal matrix: (3 : 3 : 2). Colour of the brazing metal matrix: (2 : 5 : 2). The difference in colour between the brazing metal and parent metal is only slight.

304: Same materials, same brazing conditions, same preparation and same area of the microstructure as in Fig. 303.

Photographed at 540 nm.

In monochromatic light the contrasts are so enhanced that it is easier to distinguish the brazing metal and parent metal than in the colour photograph, Fig. 303. After coating the contrast is much improved as compared with the uncoated specimen in Fig. 302. As a result of etching, the grain boundaries, faintly indicated in Fig. 301, have become clearly visible.

Figs. 305–307: **Ageing after vacuum brazing**

In order to improve the high temperature strength of the parent metal the brazed joints are aged for 16 h at 710^{0}C. During this time additional phases precipitate out in the parent metal which alter the optical properties. If the specimens are sputtered after being etched with oxalic acid, the parent metal can easily be distinguished from the brazing metal.

305: Same materials, same brazing conditions and same relative positions of parent metal and brazing metal as in Fig. 299.

Heat treatment after brazing: 710^{0}C 16 h/air.

Sputtered with an iron cathode in oxygen, photographed at 540 nm.

Essentially no different from the microstructure in Fig. 301.

306: Same materials, same brazing conditions and heat treatment, and same area of the microstructure as in Fig. 305.

Etched electrolytically with 5% oxalic acid, sputtered with an iron cathode in oxygen. Differential interference contrast.

The brazing metal (matrix: colour 3 : 6 : 2) differs clearly in colour from the parent metal (colour 6 : 5 : 3). The etch shows up the grain boundaries, even in the brazing metal.

307: Same materials, same brazing conditions, same preparation and same area of the microstructure as in Fig. 306.

Photographed at 540 nm.

The contrast between brazing metal and parent metal is noticeably greater than in the brazed joint that has not been treated, Fig. 304.

Figs. 308–313: **NiCr20TiAl brazed with BNi-1**

In this brazing metal some of the precipitates are so fine that they can be seen in the light microscope only after very careful preparation and coating (c.f. Fig. 312). Etching results in falsification of the areal proportions (c.f. Fig. 313).

308: Parent metal NiCr20TiAl (right hand side), brazing metal BNi-1 (left hand side). Vacuum brazed at $5 \cdot 10^{-5}$ mb and 1175^0C, polished only.

The individual phases in the brazing metal and the transition between brazing metal and parent metal are only faintly visible.

309: Same materials, same brazing conditions and same relative positions of parent metal and brazing metal as in Fig. 308. Beraha etch.

The individual grains in the brazing metal and parent metal are very clearly visible but individual phases within the brazing metal cannot be differentiated.

310: Same materials, same brazing conditions and same area of the microstructure as in Fig. 308.

Sputtered with an iron cathode in oxygen.

Photographed at 540 nm.

The individual phases within the brazing metal can clearly be differentiated. The grain boundaries are visible where they are occupied by precipitates.

Figs. 311–316: See page 168

311: Brazing metal BNi-1, brazing conditions as in Fig. 308. Sputtered with an iron cathode in oxygen, differential interference contrast.

The colour photograph shows the marked differences in colour between the individual phases, the exact composition of which has not yet been determined.

312: Same materials, same brazing conditions and preparation as in Fig. 311.

Photographed at 540 nm.

Very careful preparation and coating is necessary if the fine structure within the brazing metal is to be resolved at high magnifications.

313: Same materials and brazing conditions as in Fig. 308.

Etched with 5% oxalic acid, sputtered with an iron cathode in oxygen, differential interference contrast.

Photographed at 540 nm.

The resolution of the fine structure of the brazing metal is worsened by etching.

Figs. 314–316: **X 2 NiCoMo 18 8 5 brazed with BNi-5 or BNi-1**

In the case of X 2 NiCoMo 18 8 5 brazed with BNi-5 or BNi-1, the differences in composition between parent metal and brazing metal is considerable. These materials have such great differences in electrochemical potential that the phases in the brazing metal cannot be etched in the presence of the parent metal. Since a relief is usually produced on polishing, the parent metal and brazing metal can easily be distinguished by differential interference contrast. Further phases can be distinguished within the brazing metal. A considerable improvement in contrast is however achieved by deposition of an interference layer.

314: Parent metal X 2 NiCoMo 18 8 5 (right and left) brazing metal BNi-5 (centre).

Vacuum brazed at 10^{-5} mb and 1190^{0}C. Unetched, differential interference contrast.

The brazing metal stands out clearly from the parent metal. Silicide precipitates stand out dark in the light coloured brazing metal matrix.

315: Same materials, same brazing conditions and same relative positions of parent metal and brazing metal as in Fig. 314. Sputtered with an iron cathode in oxygen.

Differential interference contrast.

The contrast is considerably better than in Fig. 314.

316: Same materials and same brazing conditions as in Fig. 314. Sputtered with an iron cathode in oxygen. Differential interference contrast.

The resolution of the finely dispersed silicide precipitates in the brazing metal is excellent.

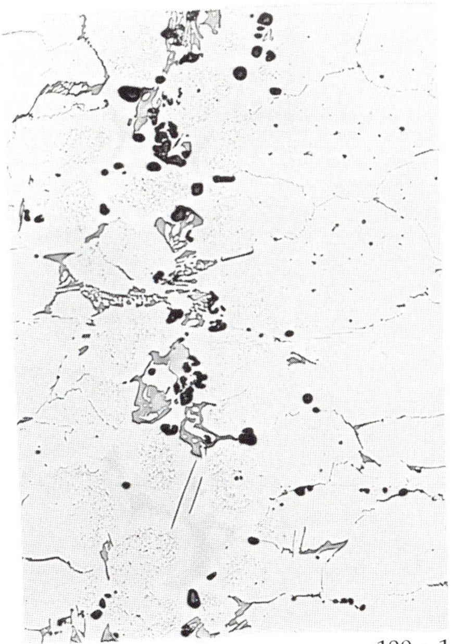

308 100 : 1

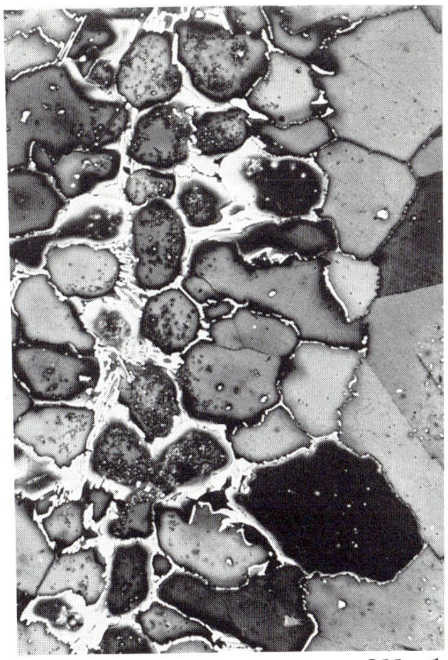

309 200 : 1

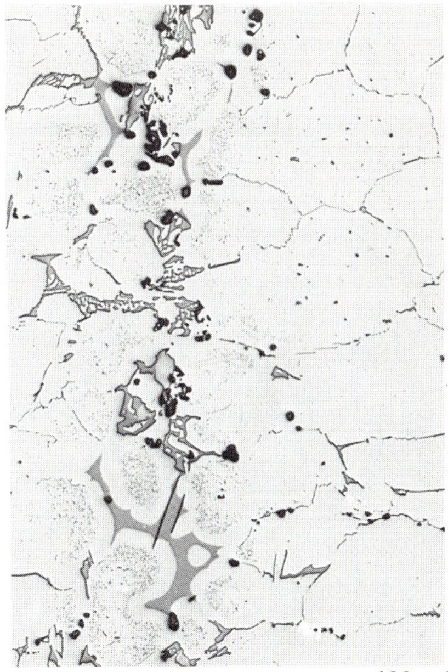

310 100 : 1

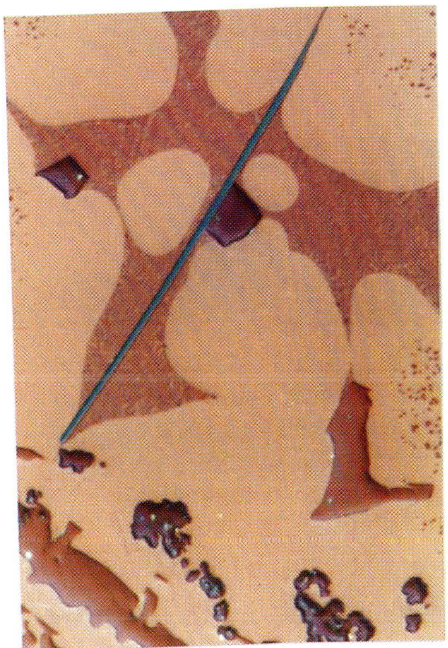

311 500 : 1

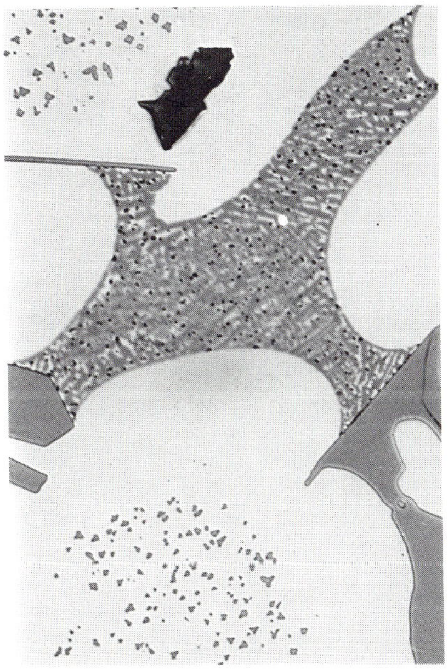

312 1000 : 1

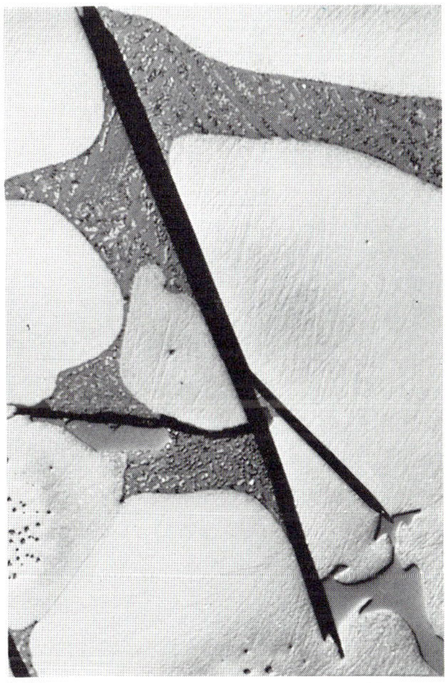

313 1000 : 1

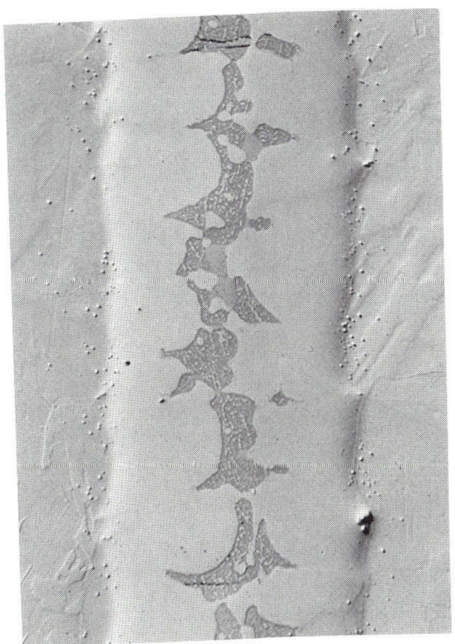

314 200 : 1

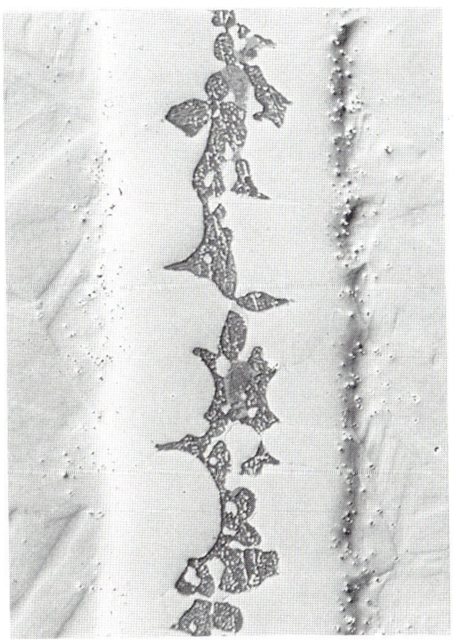

315 200 : 1

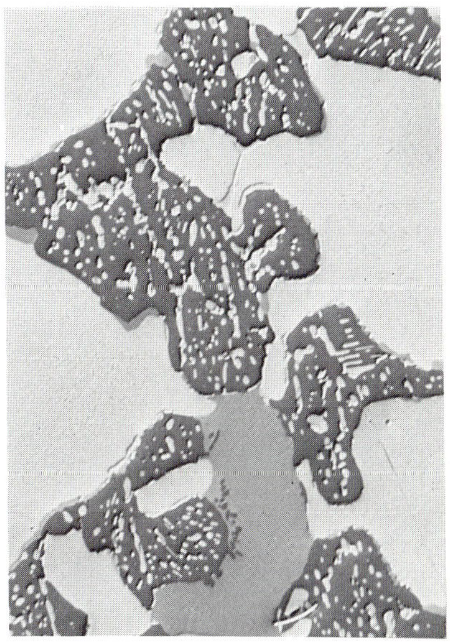

316 1000 : 1

317: Parent metal X 2 NiCoMo 18 8 5 (left and right), brazing metal BNi-1 (centre).

Vacuum brazed at $5 \cdot 10^{-5}$ mb and 1175^{0}C, unetched.

The brazing metal stands out only faintly from the parent metal.

318: Same materials, same brazing conditions and same area of the microstructure as in Fig. 317.

Sputtered with an iron cathode in oxygen. The contrast between the brazing metal and parent metal is considerably increased, the phases within the brazing metal are clearly visible.

2.1.9.4 Electron beam brazing

Materials with very different properties, such as steel and tantalum, can be joined by brazing.

Figs. 319–321: This is exemplified by electron beam brazing of a CrNi steel with tantalum.

319: Left hand side: tantalum, right hand side: austenitic CrNi-steel, centre: brazing metal consisting of 54% Ag, 25% Pd, 21% Cu. Unetched, bright field.

After polishing the three materials tantalum, brazing metal and steel can easily be distinguished and a second phase can be made out faintly in the brazing metal.

320: Same materials and same relative positions as in Fig. 319.

Etched with V2A-pickle 10), bright field.

The etch reveals the grain boundaries in the CrNi steel. The tantalum and the brazed seam are not affected.

321: Same materials and relative positions as in Fig. 319. Sputtered with a copper cathode in oxygen.

Good colour contrast is achieved between both parent metals and the brazed seam. Tantalum (left) is blue (19 : 0..1 : 6), the CrNi steel (right) is yellow (3 : 6 : 2). A red Cu-rich solid solution (7 : 5 : 3) and an orange Ag-rich region (5..6 : 4 : 2) are visible within the brazed seam. Cu solid solutions have become concentrated in both parent metal/brazing metal transition zones.

1.9.5 Composite foil (sandwich) consisting of Ag and PdCu

The layers lie alternately. Ag-Cu brazing metal has been applied at one side. The foil has been completely coated with copper to facilitate the metallographic preparation.

Figs. 322–324: Composite foil Ag-PdCu-Ag-PdCu-AgCu brazing metal.

322: Specimen etched chemically, bright field.

Sequence of phases from left to right:

copper: black

silver layer: white

PdCu: light grey

Then follow two layers of silver and PdCu.

AgCu brazing metal: medium grey, finely crystalline

copper: black

No diffusion zone is visible between the silver layer and the PdCu.

Figs. 223 and 224: See page 172

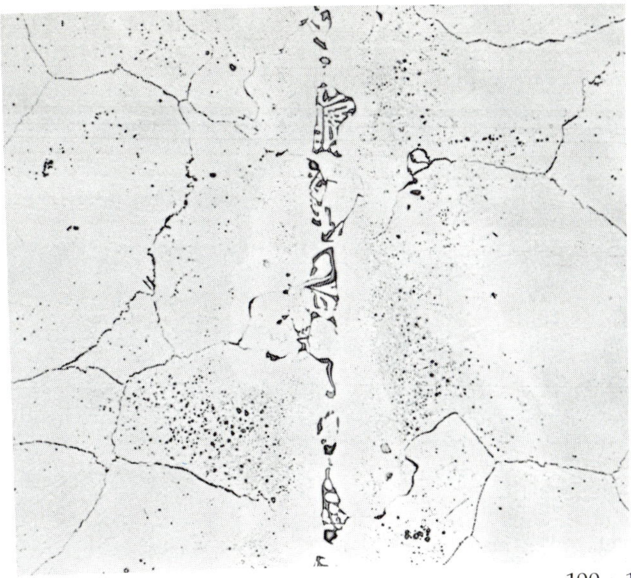

317 100 : 1

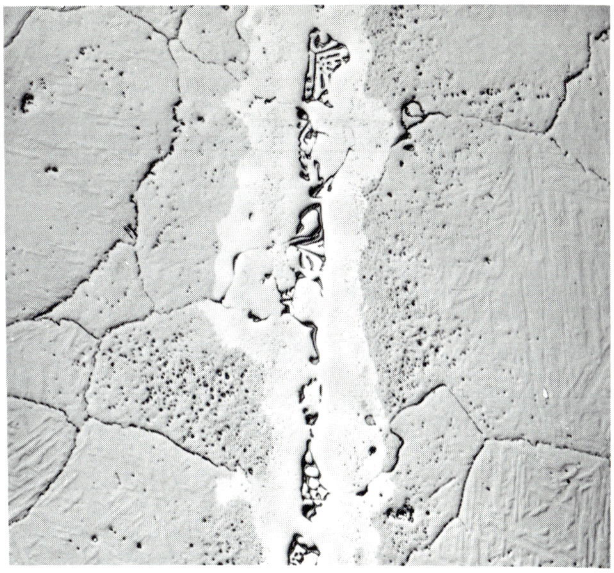

318 100 : 1

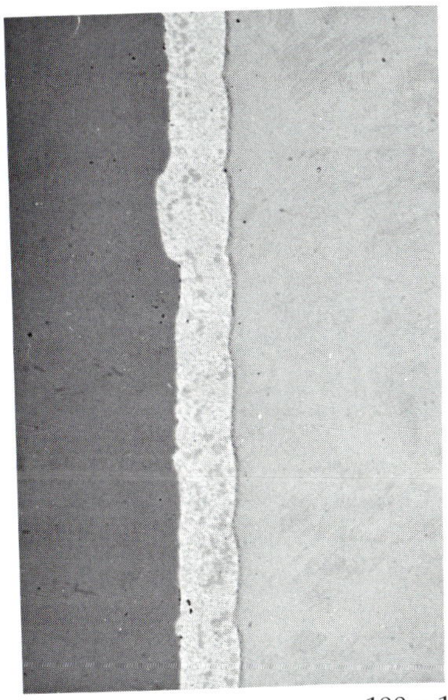

319 100 : 1

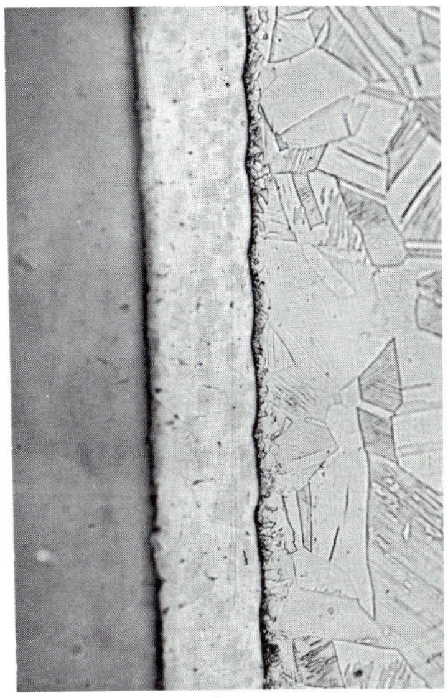

320 200 : 1

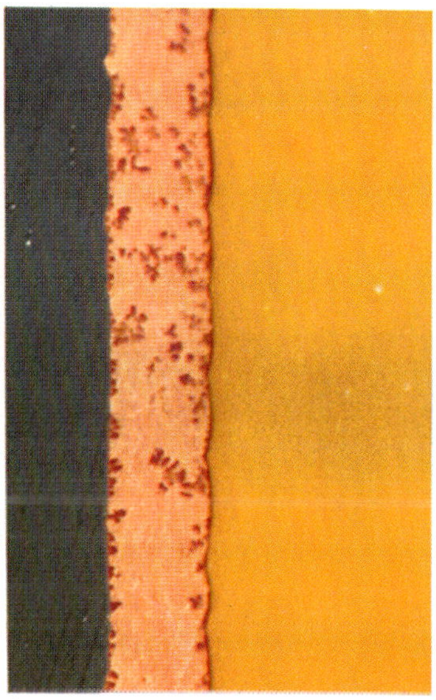

321 200 : 1

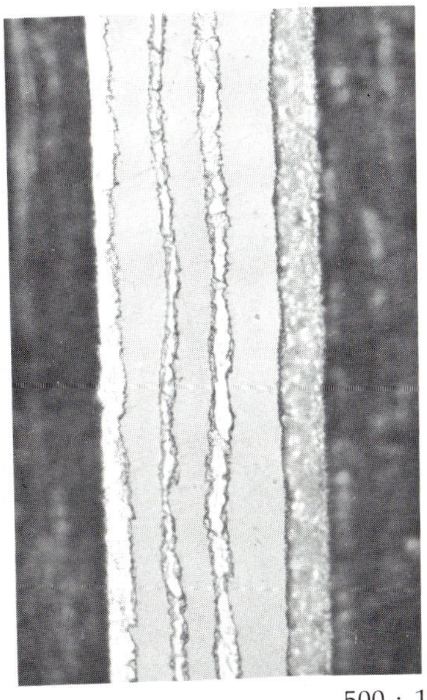

322 500 : 1

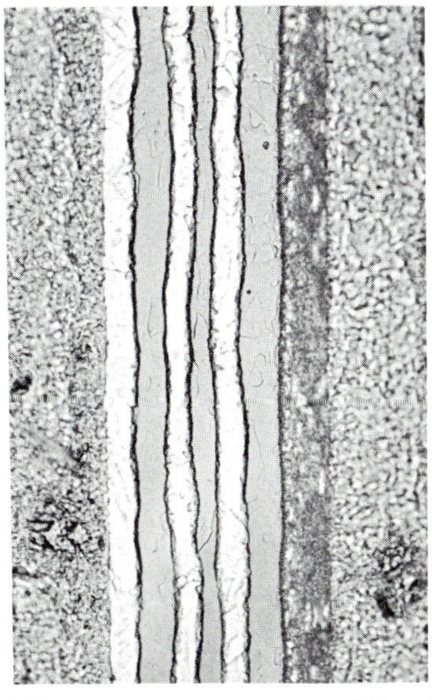

323 500 : 1

324 500 : 1

323: Same specimen, same area of the microstructure as in Fig. 322. Ion etched, bright field.

The metal surface is clearer as a result of the ion etch. The phases can be differentiated more easily. The black and white contrasts are similar to those in Fig. 322.

324: Same specimen, same area of the microstructure as in Fig. 322, sputtered with an iron cathode in oxygen.

Silver layer: greyish blue, PdCu: light beige (5 : 5 : 2), AgCu brazing metal and protective copper layer both brown (4 : 5 : 2).

A light brown diffusion zone is visible between the silver and the palladium-copper solid solution.

2.1.10 Ceramic and metal-ceramic materials and plastics

2.1.10.1 General introduction

In contrast to metallic materials, ceramics and plastics generally reflect incident light badly, i.e. the reflectivities of oxidic materials or polymers are considerably lower than those of metallic materials, see Table 9. Thus a transparent silicate with a refractive index of $n_p = 1.5$ has a reflectivity of only 4%. Microstructural examinations in the reflection microscope on polished or etched sections are made very difficult on account of the small differences in optical properties of the microstructural constituents. Using interference layer microscopy, however, the different phases can be differentiated much more clearly in light-dark contrast than in the uncoated state. Different colours are scarcely to be expected because of the small phase differences of these materials.

In **multiphase ceramics** in which the optical properties of the phases are very different, contrast produced by deposition of interference layers is very good. The same is true of metal-ceramic materials (c.f. Figs. 277 to 284) and of ceramics with metallic inclusions. In the case of metal-ceramic materials the contrast on the uncoated section is in many cases adequate for microstructural investigations.

Interference layer microscopy is particularly applicable to ceramics and metal-ceramic materials when, due to the porosity of the specimen material, a conventional chemical etch would be difficult (c.f. chapter on the preparation of metallographic sections 1.2.1).

Literature to 2.1.10

1) Kopp, W.-U. and Bühler, H.-E.: Metallographie und Keramographie – Fortschritt in der Präparationstechnik. Special Vol. **9** of Prakt. Metallographie, Dr. Riederer-Verlag GmbH, Stuttgart (1978)

2) Petzow, G., Claussen N. and Röser, H.: Gefüge und Eigenschaften metallkeramischer Werkstoffe mit keramischer Matrix. Metallographie-Schadenfallanalyse und Werkstoffentwicklung. Special Vol. **6** of Prakt. Metallographie, Dr. Riederer-Verlag GmbH, Stuttgart (1976) 225

3) Elssner, G., Pabst, R. and Aldinger, S.: Gefügeausbildung bei Keramik-Metall-Verbindungen, Ref. 2), 274

4) Elssner, G. and Petzow, G.: Metallographie von keramischen Werkstoffen und Metall/Keramik-Verbundsystemen. In: Metallographie und Keramographie – Fortschritt in der Präparationstechnik. Special Vol. **9** of Prakt. Metallographie, Dr. Riederer-Verlag GmbH, Stuttgart (1979) 207

5) Müller, E. and Schwarz, M.: Kunststoffe – Metallografisch untersucht Prakt. Metallographie **14** (1977) 151

6) Jestädt, M., Henkhaus, R., Hessel, J. and Bühler, H.-E.: Vergleichende Bewertung verschiedener Präparationsverfahren zur Sichtbarmachung der Morphologie teilkristalliner Polymerwerkstoffe. Prakt. Metallographie **16** (1979) 268

2.1.10.2 Special ceramics

Figs. 325 and 326: **Al_2O_3 + TiC ceramic**

325: Polished, bright field.

Al_2O_3: black

TiC: light grey

In this hot-pressed material, large regions of the Al_2O_3 matrix are enveloped in TiC. In the polished section, the oxide phase appears black because of its low reflectivity and the carbide phase appears bright because of its higher reflectivity.

326: Same material, same area of the microstructure, coated by sputtering with an iron cathode using oxygen as reaction gas, bright field.

Al_2O_3: light brown (3 : 1 : 2)

TiC: dark brown (3 : 5 : 3)

With this choice of coating the reflectivity of the Al_2O_3 is increased. Both components appear in a similar shade of colour, but the saturation and brightness are different. The two components can easily be differentiated by colour in the coated state.

Figs. 327 to 329: **Glass fibre reinforced resin (laminate), copper as conducting strips and sheets, protective resin layer, adhesive layer and plastic frame (plastic composite).**

327: Material as described above, polished, bright field.

copper: white

glass fibres: black (scarcely distinguishable from the resin and adhesive layers or plastic frame)

328: Same material, same area of the microstructure as in Fig. 327 but sputtered with a platinum cathode and oxygen as reaction gas, bright field.

copper: white

glass fibres: light grey

plastic frame: dark grey (upper half of picture)

adhesive layer: black (in the plane of the upper conducting strip)

resin layer: dark grey (about the same brightness as the plastic frame)

329: Same material, same area of the microstructure and same coating as in Fig. 328.

In this magnified section the orientation of the glass fibres transverse and perpendicular to the plane of section is more clearly visible than in Fig. 328. Some details may be seen which are not visible in Fig. 327.

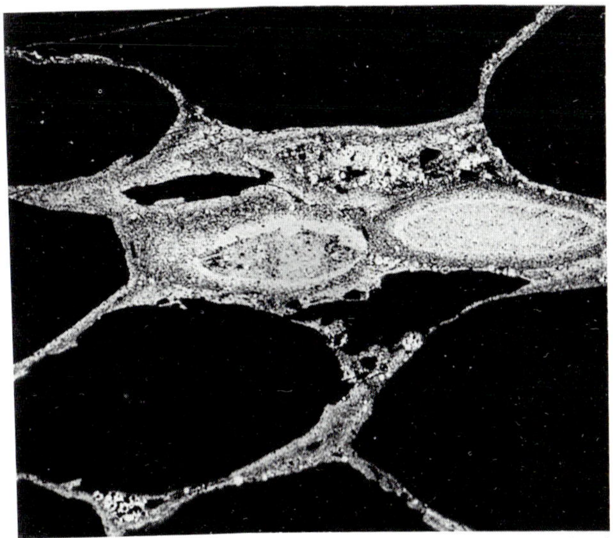

325 200 : 1

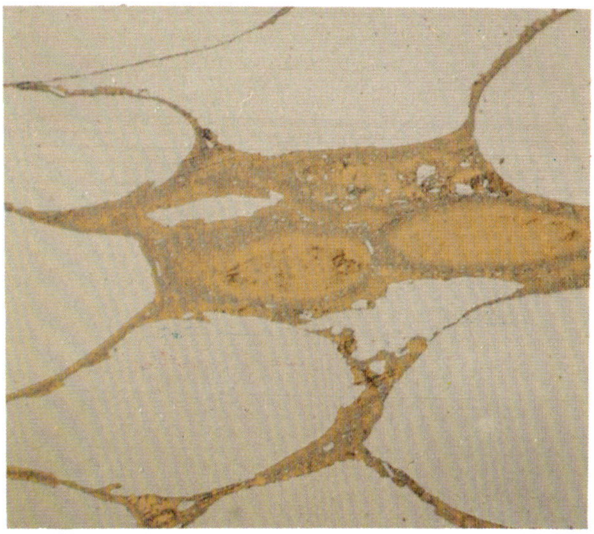

326 200 : 1

327 200 : 1

328 200 : 1

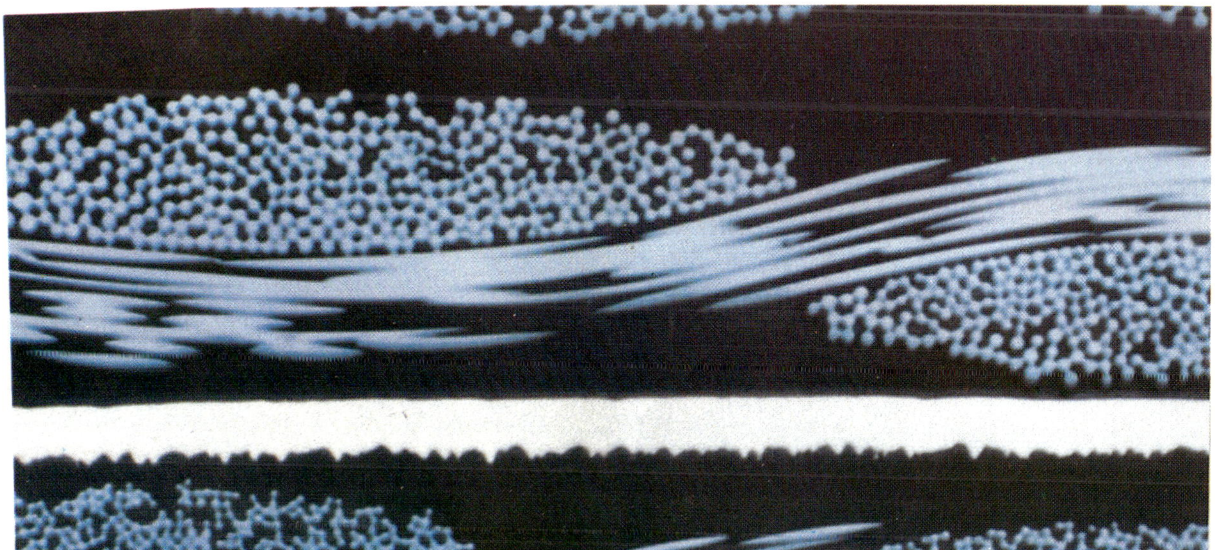

329 300 : 1

2.1.11 Magnetic materials

2.1.11.1 General introduction

The collective term "magnetic materials" covers all those materials which exhibit technologically useful magnetic polarisation. A distinction dependent upon the coercive force is made between magnetically soft (< 100 A/m) and magnetically hard (> 1000 A/m) materials 1–8).

Three different categories can be defined according to the nature of the material:

1. Metallic materials (cast alloys, wrought alloys, powder metallurgically produced materials)
2. Oxidic materials
3. Compose materials

In view of the wide variety of substances that can be used and microstructures that may arise it is clear that there can be no general rules for the metallographic preparation of magnetic materials. The examples given in this atlas of contrast production on barium ferrite microstructures are representative of most of the materials in groups 2 and 3.

No data are yet available concerning contrast production by means of interference layers on magnetic oxides. Investigations have only just begun. From among the preliminary results, reference is made to the example of multi-phase barium ferrites.

On the other hand an extensive body of literature exists on the metallography of the metallic magnetic materials and the revealing of ferromagnetic domain structures (c.f. Section 2.2.5) 6–10). In addition to the barium ferrites, which serve as examples of oxidic magnetic materials, the system samarium-cobalt will be discussed as representative of magnetically hard, powder metallurgically manufactured or cast, metallic materials.

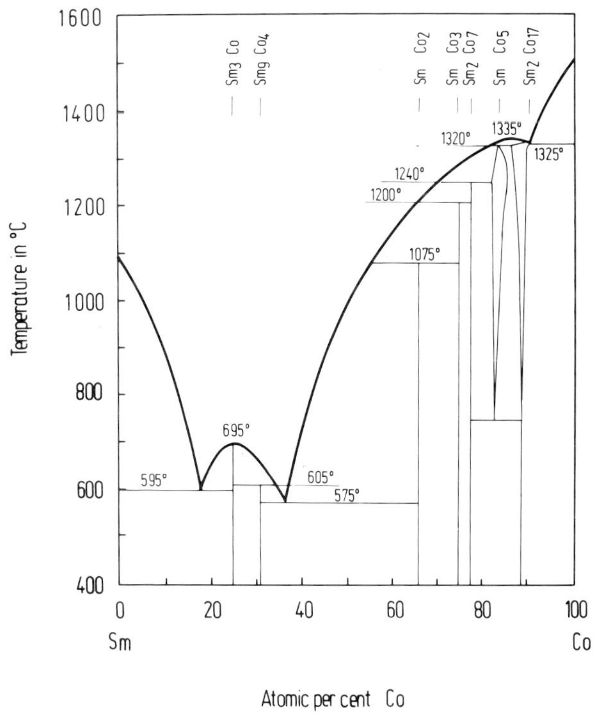

Fig. 330: The binary samarium-cobalt phase diagram.

2.1.11.2 Microstrutures of the barium ferrites and the samarium-cobalt system

In the case of the oxidic magnetic materials, the same difficulties are encountered in the preparation of sections for microstructural observation as in the production of contrast on weakly reflecting and weakly to strongly absorbing non-metallic materials (c.f. sections 1.1.2, 1.1.3, 1.2.3, 2.1.7 and 2.1.9).

From the wide range of metallic magnetic materials, the group of permanently magnetic alloys based on cobalt with the rare earths will be described in this atlas.

By consideration of the binary phase diagram for samarium-cobalt in Fig. 330 it is possible to appreciate the difficulties associated with the metallographic preparation of permanent alloys of cobalt with the rare earths. On the cobalt side of the binary system samarium-cobalt, the phases Sm_2Co_{17}, $SmCo_5$, Sm_2Co_7 and $SmCo_3$ exist in very close stoichiometric proximity. The diagram also shows that a series of peritectic reactions occur at high temperatures and that at about 750^{0}C, the compound $SmCo_5$ decomposes into Sm_2Co_7 and Sm_2Co_{17}.

The very similar optical properties of the intermetallic compounds of the rare earth metals with cobalt, and the very difficult specimen preparation of these systems of compounds (high affinity for oxygen of the rare earths) mean that it is very difficult to describe the phases and to reveal the magnetic domain structures.

Differentiation of the various phases is considerably facilitated by interference layer metallography. This is also true of the accentuation of the magnetic-optical Kerr effect by deposited layers. This accentuation will be described in the section on states of magnetic order (c.f. section 2.2.5) but it should nevertheless be mentioned here that a large Kerr-rotation in the uncoated condition due to the high magnetic flux density enables the ferromagnetic domains to be seen very clearly. This is especially true for the compounds $SmCo_5$ and Sm_2Co_{17}.

In a description of the phases in these permanently magnetic materials, therefore, the ferromagnetic domains are shown in polarised light in the uncoated as well as in the coated condition.

Literature to 2.1.11

1) Lemaire, R.: Magnetische Eigenschaften intermetallischer Verbindungen des Kobalts mit seltenen Erdmetallen und Yttrium.
Kobalt **32** (1966) 117 and **33** (1966) 175

2) Strnat, K. J.: Legierungen des Kobalts mit seltenen Erdmetallen, eine neue Gruppe aussichtsreicher Dauermagnetwerkstoffe.
Kobalt **36** (1967) 119

3) Bachmann, H., Bischoffberger, A. and Hofer, F.: Magnetic Domain Pattern in a Small Single Cristal $SmCo_5$ Particle
Journal of Material Science **6** (1971) 169

4) Herget, C. and Domazer, H.-G.: Methods for the Production of Rare-Earth-3d Metal Alloys with Particular Emphasis on the Cobalt-Alloys.
Goldschmidt informiert, No. 4, (1975), 3

5) Strnat, K. J. and Ray, A. E.: Permanent Magnets Based on $R_2\,(Co,Fe)_{17}$ Phases: Their Potential and Present State of Development.
Goldschmidt informiert, No. 4, (1975), 47

6) Herget, C. and Altenhoff, H.: Prakt. Metallographie der Seltenerd-Kobalt-Legierungen /II. Zur Präparationstechnik der Seltenerdmetalle und Seltenerd-Kobalt-Legierungen.
Prakt. Metallographie **10** (1974) 599

7) Feldmann, D. and Schulz, I.: Sichtbarmachung magnetischer Bereichstrukturen in Seltenerd-Kobalt-Legierungen durch den magnetooptischen Kerr-Effekt.
Prakt. Metallographie **8** (1973) 462

8) Pepperhoff, W.: Sichtbarmachung ferromagnetischer Elementarbereiche im polarisierten Licht.
Arch. Eisenhüttenwes. **34** (1963) 767

9) Bühler, H.-E. and Horn, E.: Magnetooptik der Metalle.
Special Vol. **6** of Prakt. Metallographie (1976) 137
Dr. Riederer-Verlag GmbH, Stuttgart

10) Bühler, H.-E., Schorr, M. and Velicescu, M.: Über Anwendungsmöglichkeiten der magnetooptischen Kerr- und Farady-Effekte zur metallographischen Untersuchung magnetischer Werkstoffe.
Special Vol. **10** of Prakt. Metallographie (1979) 607
Dr. Riederer-Verlag GmbH, Stuttgart

Figs. 331 to 334: **Microstructure of barium ferrites**

331: Polished, unetched, bright field.

Three phases are present in the section, magnetite (Fe_3O_4), Y-phase ($Ba_2Fe_{14}O_{22}$) and monoferrite ($BaFe_2O_4$) but they can scarcely be differentiated in the unetched condition.

magnetite and Y-phase: light grey

monoferrite: dark grey

pores: black

332: Same material and same area of the microstructure, etched, bright field.

There is no improvement as compared with Fig. 331. The monoferrite phase has been etched out of the microstructure leaving a pore.

333: Same material and same area of the microstructure, sputtered with an iron cathode using nitrogen/oxygen as reaction gas, bright field.

magnetite (Fe_3O_4): light grey (lightly graded shading)

Y-phase ($Ba_2Fe_{14}O_{22}$): dark grey

monoferrite ($BaFe_2O_4$): light grey to white

pores: black

334: Same material, same area of the microstructure and same preparation as in Fig. 333.

magnetite (Fe_3O_4): greenish blue (19 : 1 : 2)

Y-phase ($Ba_2Fe_{14}O_{22}$): orange (6 : 1 : 3)

monoferrite ($BaFe_2O_4$): light beige (2 : 2 : 2)

pores: black

After coating, the phases can be distinguished better than in Fig. 331 and 332, both in the coloured and black and white micrographs.

Figs. 335–340: **Cobalt-Samarium system, alloy containing 34% Sm**

The section contains a cobalt crystal surrounded by the compound Sm_2Co_{17}. Residual eutectic composed of cobalt and Sm_2Co_{17} is embedded in the cobalt crystallite.

335: Polished, bright field.

There is no light-dark contrast between the cobalt and the samarium-cobalt compound.

336: Same material and same area of the microstructure as in Fig. 335, specimen coated with ZnSe, reflection minimum of cobalt lies at 540 nm, bright field.

cobalt: light grey border round the cobalt crystal

Sm_2Co_{17}: dark grey

Sm_2Co_{17} in eutectic: also dark grey

Figs. 337–340: See page 178

331 200 : 1

332 200 : 1

333 200 : 1

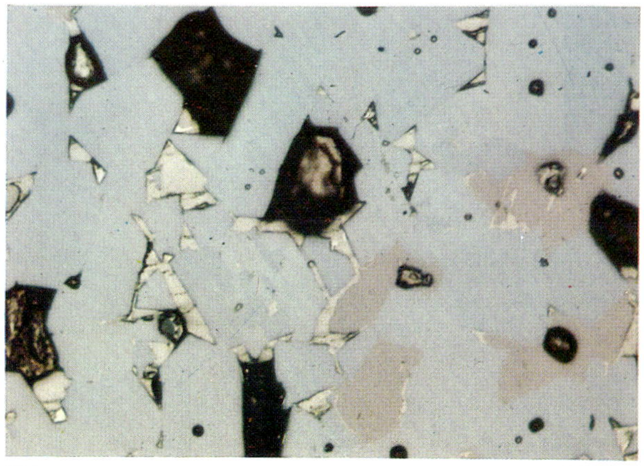

334 200 : 1

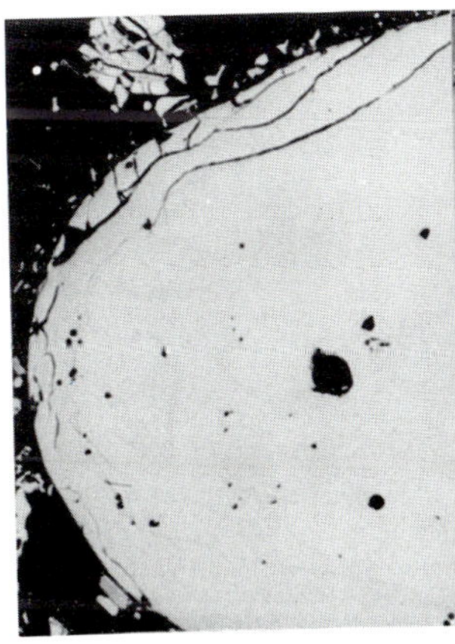

335 200 : 1

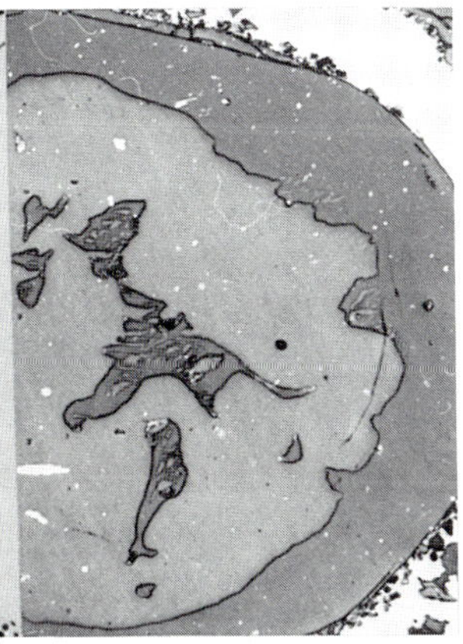

336 200 : 1

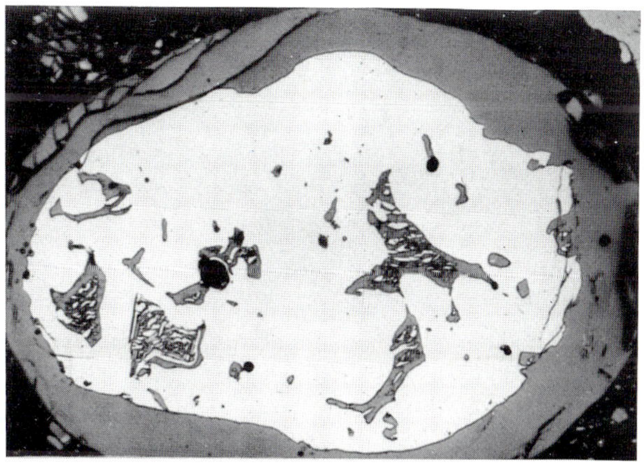

337 200 : 1

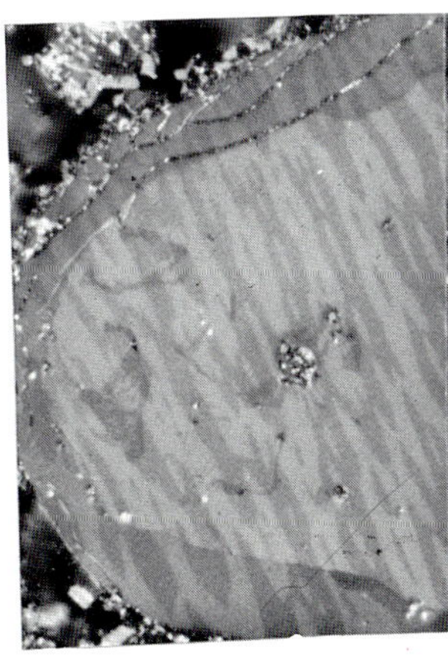

338 200 : 1

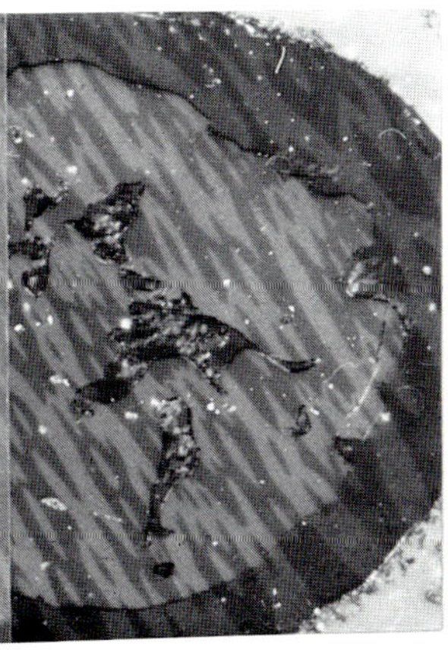

339 200 : 1

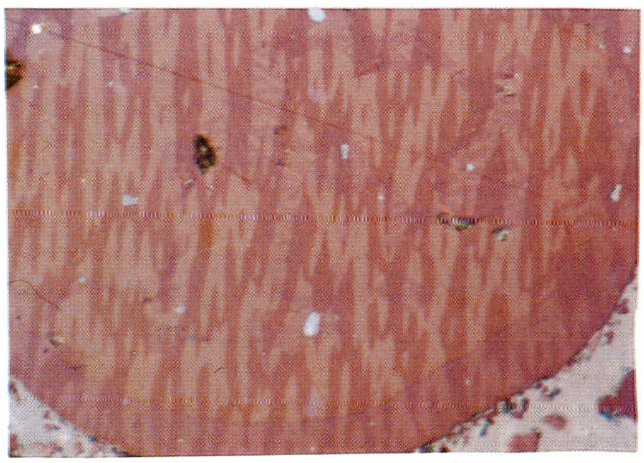

340 200 : 1

337: Same material and same area of the microstructure as in Fig. 335, etched for 5 s in 3% alcoholic nitric acid, bright field.

cobalt: white

Sm_2Co_{17}: dark grey

After etching, the cobalt – Sm_2Co_{17} eutectic stands out clearly from the cobalt. The faithfulness of the geometric reproduction is only minimally poorer than in Fig. 336.

338: Same material, same area of the microstructure and same preparation as in Fig. 335, observed in polarised light.

Even in the uncoated condition, the magnetic order can be seen clearly in both the cobalt and the Sm_2Co_{17} when the specimen is observed in polarised light.

The magnetic domain structures appear in various degrees of light-dark contrast, and in parts the Bloch walls intersect the cobalt/Sm_2Co_{17} phase boundary.

339: Same material, same area of the microstructure, same coating as in Fig. 336, same mode of observation as in Fig. 338.

The evaporated ZnSe layer serves to reinforce the magneto-optic Kerr effect. The light-dark contrast between the magnetic domains is therefore stronger than in Fig. 338.

340: Same material, same area of the microstructure, same type of coating and same observation conditions as in Fig. 339.

The colour photograph clearly shows that there is a noticeable colour contrast between the ferromagnetic domains when the nicols are not completely crossed. As in Fig. 339, the intersection of the magnetic domains with the cobalt/Sm_2Co_{17} phase boundary is particularly distinct.

2.2 APPLICATIONS INDEPENDENT OF THE MATERIAL

2.2.1 Concentration differences in solid solutions (diffusion, segregation)

2.2.1.1 General introduction

Diffusion processes and also segregations (concentration differences in solid solutions) can in many cases be revealed by means of interference layers. Accentuation of both light-dark and colour contrast can be achieved.

Steel-aluminium and steel-zinc composites are particularly impressive examples of how **diffusion processes** can be revealed 1,2).

Figs. 343 to 349 show a reaction front between steel and aluminium in which the phases steel, α-Fe-Al solid solution, Fe_3AlC_x, $FeAl_2$ and Fe_2Al_5 are visible. The iron-aluminium solid solution, which is shown particularly clearly in Figs. 343 and 344, shows a pronounced change in optical properties with Al-content. The same situation obtains in the steel-zinc system (Figs. 350 to 355).

If a quantitative evaluation is made (Fig. 341) the reflection minimum is observed to shift to longer wavelengths as the aluminium content of the solid solution increases (if the α-iron is purple, the saturated α-iron-aluminium solid solution appears violet) (Fig. 346). Using layer materials with little variation in refractive index over the visible range (above all ZnS), the reflectivity and the electrical conductivity suffer a discontinuity in the zone of crystallographic order (Fe_3Al).

As the aluminium content of the solid solution increases, the reflectivity at the interference minimum decreases (Fig. 346). In Fig. 341, the reflectivity at the minimum was measured using ZnTe. In this case an increase in the aluminium content of the solid solution is accompanied by an increase in reflectivity. This is a consequence of the marked decrease in the refractive index of ZnTe at long wavelengths. In addition ZnTe exhibits noticeable self-absorption over the whole visible range (cf. section 1.2.3.1).

As well as for materials with very different optical properties, interference layers can also be used to reveal concentration differences in the case of transition metals which occur close together in the periodic table of elements. Examples of this are diffusion chromed steels (Fe-Cr solid solutions) or also directly white enamelled sheet in which, as a result of the manufacturing process, Fe-Ni solid solutions can be detected in the phase boundary region 3).

Particularly good examples of crystal segregations are provided by binary or ternary systems in which the liquidus and solidus lines are well separated over a broad range of concentration.

The gold-platinum system has been selected as an example of the use of interference layers to reveal crystal segregations (Fig. 342). An alloy with about 40% Pt starts to solidify at about 1500^{0}C forming a crystal segregation of between 20 and about 85% Pt under non-equilibrium conditions. The first solid solutions to solidify contain 85% Pt, the last less than 20% Pt. Considering these great differences in concentration it is clear that a long time is needed to establish equilibrium by diffusion annealing. Figs. 356 to 361 illustrate different degrees of crystal segregation in the Au-Pt system.

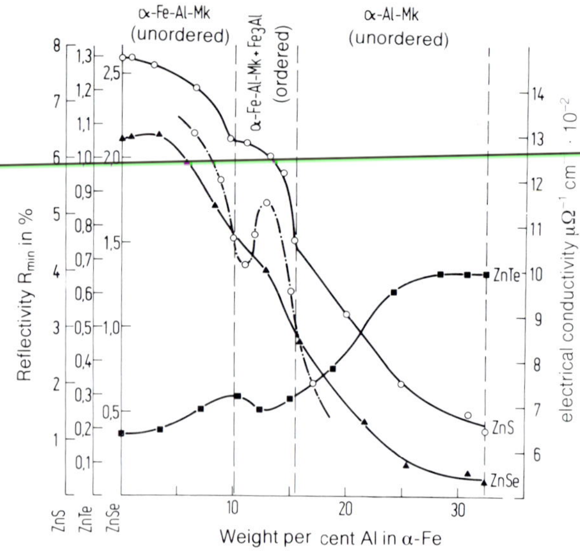

Fig. 341: Influence of the optical properties of the iron-aluminium solid solution in the iron-aluminium system on the reflectivity and on the position of the reflection minimum after coating with zinc sulphide, zinc selenide and zinc telluride. The dependence of the electrical conductivity on the aluminium content is also shown. There is a discontinuity in the zone of crystallographic order (Fe_3Al) which is confirmed by optical measurements.

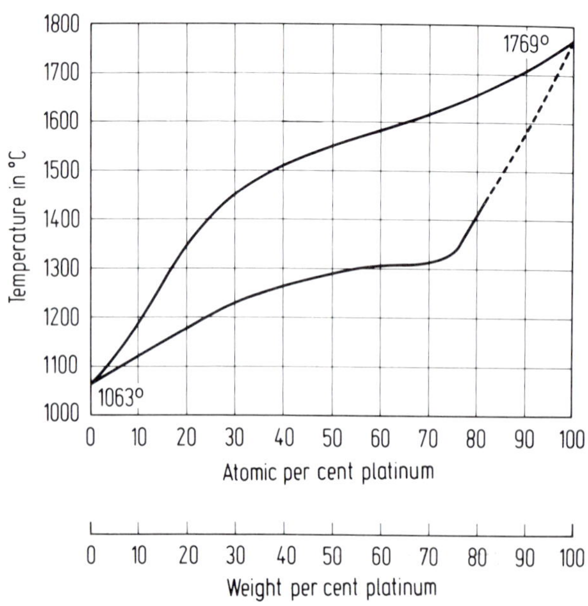

Fig. 342: The binary system gold-platinum.

Literature to 2.2.1

1) Pepperhoff, W. and Ettwig, H. H.: Interferenzschichten-Mikroskopie. Dr. Dietrich Steinkopff-Verlag Darmstadt 1970

2) Jackel, G., Bühler, H.-E. and Meyer, L.: Über Bildung und Eigenschaften von Phasen im Dreistoffsystem Eisen-Zink-Kohlenstoff im Temperaturbereich von 500°–1100° C.
Proceedings of 9th Int. Galvanising Conference, Düsseldorf 1970

3) Jackel, G. and Bühler, H.-E.: Interferenzschichten-Mikroskopie in der Metallographie. Fortschritte in der Metallographie. Special Vol. **3** of Prakt. Metallographie (1972) 107, Dr. Riederer-Verlag, Stuttgart

4) Meyer, L. and Bühler, H.-E.: Über den Aufbau von Legierungsschichten zwischen Stahl und Aluminium bei Temperaturen unter 1300° C. Aluminium **43** (1967) 733

2.2.1.2 Diffusion processes in the neighbourhood of the steel-aluminium interface

Figs. 343 and 344: **Diffusion zone between steel containing 0.06% C and aluminium,** 1 h/1100° C followed by accelerated cooling, vacuum deposit of ZnSe, zero order, bright field.

343: Material and treatment as described above.

The sequence of phases from left to right is as follows: $FeAl_2$ + Fe-Al eutectoid, Fe-Al solid solution with decreasing aluminium content, Fe_3AlC_x, steel The reduction in aluminium content of the solid solution is characterized by a gradual lightening in colour in the steel direction. At the boundary between the eutectoid and the solid solution, the Fe-Al solid solution contains 33% Al, the maximum amount soluble at the annealing temperature of 1100° C.

As the distance from the phase boundary increases the aluminium content decreases. Needle-like precipitates of $FeAl_2$ have formed from the supersaturated solid solution in the region of the diffusion zone. The dark, needle-shaped carbides of type Fe_3AlC_x visible in Fig. 343 have the cubic structure and show approximately the same interference colour as the saturated iron-aluminium solid solution.

344: Same material, same area of the microstructure and same coating as in Fig. 343.

$FeAl_2$ and Fe-Al eutectoid: lamellar structure, similar to pearlite, light brown-violet (4 : 5 : 2 = $FeAl_2$ and 10 : 3 : 4 = Fe-Al)

iron-aluminium solid solution: with decreasing aluminium content dark violet to light violet (11 : 2 : 4 to 8 : 1 : 3)

Fe_3AlC_x carbide: needle-shaped, violet (11 : 3 : 4)

steel: light brown (3 : 3 : 2)

Figs. 345, 347–349: **Diffusion zone between steel containing 0.06% C and aluminium,** 1 h/900^{0}C followed by accelerated cooling, vacuum deposit of ZnSe, zero order.

345: Schematic representation of the arrangement of the phases in the diffusion zone in Figs. 347 to 349.

346: **Influence of the aluminium content of the α-Fe-Al solid solution on the position of the reflection minimum.**

As in Fig. 344 it is clear that at the $FeAl_2$/Fe-Al-solid solution boundary the reflection minimum of the Fe-Al solid solution lies in the long wave range of the spectrum (600 nm). The interference colour is violet (11 : 2 : 4). As the distance from the boundary increases and the aluminium content of the solid solution correspondingly decreases the reflection minimum in the case of a ZnTe vacuum evaporated layer shifts to shorter wavelengths until finally a value of 530 nm is measured for pure α-iron (steel).

Figs. 347–349: Same material, same area of the microstructure as in Fig. 345, vacuum deposited ZnTe layer, photographed at 550 nm, zero-order, with different layer thicknesses.

347: Same material, same area of the microstructure as in Fig. 345, photographed at 550 nm. The reflection minimum of α-iron is at 510 nm.

$FeAl_2$ and Fe_2Al_5: white, indistinguishable.

Fe-Al solid solution: black and, as the reflectivity increases in the direction of the steel, grey to white.

Fe_3AlC_x: black

steel: white

348: Same material, same area of the microstructure as in Fig. 347. The reflection minimum of α-iron is at 560 nm.

Fe_2Al_5: light grey

$FeAl_2$: medium grey

Fe-Al solid solution: white when the Al content is high, dark grey to black when it is low.

Fe_3AlC_x: white

steel: dark grey

349: Same material, same area of the microstructure as in Fig. 347. The reflection minimum of α-iron is at 610 nm.

Fe_2Al_5: dark grey to black

$FeAl_2$: dark grey

Fe-Al solid solution, Fe_3AlC_x and steel: all white, indistinguishable.

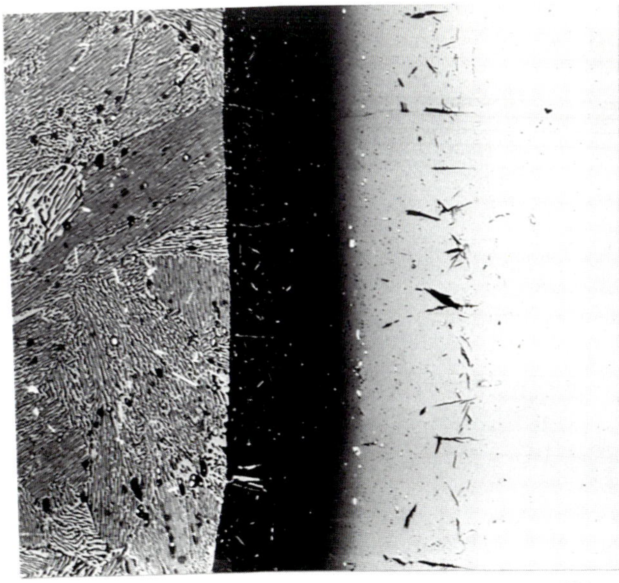

343 200 : 1

344 200 : 1

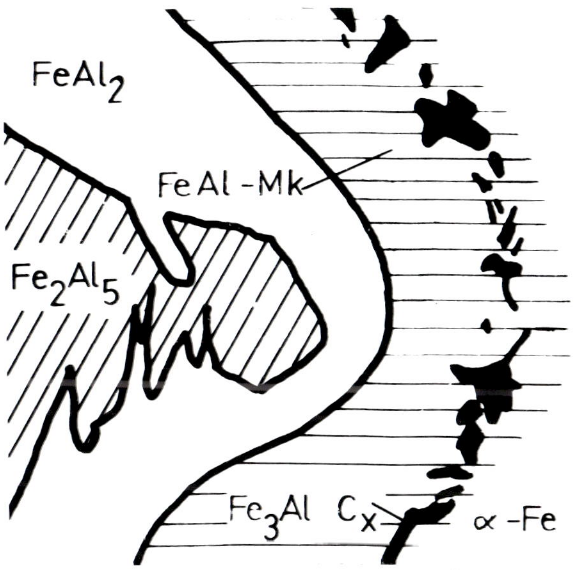

345

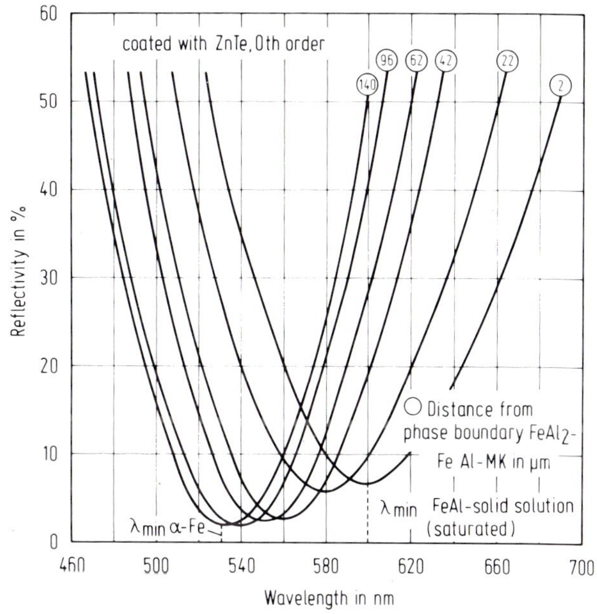

346

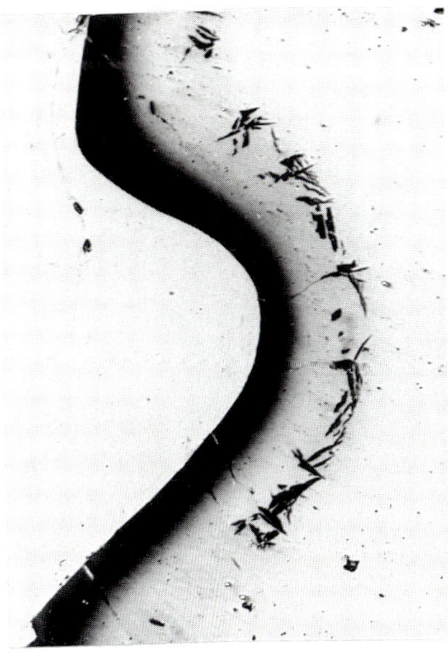

347 200 : 1

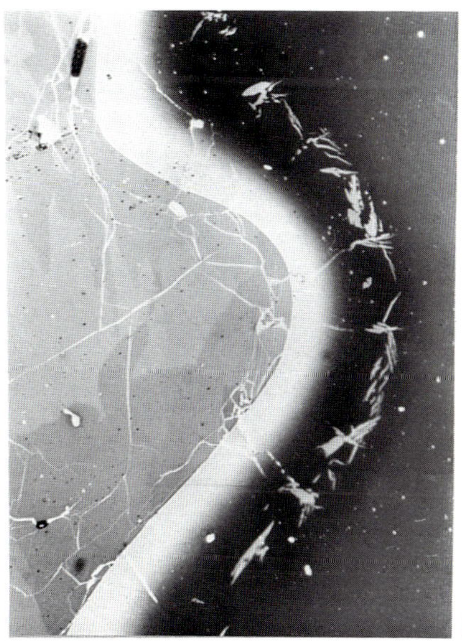

348 200 : 1

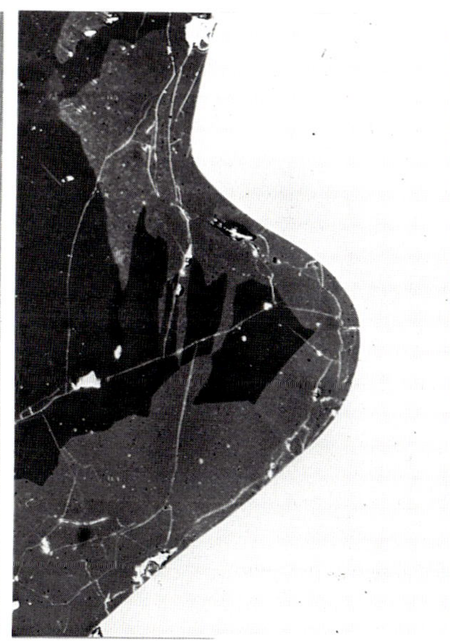

349 200 : 1

2.2.1.3 Diffusion processes in the steel-zinc system

Fig. 350: **Steel with 0.45% C, annealed at 700°C with pure zinc,** accelerated cooling, coated with ZnSe, bright field.

steel (lower half of picture): red (8 : 4 : 3 to 9 : 4 : 3)

Fe_3ZnC_x: carbides: dark violet platelets (11 : 4 : 4)

δ_1-phase (upper half of picture): ochre (4 : 2 : 2)

Γ-phase (between steel and δ_1-phase): light brown (5 : 4 : 3)

Together with the α-iron (α-iron-zinc solid solution) the mixed carbides lie in an ordered configuration in the phase boundary zone and randomly distributed in the δ_1-phase. As well as solid solution-carbide and δ_1, Fig. 350 also clearly shows Γ-phase lying between the iron-zinc solid solution and the δ_1 and coloured a darker shade of brown than the δ_1-phase.

Figs. 351–355: **Steel with 0.06% C was annealed with pure zinc for a prolonged period at temperatures between 700 and 1100°C** so that a broad diffusion zone could build up between the two components. The specimens were cooled rapidly after the anneal. All specimens were coated with ZnSe.

351: Specimen annealed at 800°C, accelerated cooling, coated with ZnSe, bright field.

The transition from steel low in zinc to the zinc saturated iron-zinc solid solution is characterized in Fig. 351 by a sudden colour change from brownish-red to violet. Judging from microprobe measurements this sharp transition marks an increase in the zinc content of the solid solution to 43%.

steel: reddish brown (6 : 5 : 2)

iron-zinc solid solution: dark reddish brown (9 : 5 : 3)

intermetallic iron-zinc compounds (δ_1 and ξ): brownish red (5 : 5 : 3) and yellow (3 : 5 : 1)

352: Steel with 0.06% C annealed with pure zinc at 700°C, accelerated cooling, coated with ZnSe, bright field.

steel matrix (lower half of picture): light grey

zinc-rich solid solution: somewhat darker grey

Fe_3ZnC_x carbide: black (preferential growth direction perpendicular to the advancing reaction front in the zinc-rich solid solution).

δ_1-phase: white

353: Same material, same area of the microstructure and same coating as in Fig. 351, bright field.

steel matrix: medium grey

iron-zinc solid solution (saturated with zinc): black

intermetallic iron-zinc compounds (δ_1 and ζ): light grey and white

The phases may be distinguished well also in the black and white micrograph.

354: Steel with 0.06% C annealed with pure zinc at 1100°C, accelerated cooling, coated with ZnSe, bright field.

steel matrix (lower half of picture): dark grey

iron-zinc solid solution saturated with zinc: black

intermetallic iron-zinc compounds: light grey to white.

Fig. 355: See page 190

350 1000 : 1

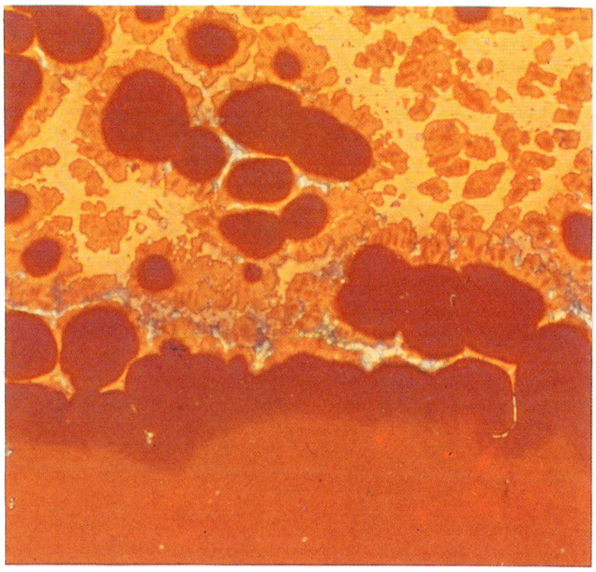

351 350 : 1

352 500 : 1

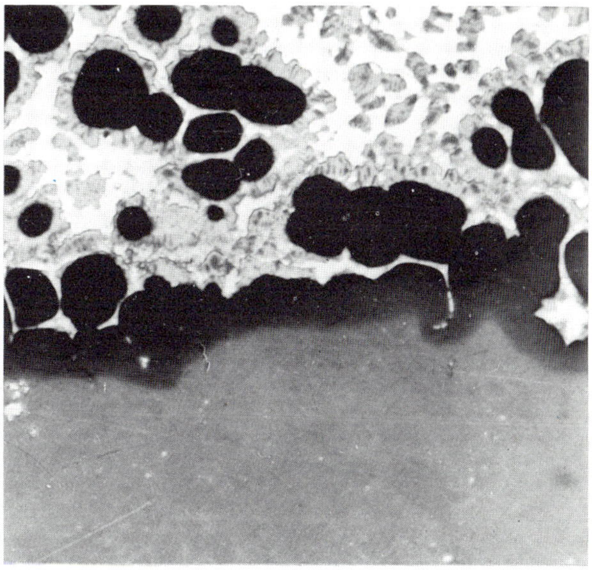

353 300 : 1

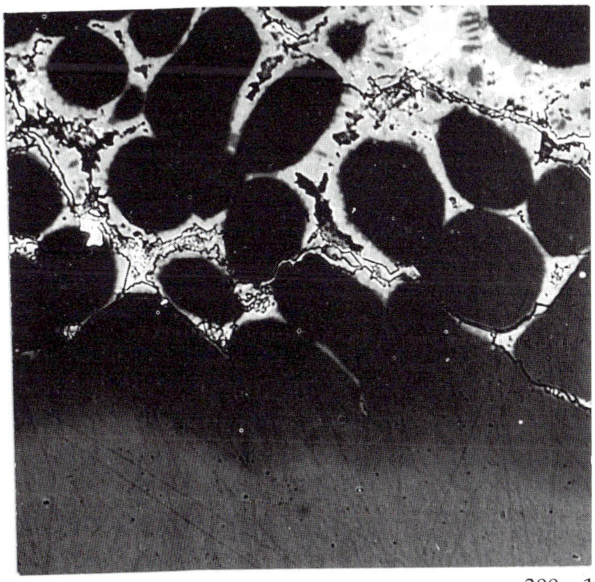

354 300 : 1

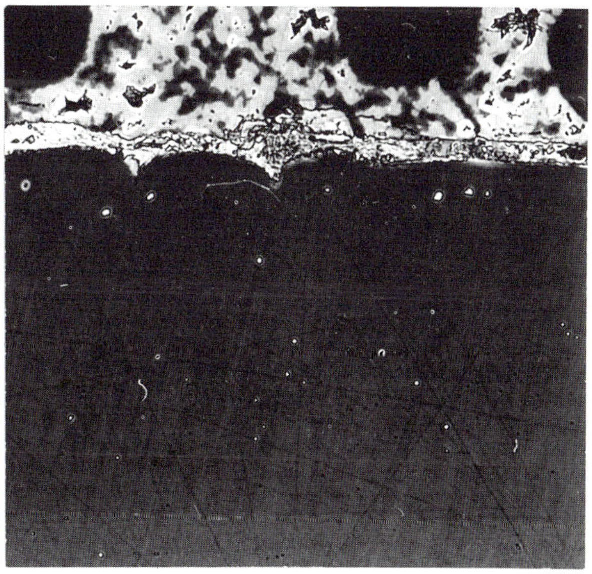

355 300 : 1

355: Same material, same treatment, same area of the microstructure as in Fig. 354 but at a higher magnification.

steel matrix (lower half of picture): dark grey to black

iron-zinc solid solution saturated with zinc: black

The sharp transition in the concentration gradient to zinc is still clearly visible even at this high magnification.

intermetallic iron-zinc compounds (upper half of picture): light grey to white.

2.2.1.4 Segregations in the gold-platinum system

Figs. 356–361: **Gold-platinum alloy with 40% Pt.**

356: Specimen as cast, ion-etched, bright field.

The differences in light-dark contrast between solid solutions rich or deficient in platinum are relatively slight. The platinum-rich dendrites, which were first to solidify, have a high reflectivity and appear light grey [1].

357: Same material as in Fig. 356, diffusion annealed 20 h at 1150^{0}C, ion-etched, bright field.

The diffusion anneal has resulted in extensive homogenisation of the material. Equilibrium is not fully established, however, and light grey areas of platinum-rich solid solution are still visible [1].

358: Same material, same area of the microstructure as in Fig. 356, same treatment (ion-etched), sputtered with an iron cathode using oxygen as reaction gas, bright field.

Pt-rich solid solution: yellow (2 : 6 : 1)

Pt-deficient solid solution: brown (3 : 4 : 3)

359: Same material as in Fig. 356, same heat treatment as in Fig. 357, same coating layer as in Fig. 358, bright field.

Pt-rich solid solution: yellow (2 : 5 : 1)

Pt-deficient solid solution: brown (3 : 3 : 3)

360: Same material, same area of the microstructure and same treatment as in Fig. 358, bright field.

Pt-rich solid solution: light grey

Pt-deficient solid solution: dark grey

After sputtering, the differences in concentration between the various solid solutions are more clearly visible in light-dark contrast than in Fig. 356.

361: Same material as in Fig. 356, same treatment and same coating layer as in Fig. 359, bright field.

Pt-rich solid solution: light grey

Pt-deficient solid solution: medium grey

As in Fig. 360 the contrast on the coated specimen is considerably better than on the uncoated specimen.

356 500 : 1

357 500 : 1

358 750 : 1

359 750 : 1

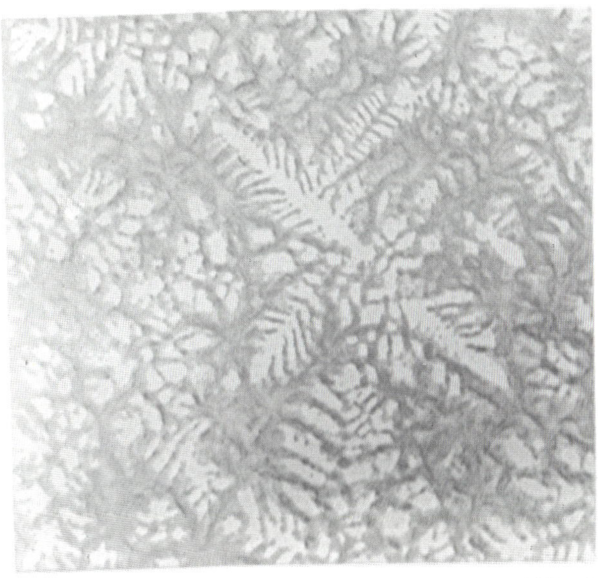

360 750 : 1

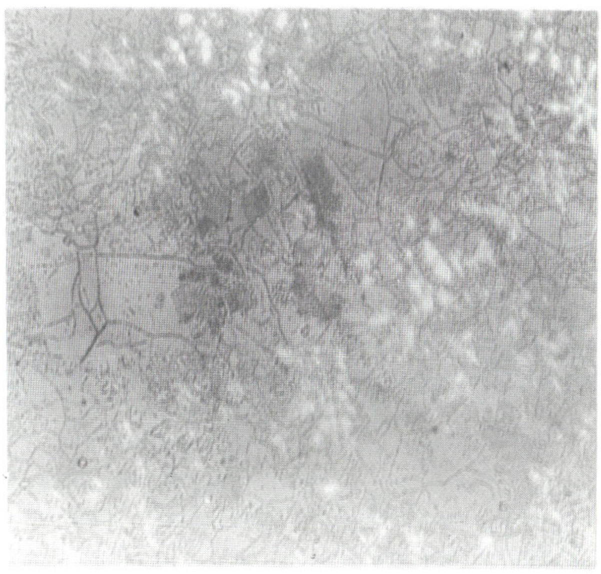

361 750 : 1

2.2.2 Reinforcement of anisotropy effects

General introduction

Contrast enhancement by interference layers can also be used for the observation of only slightly anisotropic crystals in the polarizing microscope (between crossed nicols). Just as the technique can be used to accentuate the differences in the black and white and colour contrast between different microstructural constituents in bright field it also serves to reinforce differences based on the crystallography. A quantitative consideration of this reinforcement effect will not be given here. A thorough discussion can be found in the literature 1,2).

For the metallographer working in the field, the examples that follow illustrate the diverse ways in which optical anisotropy can be used to characterize a microstructure.

As a rule of thumb, interference layers should be used in the characterization of optically uniaxial crystals when:

a) a distinction must be made between cubic and non-cubic crystals. It frequently happens that two phases differ only very slightly in interference colour or reflectivity. If one of the two phases is optically uniaxial, i.e. not cubic, the two microstructural constituents can easily be differentiated by observation under the polarizing microscope.

b) The technique can be employed with great success in quantitative work (determination of anisotropy parameters). Data is available showing that like the optical constants (refractive index and absorption coefficient) the anisotropy parameters can be measured very accurately using the interference layer technique 1).

The examples given in section 2.2.2 refer only to case (a). They serve to illustrate the advantages of observation in the polarizing microscope with the simultaneous use of the interference layer technique as compared with uncoated specimens.

Literature to 2.2.2

1) Pepperhoff, W. and Ettwig, H.-H.: Polarisations-Mikroskopie durch aufgedampfte Interferenzschichten.
Radex-Rundschau (1967) No. 3/4, 667

2) Pepperhoff, W. and Bühler, H.-E.: Lichtmikroskopische Untersuchungen an martensitischen Gefügen.
Arch. Eisenhüttenwes. **34** (1963) 839

Figs. 362 and 363: **Anisotropy of tetragonal martensite in an iron-carbon alloy**

362: Iron-carbon alloy containing 1% C, quenched in water from temperatures around 1000^0C. Vacuum deposit of TiO_2, zero order, bright field.

martensite: black needles

transformation microstructure (bainite, pearlite): medium grey

363: Same material, same area of the microstructure and same coating as in Fig. 362, observed in polarized light.

martensite: black or white according to orientation

transformation microstructure (bainite, pearlite): medium grey

The optical anisotropy of the tetragonal martensite needles is clearly visible from the different grey levels (in this case better black-white contrast).

Figs. 364 and 365: **Reinforcement of the anisotropy of martensite in an iron-carbon alloy**

364: Iron-carbon alloy with about 1% C, quenched in water from 1000^0C, polished, observed in polarized light.

martensite: dark grey or light grey depending on orientation

transformation microstructure (bainite, pearlite): medium grey

365: Same material, same area of microstructure as in Fig. 364, vacuum deposit of TiO_2, observed in polarized light.

martensite: black or white depending on orientation

transformation microstructure (bainite, pearlite): medium grey

The light-dark contrasts between the differently oriented martensite crystals are more clearly visible after coating with an interference layer.

Figs. 366 and 367: **Optical anisotropy of the carbide M_7C_3**

366: **Iron-chromium alloy containing 7.9% Cr,** 250 h / 500^0C, vacuum deposit of TiO_2, zero order, bright field.

M_7C_3: medium grey, extensive areas of carbide with a hexagonal shape.

Residual melt consisting of α-iron-chromium solid solution, σ-phase and the carbide $M_{23}C_6$: white with black inclusions.

367: Same material, same area of the microstructure and same coating as in Fig. 366 but observed in polarized light.

The optical anisotropy of the primary carbides solidified from the melt is clearly visible, and the growth twins show up particularly well.

Figs. 368–372: See page 194

Figs. 368 and 369: **Optical anisotropy of carbosulphides in steel of type St 52.**

368: Titanium carbosulphide ($Ti_4C_2S_2$) in a steel of type St 52 with the addition of 0.1% Ti, rolled, vacuum deposit of ZnSe, zero order, bright field.

steel matrix: medium grey

titanium carbosulphide ($Ti_4C_2S_2$): medium to dark grey

369: Same material, same area of the microstructure and same coating as in Fig. 368, but observed in polarized light.

steel matrix: medium grey

titanium carbosulphide ($Ti_4C_2S_2$): white or medium grey to black depending on the orientation of the grains.

The titanium carbosulphides have a hexagonal lattice so that, like the M_7C_3 in Fig. 367, their light-dark contrast varies in polarized light. In steels with the above basic composition, titanium carbosulphides cannot easily be distinguished in bright field from titanium carbonitrides because their interference colours are virtually identical. Ti (C,N) has the cubic structure and so the two phases can be differentiated in polarized light.

Figs. 370–372: **Optical anisotropy of compounds in the iron-aluminium system**

370: A section of the steel/aluminium interface region in a hot aluminized steel specimen after annealing the aluminium coating at high temperatures. The sequence of phases from left to right is: Fe_2Al_5, $FeAl_2$, eutectoid consisting of $FeAl_2$ and iron-aluminium solid solution, iron-aluminium solid solution. Coated with ZnSe, zero order, interference colour of steel purple, bright field.

Fe_2Al_5: medium grey (left hand side of micrograph)

$FeAl_2$: medium grey (varing grey levels)

eutectoid: pearlite-like configuration of iron-aluminium solid solution (black) and $FeAl_2$ (medium grey)

iron-aluminium solid solution (lower right hand edge of picture): black

371 and 372: Same material, same area of the microstructure and same coating as in Fig. 370 but observed in polarized light at different angles to the polarizing prisms.

It is clear that all iron-aluminium alloys or compounds are optically anisotropic.

At different angles to the polarizing prisms:

a) Fe_2Al_5: medium grey in Fig. 371, white in Fig. 372

b) $FeAl_2$: various shades of light grey in Fig. 371, dark grey to black (again in various shades) in Fig. 372.

c) In the case of the eutectoid consisting of $FeAl_2$ and iron-aluminium solid solution, the different grains with different orientations can be recognized by virtue of the different light-dark contrasts.

362 200 : 1

363 200 : 1

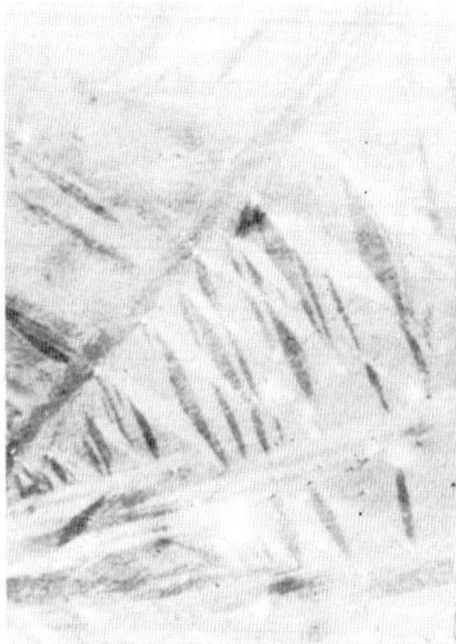

364 200 : 1

365 200 : 1

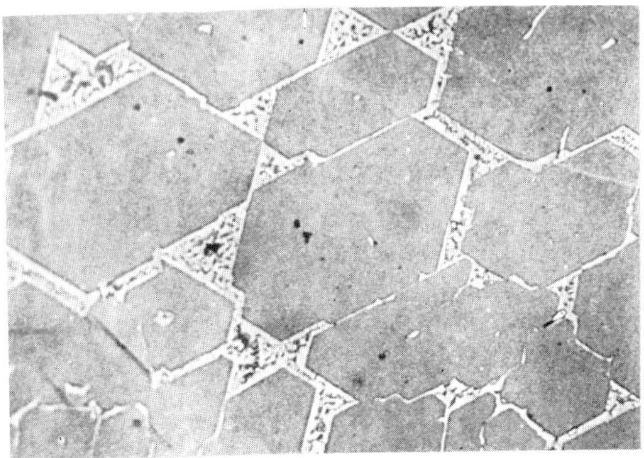

366 100 : 1

367 100 : 1

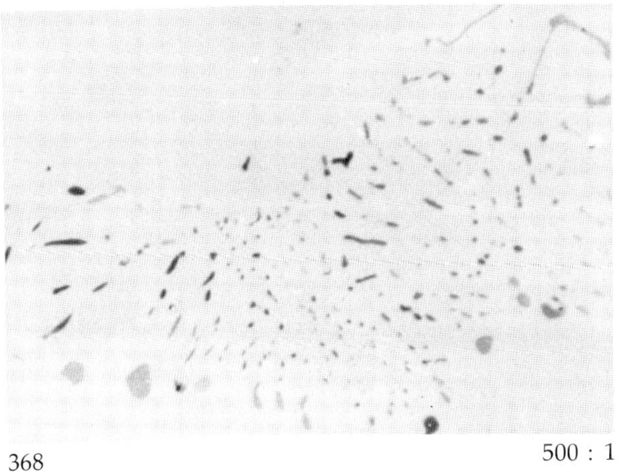

368 500 : 1

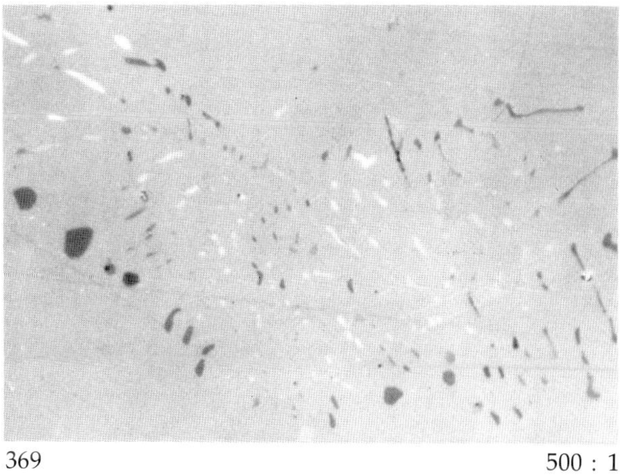

369 500 : 1

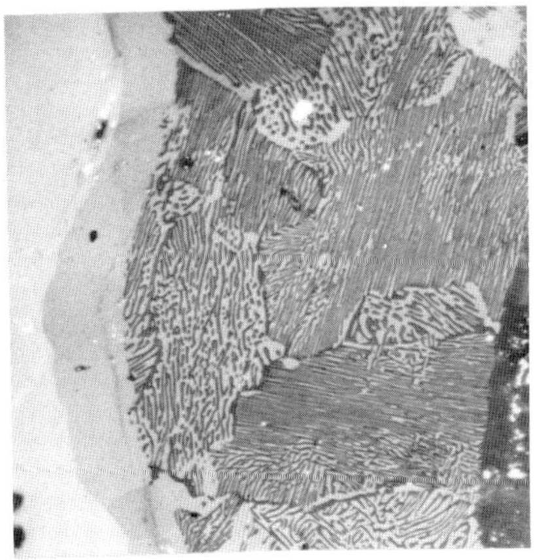

370 50 : 1

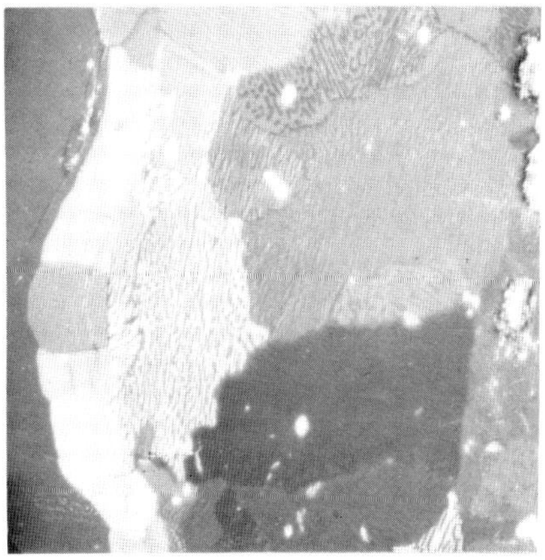

371 50 : 1

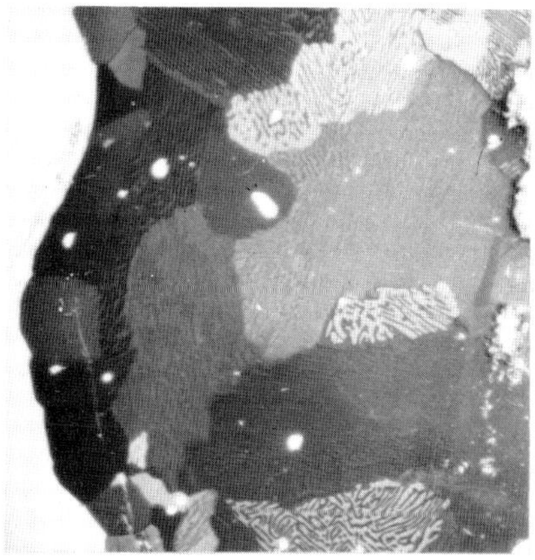

372 50 : 1

2.2.3 Pore volume
Investigation of the pore geometry in porous materials

2.2.3.1 General introduction

Important technical materials (e.g. sintered steels and other sintered metals, materials for catalysis and filtration) contain pores, desired or undesired, the geometry of which has a decisive influence on the properties. A faultless metallographic preparation of such materials is difficult because in soft materials the pores become smeared over and in hard materials they tend to break out. These preparation faults can be avoided by infiltration of the pores. For this to be successful the pores must be interconnecting and open. Only a fraction of the closed pores can be opened by grinding and etching.

Infiltration with viscous synthetic resins (e.g. epoxyresins) is often unsatisfactory because the penetration depth is too shallow and only truly reflects the porosity of the surface zone. Less viscous synthetic resins have poor adhesion or are difficult to handle (e.g. bubbles often form in methyl methacrylate). Metallic materials (low melting solders) are often more suitable because if the choice is favourable the wetting of one metal by the other is very good, with the result that complete infiltration and perfect adhesion are guaranteed.

After polishing, however, the contrast between the solder and the parent materials is often very poor. Quantitative investigation of the pore structure with automatic instruments (television microscopes) is for example impossible without further treatment. By deposition of suitable interference layers the contrast can be improved to such an extent that automatic phase detection presents no problems.

Brazing metal L-Ag 40 Cd (see DIN 8513) can be used as infiltration agent for sinter steel 1).

Another example of the importance of the exact identification of pores and the differention between such pores and oxides, is provided by a W-Ta-Ni alloy (Figs. 374 and 375). Fine oxide inclusions and pores of about the same size were visible over the entire cross section of this material. A definite distinction between oxides and pores could not always be made when a polished section was observed under the microscope. The deposition of a ZnSe layer, however, allowed the pores and oxides to be differentiated easily.

In the field of surface modification, a knowledge of pore volume and pore geometry is often of interest in the case of enamelled materials. A well-contrasted picture of the pores, also in the case of double-layer enamelling, is often difficult because the glassy enamel matrix has only a low reflectivity. By coating with suitable materials the reflectivity can be increased (c.f. sections 1.2.3 and 2.1.10) and the quantity and size of the pores can be evaluated automatically.

Literature to 2.2.3

1) Metzler, H. and Nitsche, D.: Schliffherstellungsmethode für poröse Preßlinge und Sinterteile auf Eisenbasis zur vollautomatischen Gefügeanalyse.
Prakt. Metallographie 7 (1972) 408

2) Elssner, G. and Petzow, G.: Metallographie von Keramischen Werkstoffen und Metall/Keramik-Verbundsystemen.
In: Metallographie und Keramographie – Fortschritte in der Präparationstechnik
Special Vol. **9** of Prakt. Metallographie (1978) 207
Dr. Riederer-Verlag GmbH, Stuttgart.

Fig. 373: **Sintered electrolytic iron powder**

Pressed electrolytic iron powder sintered at 1000^0C and treated with braze metal L-Ag 40 Cd (after DIN 8513), coated with ZnSe, bright field.

The silver solder used to prepare the specimen melts at low temperatures (610^0C) with the result that the shape of the pores is retained exactly. Fluxes may not be used because residues may break out on polishing. A temperature just above the melting point of the solder is sufficient for infiltration. Before coating, the contrast between the solder-filled pores and the iron matrix is very poor because of the high reflectivity of both iron and silver solder. After coating the reflectivity of the iron is considerably more reduced than that of the highly reflecting silver solder. Thus in Fig. 373, the pore-free electrolytic iron powder appears black and the pores white. Thanks to the good contrast, a quantitative determination of the pore geometry is possible with an automatic image analyser.

Figs. 374 and 375: **Differentiation between pores and oxides in a sintered W-Ta-Ni alloy**

374: Polished section, bright field.

Black, finely dispersed pores [P] and oxide inclusions [O] can be seen over the whole cross section, but they cannot be differentiated in the as-polished specimen.

375: Same material, same area of the microstructure as in Fig. 374, coated with ZnSe, zero order, bright field.

oxides: light grey to white [O]

pores: dark grey to black [P]

This method of specimen preparation shows that the proportion of pores, which remained dark, is considerably smaller than might at first have been assumed.

Figs. 376 and 377: **Double-layer enamelled steel sheet**

376: Polished specimen, bright field.

The photograph shows double-layer enamelled steel sheet in which the foundation enamel is dark grey and the surface enamel is light grey. Both enamel layers are permeated with pores of varying diameter.

377: Same material, same area of the microstructure as in Fig. 376, coated with ZnTe, the interference minimum of the α-iron lies at 550 nm, photographed at 530 nm.

The reflectivity of the vitreous surface enamel has been increased by the deposition of a highly refracting interference layer. Both layers of enamel now appear to be an almost uniform shade of grey. The pore volume and pore geometry are now clear in both layers. A quantitative evaluation made for different baking conditions (different baking temperatures and times) yielded reproducible results.

373 500 : 1

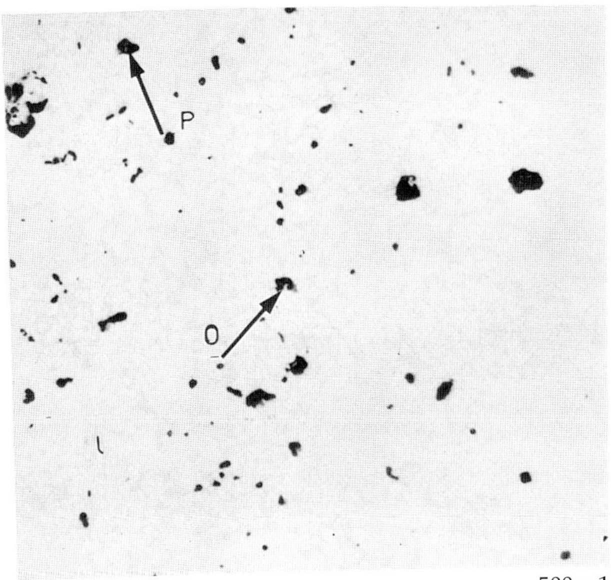

374 500 : 1

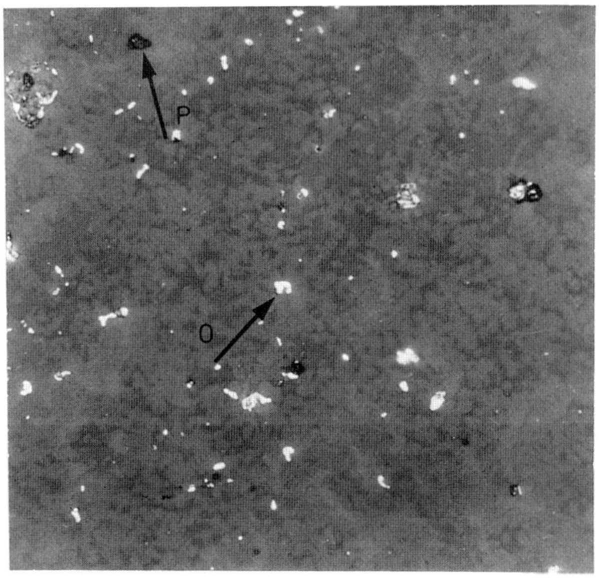

375 500 : 1

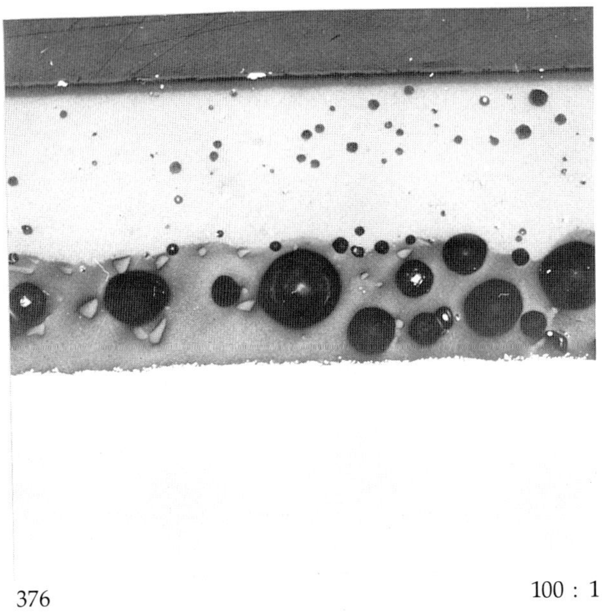

376 100 : 1

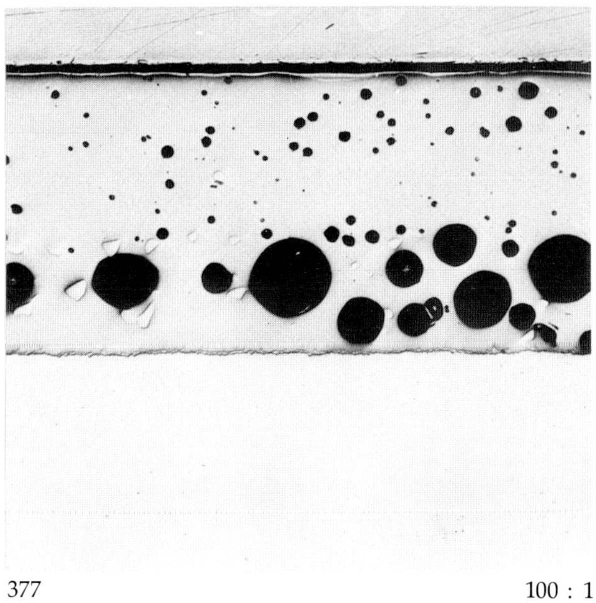

377 100 : 1

2.2.4 Corrosion and Oxidation

2.2.4.1 General Introduction

At the high operational temperatures of industrial plant heavy demands are made on the high temperature strength and scale resistance of metallic and non-metallic materials. Failures occur as a result of excessively high temperatures, changes in the composition of the medium (air, gases) and in very many cases as a result of an inappropriate choice of material. The types of plant mainly affected are high pressure steam boilers, process furnaces and gas turbines.

The most important types of high temperature corrosion are: scaling, internal oxidation, carburization, attack by hydrogen or nitrogen and attack by molten salts and molten metals. Whereas failure due to wet corrosion can be revealed and identified effectively by the normal metallographic techniques, the appearance of the failure due to high temperature corrosion is particularly interesting for interference metallography. In the majority of failures, high temperature corrosion (oxidation, carburization, attack by nitrogen) has resulted in the formation of surface layers consisting of reaction products as well as of carbides or nitrides often of unknown chemical composition and structural properties. In most cases these can not be identified more exactly by conventional metallographic techniques.

The reaction products, whether oxides, sulphides or carbides, are characterized by a reflectivity which is lower than that of the metallic material. In the case of oxides, the layers are frequently inhomogeneous or porous so that a high degree of care in specimen preparation is necessary to produce a surface which is physically well-defined.

In order to obtain the best possible light-dark contrasts on substrates with a low reflectivity, layer materials are needed with lower refractive indices than those used on metallic materials (c.f. section 1.2.3 and Fig. 120).

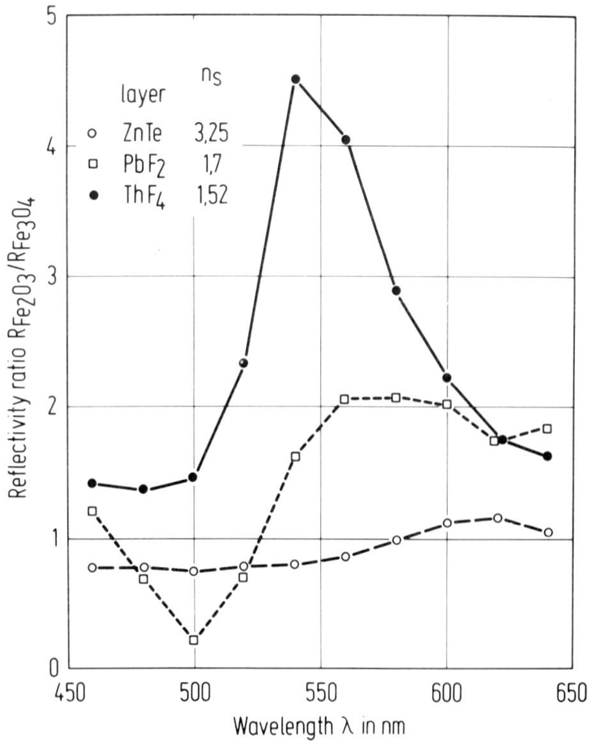

Fig. 378: Influence of the observation wavelength on the ratio of the reflectivities of Fe_2O_3 and Fe_3O_4 after coating with various evaporated layers.

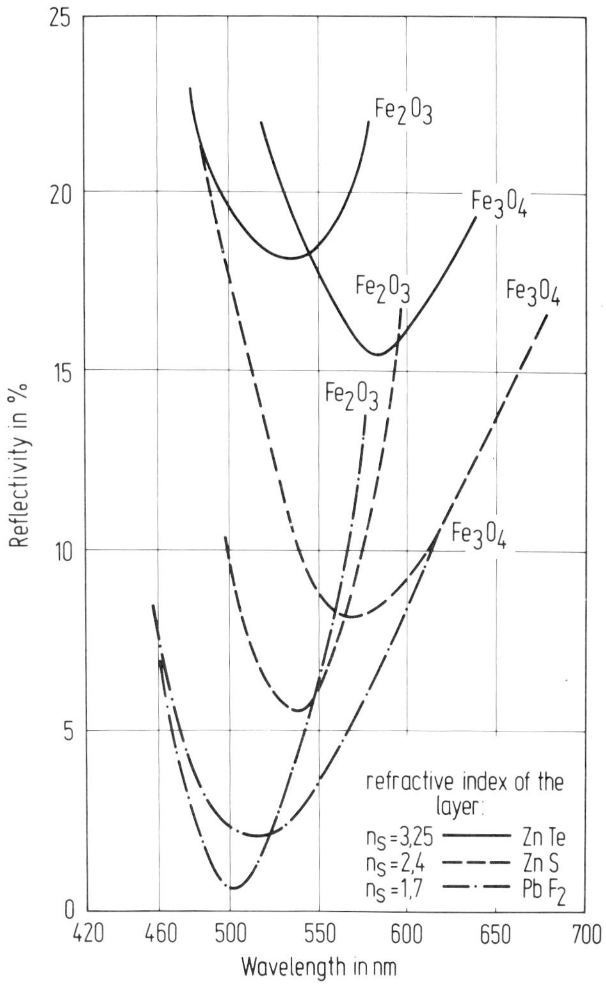

Fig. 379: Spectral reflectivity of Fe_2O_3 and Fe_3O_4 after coating with evaporated layers of various materials and thicknesses.

On the other hand, the use of very highly refractive layers increases the difference in colour between different microstructural constituents (c.f. Figs. 120 and 379).

In investigations of the microstructure of layers formed by oxidation, carburization or nitriding the metallographer is frequently obliged to choose between the best possible colour reproduction and the production of an optimum light-dark contrast. Examples have already been given of the possibilities of revealing the layers formed as a result of high temperature corrosion with a good degree of contrast using interference layer microscopy 1–3).

In order to illustrate the potential of interference layer metallography for describing high temperature corrosion damage, the following examples have been selected for inclusion in the present volume:

a) influence of the refractive index (of the layer material) on the light-dark and colour contrast between the various oxides of iron (FeO, Fe_3O_4 and Fe_2O_3) and

b) investigation of hot gas corrosion of nickel alloys with and without a protective surface layer by flowing exhaust gases.

Literature to 2.2.4

1) Kiesheyer, H.: Die Unterscheidung von Gefügebestandteilen mittels Interferenzbedampfung zur gezielten Mikrosondenanalyse. DEW-Technische Berichte **12** (1972) 70

2) Semlitsch, M.: Interferenz-Aufdampfschichten bei Mikroanalysengefüge von Werkstoffen mit der Elektronenstrahl-Mikrosonde. Prakt. Metallographie **8** (1971) 511

3) Geiger, T. and Semlitsch, M.: Hochtemperaturkorrosion von Gichtgas-Turbinenschaufeln. Tech. Rundschau SULZER, Forschungsheft (1968) 31

4) Semlitsch, M.: Untersuchungen von Hochtemperatur-Korrosionsschäden an Gasturbinenschaufeln mit der Elektronenstrahl-Mikrosonde. Mikrochimica Acta (Vienna), Suppl. III (1968) 65

5) Bühler, H.-E., Jackel, G. and Thiemann, E.: Möglichkeiten der quantitativen Beschreibung von oxidischen Phasen mit Hilfe der ätzfreien Gefügeentwicklung nach Pepperhoff. Arch. Eisenhüttenwes. **41** (1970) 409

2.2.4.2 Influence of the refractive index (of the layer material) on the light-dark and colour contrast between the oxides of iron (FeO, Fe_3O_4 and Fe_2O_3)

Figs. 380–386: Armco iron, 10 h 1200°C/air, cooled in the furnace.

The oxide layer was coated with various materials with different refractive indices, measured photometrically, and observed in monochromatic light.

380: Specimen coated with ZnTe, first order, photographed at 460 nm.

Armco iron: medium grey (lower half of picture)

Fe_2O_3 and Fe_3O_4: medium grey (Fe_2O_3 is a slightly darker shade of grey than Fe_3O_4 in the upper half of the picture)

381: Same material, same specimen preparation as in Fig. 380, photographed at 620 nm.

Armco iron: dark grey

Fe_3O_4: medium grey

Fe_2O_3: white

At the high refractive index of ZnTe, the light-dark contrast between the two oxides Fe_3O_4 and Fe_2O_3 is visible but not optimum. The reader is referred to Fig. 378, in which the ratio of the reflectivities $R_{Fe_2O_3}/R_{Fe_3O_4}$ is plotted as a function of wavelength for the layer materials ZnTe, PbF_2 and ThF_4. It can be seen that for ZnTe the ratio fluctuates about the value 1 (i.e. equal reflectivities) over the entire spectral range. Better light-dark contrasts can be obtained using layer materials with lower refractive indices than that of ZnTe. According to Fig. 378 PbF_2 should give a better contrast. This can also be expected on the basis of the measured interference curves in Fig. 379. Here, the reflectivity at the minimum shows a distinct shift to lower values for materials with lower refractive indices.

382: Same material as in Fig. 380, coated with PbF_2, first order, photographed at 500 nm.

Armco iron: medium grey (lower half of picture)

Fe_3O_4: medium grey

Fe_2O_3: black

A narrow crack has formed between the iron and the oxide layer. Pores are also visible. Both regions appear white in the photographic reproduction.

Figs. 383–385: See page 204

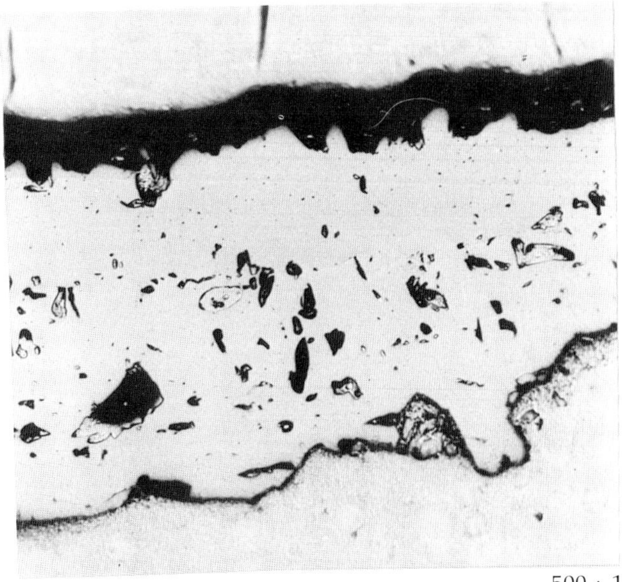

380 500 : 1

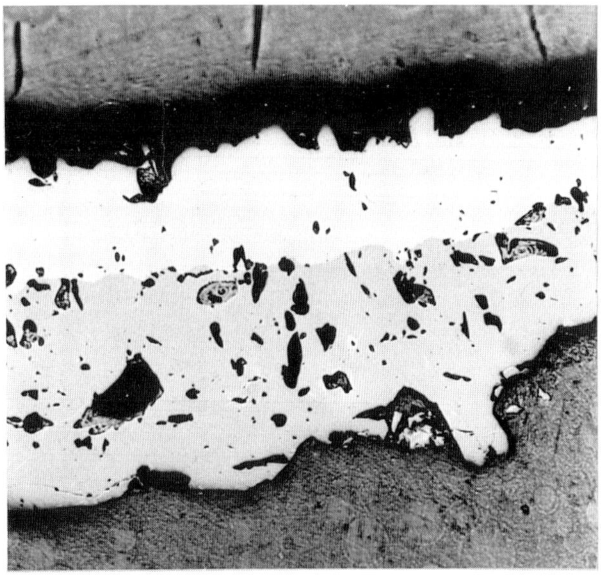

381 500 : 1

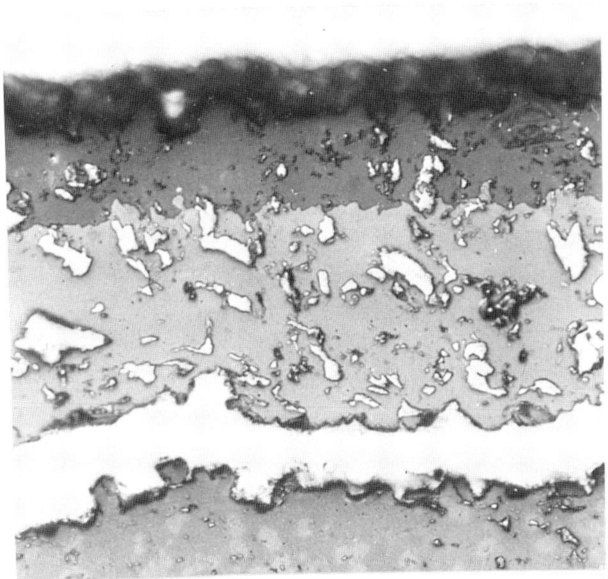

382 500 : 1

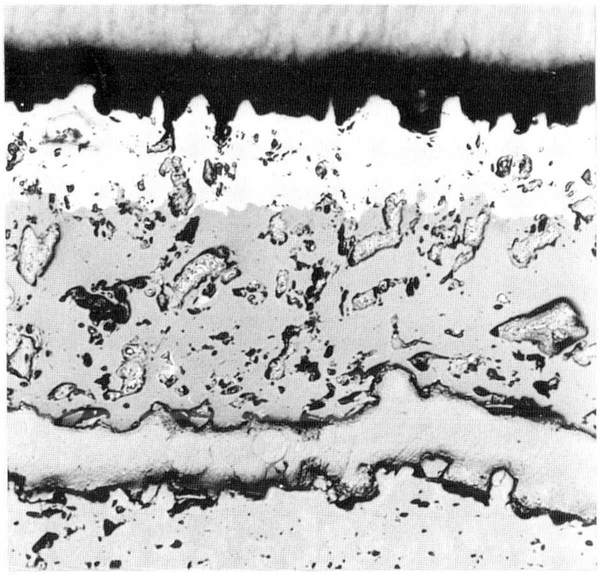

383 500 : 1

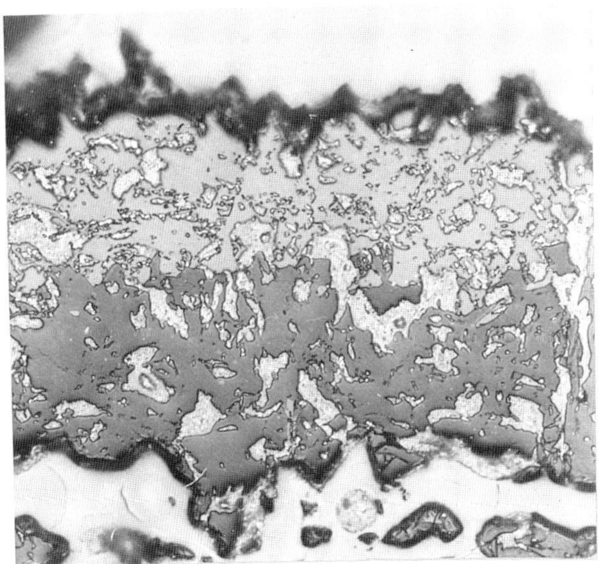

384 500 : 1

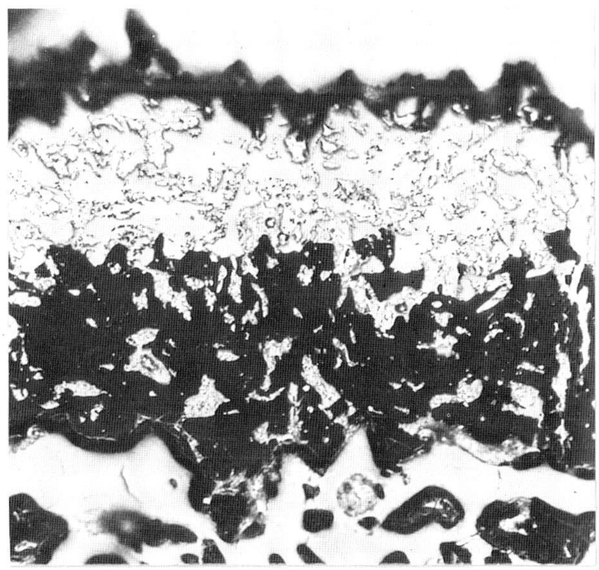

385 500 : 1

383: Same material, same coating layer as in Fig. 380, photographed at 640 nm.

Armco iron: medium grey

Fe_3O_4: medium grey

Fe_2O_3: white

Cracks and pores appear in a somewhat different shade of grey from the oxide Fe_3O_4.

384: Same material as in Fig. 380, coated with ThF_4, first order, photographed at 460 nm.

Fe_3O_4: dark grey

Fe_2O_3: light grey

pores and cracks: light grey to white

385: Same material, same coating layer as in Fig. 384, photographed at 540 nm.

Fe_3O_4: black

Fe_2O_3: white

cracks and pores: white

This sequence of photographs shows, in agreement with Figs. 378 and 379, that in the case of layer materials with low refractive indices, the reflectivity at the interference minimum decreases for both oxides. In consequence of this, the reflection ratio using monochromatic light can be substantially increased.

386: Same material, same heat treatment as in Fig. 380, specimen coated with ZnS, first order, bright field.

FeO: dark blue (16 : 4 : 3)

Fe_3O_4: blue (15..16 : 2 : 3)

Fe_2O_3: pale violet (11..12 : 1 : 2)

Fig. 386 demonstrates that the reflection minimum shifts from short to long wavelengths in the order Fe_2O_3, Fe_3O_4 to FeO.

2.2.4.3 Hot gas corrosion of nickel alloys with and without a protective surface layer in streaming exhaust gases

Fig. 387: **Nickel base alloy Nimonic 80 A**

Heat treatment: 8 h 1080°C/air + 46 h 700°C/air, corroded in blast furnace turbine 36,000 h 710°–720°C.

Coated by evaporation with ZnSe, zero order, bright field.

Matrix, depleted in chromium: light blue (17 : 1 : 2)

$(Cr, Ni, Ti)_3 S_4$: dark blue (17 : 4 : 5)

Cr_7C_3: red platelets (8 : 4 : 3)

Figs. 388–392: **Platinaluminized protective layer on a nickel base alloy** (Material IN 738, Material No. 2.4990).

A duplex protective layer was deposited on the base material IN 738 in a two-stage process consisting of an electrolytic deposition of several μm of platinum followed by powder aluminization.

After this pretreatment, the duplex layer consists of three characteristic strata. The outermost layer is a platinum-aluminium phase $PtAl_2$ containing some nickel. The middle layer consists of NiAl containing cobalt in which chromium rich solid solution and titanium carbonitrides have precipitated. Between the NiAl layer and the unchanged base material,

Figs. 388–392: See page 206

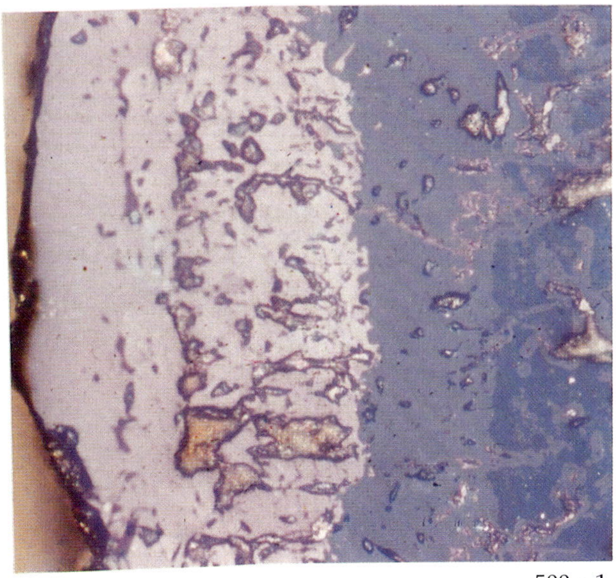

386 500 : 1

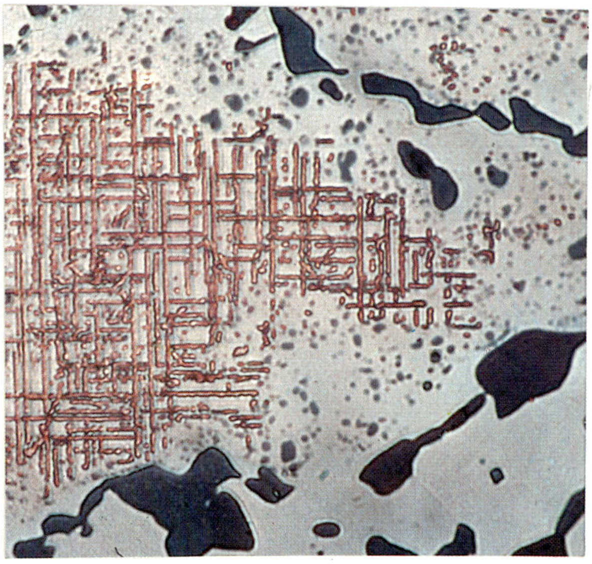

387 1000 : 1

388 1000 : 1

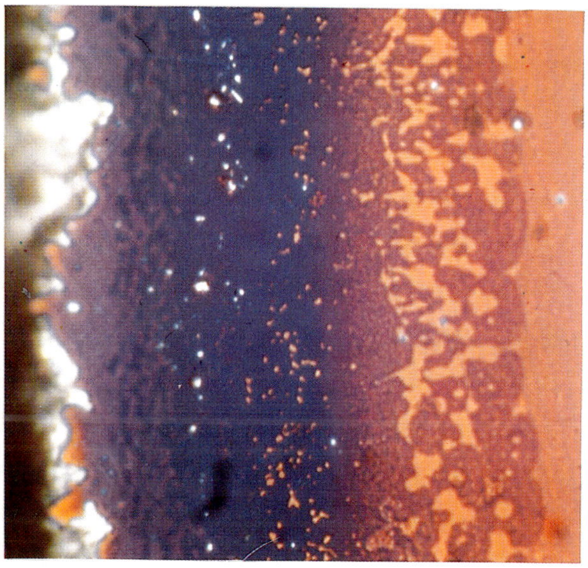

389 1000 : 1

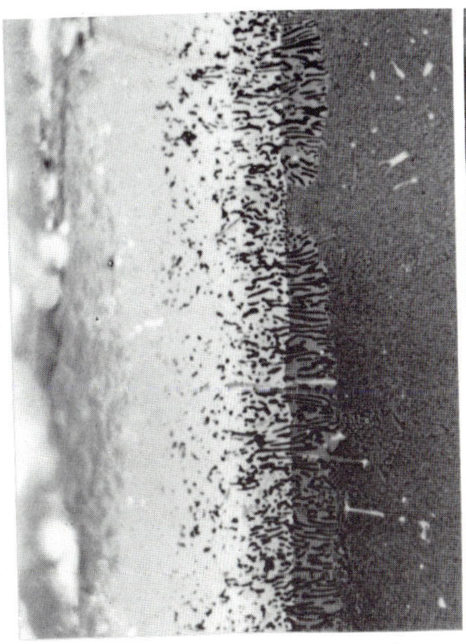

390 500 : 1

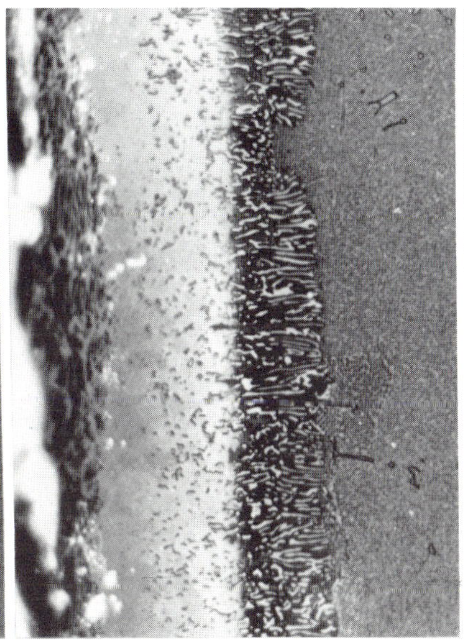

391 500 : 1

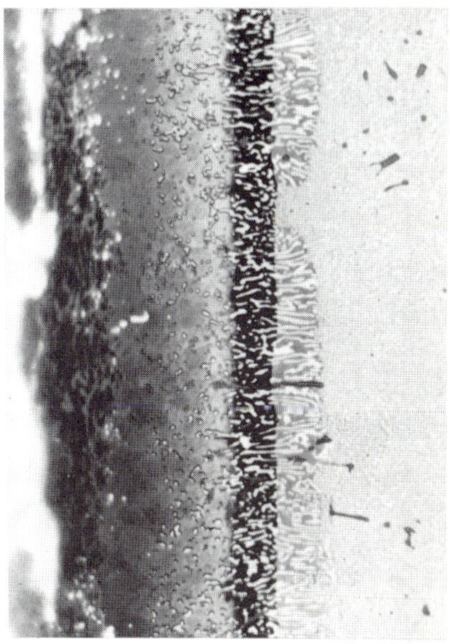

392 500 : 1

there is a narrow zone consisting of the intermetallic compound Ni_3Al, chromium rich solid solution and titanium carbonitrides. It can be seen from the matrix of the material IN 738 that very fine precipitates of the γ'-phase have formed.

388: Material as described above, coated with CdSe, zero order, reflection minimum of the nickel base alloy lies at 490 nm.

$PtAl_2$: light blue (16 : 1 : 3)

NiAl (containing cobalt): dark blue (16 : 4 : 4)

Ni_3Al: violet (12 : 2 : 4)

chromium-rich solid solution: reddish yellow (10 : 1 : 2)

nickel base alloy IN 738: pale violet (11 : 2 : 2)

titanium carbonitrides: Finely distributed between the Ni_3Al and chromium-rich solid solution (dark brown colour).

389: Same material, investigated after an experimental period of 400 h at 850^0C in hot gases produced from light oil containing 0.4% sulphur and with an addition of 14 ppm sodium.

Same coating layer as in Fig. 388.

After the prolonged attack at 850^0C the diffusion zone consisting of the chromium rich solid solution and the intermetallic compound Ni_3 (Al, Ti) has extended both in the direction of the base material and towards the NiAl phase of the protective layer. The outermost platinum-aluminium layer $PtAl_2$ is still completely intact after this period of exposure.

$PtAl_2$: reddish violet (11 : 2 : 4)

NiAl: dark blue (16 : 4 : 4)

$Ni_3(Al,Ti)$: reddish violet (10 : 3 : 4, basis colour)

chromium-rich solid solution: yellow (3 : 4 : 2)

nickel base alloy: light brown (8 : 2 : 3)

Figs. 390–392: Same material and same heat treatment as in Fig. 389, but coated with ZnSe, zero order.

390: Photographed at 520 nm.

$PtAl_2$: dark grey

NiAl: light grey

chromium rich solid solution: black

Ni_3Al: dark grey (difficult to distinguish from the nickel base matrix)

nickel base alloy matrix: dark grey

391: Photographed at 580 nm.

$PtAl_2$: dark grey to black

NiAl: medium grey to light grey

chromium-rich solid solution: white

Ni_3Al: black

nickel base alloy matrix: dark grey

392: Photographed at 600 nm.

$PtAl_2$: dark grey to black

NiAl: light grey

chromium-rich solid solution: white

Ni_3Al: black and dark grey

Figs. 393–398: **Chromium plating on nickel base alloy (material IN 738, material no. 2.4490),** coated with CdSe, zero order. The reflection minimum of the nickel base alloy is at 490 nm.

A protective layer was applied to the nickel base alloy by diffusion chromium plating.

393: Material and preparation as given above, as-received condition:

chromium-rich outer layer (α-Fe-Cr solid solution [M]): light brown (5 : 3 : 2)

Alumina inclusions [O] in α-Cr-Fe-solid solution: light yellow (1..2 : 1 : 1)

diffusion layer [D] (transition to nickel base alloy): violet (10 : 3 : 3)

titanium carbonitrides: dark violet to light yellow [1] (4 : 1 : 1)

chromium-rich phases between the outer and inner protective layers in the diffusion zone (M_4N, c.f. Fig. 397): red (difficult to distinguish from the titanium carbonitrides, cf. also Fig. 397) (8 : 3 : 3)

After diffusion chromium plating a chromium-rich outer layer (α-Fe-Cr solid solution) containing white alumina inclusions has formed. Further, an inner diffusion zone may be seen, in which needle-shaped, violet to grey titanium carbonitrides have precipitated. At the transition between the two layers another, brown phase, hardly visible, has formed, which has a lower nickel and higher chromium content than the outer layer.

394: Same material, same position and same coating as in Fig. 393, observation at 560 nm.

chromium-rich outer layer [M]: light grey

alumina inclusions [O]: white

inner protective layer [D] with transition to nickel base alloy: dark grey

titanium carbonitrides [1] in this layer: dark grey to white

chromium-rich phase [M_4N] in the diffusion zone: dark grey (same darkness level as in Fig. 396 and same shade as in Fig. 395).

395: Same material as in Fig. 393 after testing for 400 h at 850^{0}C in hot gas.

oxide layer: brownish yellow

chromium-rich phase on the surface (Cr_2N, cf. Fig. 397): dark violet (10 : 4 : 4)

chromium plating layer [M] (α-Fe-Cr solid solution): light brown (6 : 3 : 3)

chromium-rich, nickel-depleted phase (M_4N, cf. Fig. 397): red (8 : 4 : 3)

titanium carbonitrides [1] in the form of rods: violet to light yellow (4 : 2 : 1)

matrix of nickel base alloy: violet (10 : 3 : 3)

In the specimens the beginning of scale formation can be seen on the outer surface. The diffusion chromium plating layer is made up of three phases. Compared with the original layer (cf. Figs. 393 and 394) the α-Fe-Cr solid solution has a lower chromium and a higher nickel content. On the outer surface of the layer a violet to dark red coloured zone has formed which shows high chromium concentrations and only small amounts of other elements. The brown phase noted already in Figs. 393 and 394 has formed in greater quantities and is found also in greater volume fractions in the chromium layer.

396: Same material and same position as in Fig. 395, observed at 560 nm.

oxide layer: light grey

chromium-rich phase near the surface (Cr_2N, cf. Fig. 397): black

chromium plating layer [M] (α-Fe-Cr solid solution): light grey

chromium-rich, nickel-depleted phase in the chromium plating layer (M_4N, cf. Fig. 397): dark grey

titanium carbonitrides [1]: white to dark grey

nickel base alloy: light to medium grey

The phases may be distinguished easily in the black and white micrograph.

397: Same material as in Fig. 393, but subjected to the hot gas corrosion for 600 h.

scale layer: light yellow to brownish yellow (2 : 3 : 1)

Cr_2N: dark violet (10 : 4 : 3)

nitride M_4N: brown (8 : 4 : 3)

titanium carbonitrides: yellow

matrix: violet (10 : 3 : 3)

In this specimen, the reaction of the chromium plating layer with the corrosive medium is even more apparent. Together with an increased scale layer concisting of chromium-rich oxides, the violet phase has formed in greater amounts near the surface. Further, the brown phase noted already in Figs. 393 and 395 is present in a greater proportion by volume. In this case the various microstructural constituents were large enough to permit a quantitative electron beam microanalysis of the phases in the chromium plated zone. The following results were obtained: The violet phase near the surface contained approximately 80% Cr and only 1.5% Fe and 2% Ni, but about 12% N. On the basis of this composition it may be identified as the chromium nitride Cr_2N. The brown phase clearly visible in Fig. 395 but only in small amounts in Fig. 393 contained 60% Cr, 12% Fe, 19% Ni, 3% Co and about 6% N. It may therefore be assumed that this could be the nitride M_4N. Between the chromium layer and the nickel base alloy a diffusion exchange has taken place after 600 h testing. The colour difference between the chromium diffusion layer and the nickel base alloy is no longer visible as in Fig. 393 and 395.

398: Same material and same position as in Fig. 397, observed at 560 nm.

scale layer: white to light grey

Cr_2N: black

nitride M_4N: dark grey to almost black

titanium carbonitrides: white

matrix of nickel base alloy: medium grey

The black and white micrograph again gives a satifactory distinction of the phases.

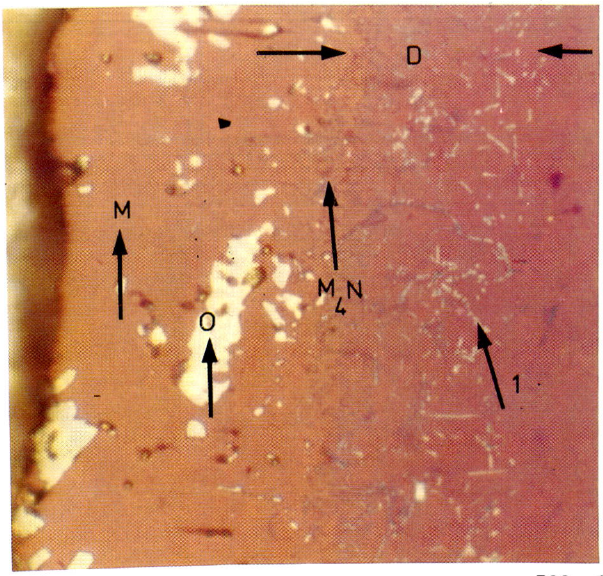

393 500 : 1

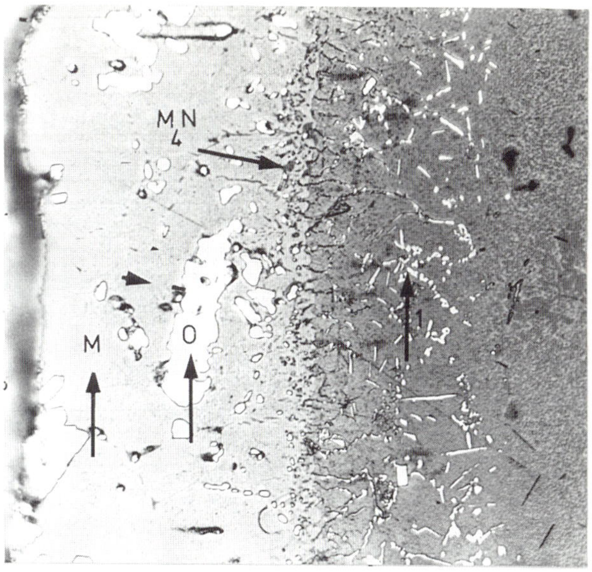

394 500 : 1

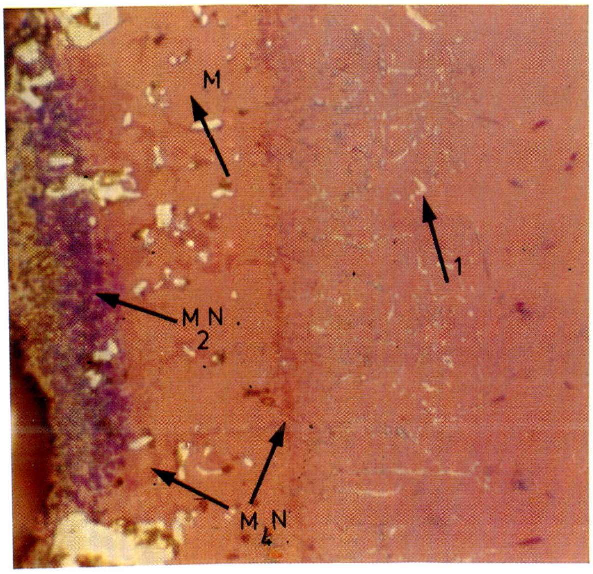

395 500 : 1

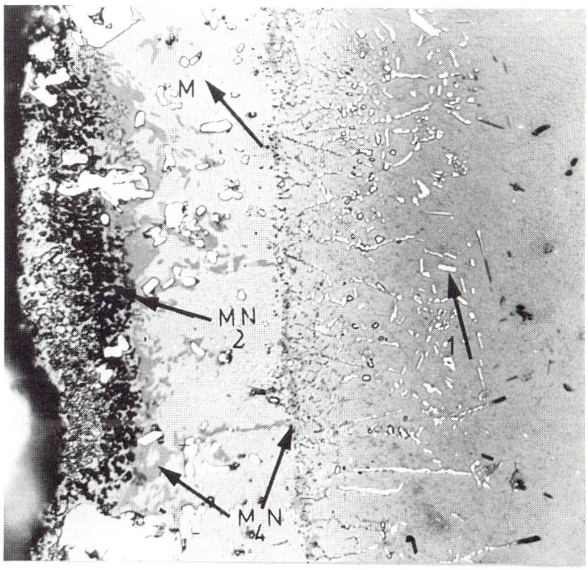

396 500 : 1

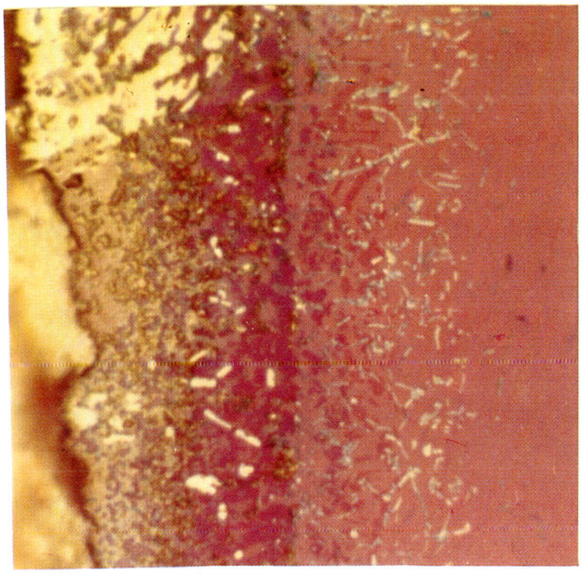

397 500 : 1

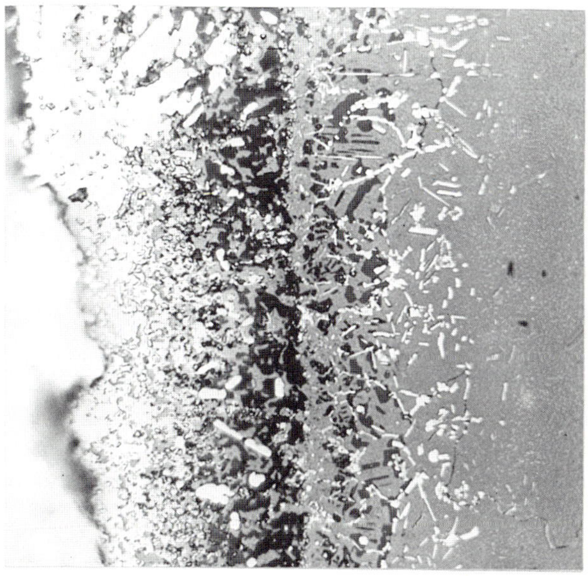

398 500 : 1

2.2.5 Observation of states of magnetic order

2.2.5.1 General introduction

In contrast to magnetically disordered crystals (paramagnetic state), ferro-, ferri- and antiferromagnetic materials are characterized by an ordered magnetic structure at lattice level. These structures can be classified in a symmetry scheme similar to the crystallographic crystal classes (point and space groups).

In this scheme the magnetically ordered crystal is always less symmetrical than the paramagnetic crystal.

As a consequence of the reduction in symmetry, optical anisotropy is possible in magnetically ordered crystals which does not exist in the paramagnetic state.

The significance of magneto-optical effects in polarised light microscopy is that they provide a means of observing the magnetic domains of which the magnetically ordered crystal is composed 1,2). The effects are normally so small that they need amplifying.

The magneto-optical Kerr Effect

The magneto-optical Kerr effect describes the rotation of the plane of polarisation of light reflected from the surface of a ferromagnetic substance. Since this rotation depends on the direction of magnetisation of the domain, zones magnetised in different directions appear in different light-dark contrast. A diagrammatic representation of the magneto-optical Kerr effect is shown in **Fig. 399.**

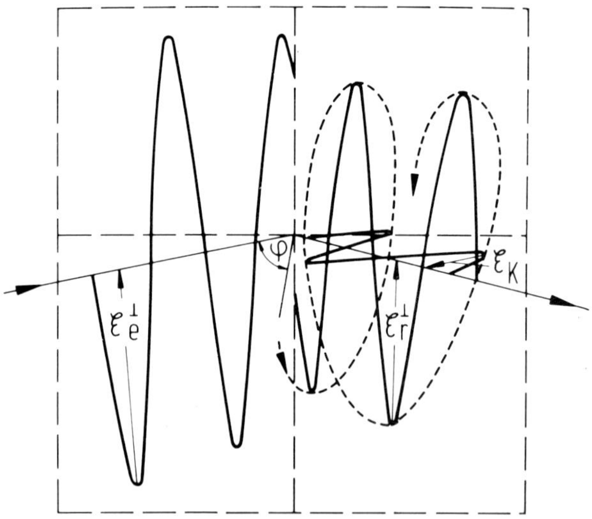

$\varepsilon_e^\perp$ = linearly polarised incident light

$\varepsilon_r^\perp$ = linearly polarised reflected light in the absence of Kerr effect

φ = angle of incidence

ε_K = component perpendicular to ε_r dashed: elliptically polarised light, superposition of ε_r and ε_K.

Fig. 399: Schematic diagram of the magneto-optical Kerr effect.

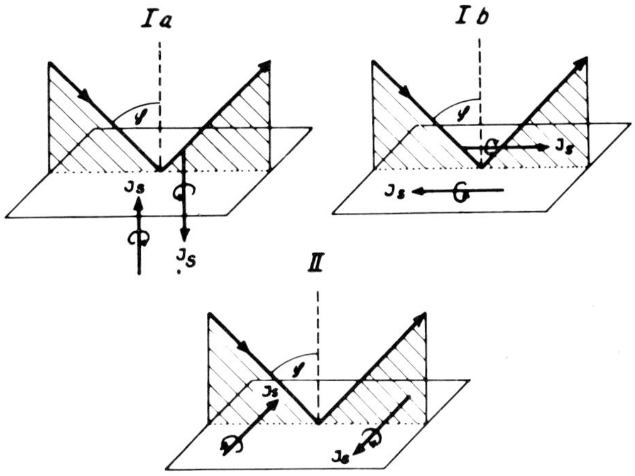

Fig. 400: The magneto-optic Kerr effect.

I a) polar Kerr effect
 b) meridional (longitudinal) Kerr effect
II equatorial (transverse) Kerr effect
(J_s = direction of magnetisation φ = angle of incidence)

If linear polarised light ($\mathcal{E}_e^{\perp}$) strikes a polished surface at an angle of incidence φ, a similarly linear-polarised reflected wave ($\mathcal{E}_r^{\perp}$) is normally produced. If however the material is ferromagnetic, the light is given a Kerr component ($\mathcal{E}_K$) perpendicular to the normally emergent vibration as a result of the Lorentz force. Both components $\mathcal{E}_r$ and $\mathcal{E}_K$ combine to given elliptically polarised light indicated by the spiral in Fig. 399. The Kerr effect thus results in a change of brightness, whereby the nature and intensity of this change are dependent on the degree and direction of the spontaneous magnetisation of the individual ferromagnetic domains. Details of the possibilities of exploiting the various types of Kerr effect can be found in the literature 2).

Accentuation of the magneto-optical Kerr effect.

Only in the case of uniaxial crystallites, e.g. Fe_3C or the intermetallic compound MnBi, and after very careful metallographic preparation, can a sufficiently well contrasted optical micrograph of magnetic order be obtained using the Kerr effect. In these cases the Kerr rotation is large. Contrast and reproduction can however be vastly improved by the deposition of interference layers. As well as the attenuation ot the normally reflected wave (reduction in reflectivity) already described, these layers bring about an additional increase in the Kerr rotation.

In principle, the same coating materials and procedures can be used as described earlier.

The different forms of the Kerr effect are illustrated in **Fig. 400:** a polar effect, a meridional (longitudinal) effect and an equatorial (transverse) effect.

In the first two cases the direction of magnetisation is parallel or anti-parallel to the plane of incidence of the light. In the polar effect, the direction of magnetisation is perpendicular to the surface, and in the meridional effect parallel. The Kerr component already described in Fig. 399 appears in both cases in a direction given by the circular arrows in Fig. 400. The plane of the arrows gives the plane of the electronic vibration induced by the incident light. The equatorial effect is negligibly small and will be ignored in the following discussion.

The Kerr rotations represented in Fig. 399 are extremely small and fluctuate, according the observation conditions, between several minutes of arc and in very favourable cases up to 0.5^0. The schematic representation in Fig. 399 has therefore been exaggerated so that it is easier to understand.

The revealing of magnetic domain structures on the surface of uniaxial ferromagnetic crystals using the polar Kerr effect in a normal vertical polarising microscope with conventional ray path yields very well contrasted images superior in every way to those obtained by the powder line technique.

For the observation of magnetic domains on cubic crystals using the meridional Kerr effect, the optical arrangement must be such that the angle between the illuminating and imaging ray paths is very large. Research 1) has shown that it should be between 100^0 and 120^0. High magnifications ($> 50x$) are possible in a normal vertical polarising microscope only if a very oblique and narrow beam is used 2,3). The aperture stop in the illuminating ray path of the microscope has to be swung out of the optical axis and closed so that only a fraction of the outer peripheral rays from an objective with a high aperture are used for imaging. By adjustment of the excentric aperture stop, it is possible to determine the angle of incidence of the narrow light beam which yields the greatest possible contrast. All objectives with numerical apertures > 0.40 can be used for this observation. The use of immersion objectives permits the imaging of magnetic domains utilising the meridional Kerr effect at the full resolution of the polarising microscope.

Observation of antiferromagnetic domains and superconductivity structures

Antiferromagnetic order can also be observed in the microscope. The paramagnetic cubic oxides of the transition metals, nickel, cobalt and iron become rhombohedrally (NiO, FeO) or tetragonally distorted (cobalt oxide) below the Néel temperature 4). Only in the case of weakly absorbing and low reflecting materials is it possible to reveal these lattice distortions successfully without the help of interference layers. Investigations of antiferromagnetic order in metallic systems have been reported, for example on gold-manganese alloys 1).

Superconducting structures of the mixed state of Type I superconductors (superconducting zones alternating with normally conducting zones) have been rendered visible in lead using evaporated europium sulphide layers 1).

Literature to 2.2.5

1) Pepperhoff, W. and Ettwig, H.-H.: Interferenzschichten-Mikroskopie. Dr. Dietrich Steinkopff, Verlag Darmstadt (1970), see particularly P. 57–77

2) Pepperhoff, W. and Bühler, H.-E.: Lichtmikroskopische Untersuchungen an martensitischen Gefügen. Arch. Eisenhüttenwes. **34** (1963) 839

3) Pepperhoff, W.: Sichtbarmachung ferromagnetischer Elementarbereiche in polarisiertem Licht. Arch. Eisenhüttenwes. **34** (1963) 767

4) Bühler, H.-E. and Horn, E.: Magnetooptik der Metalle. Special Vol. **6** of Prakt. Metallographie, Dr. Riederer-Verlag, Stuttgart (1976) 137

5) Feldmann, D. and Schulz, I.: Sichtbarmachung magnetischer Bereichstrukturen in Seltenerd-Kobalt-Legierungen durch den magnetooptischen Kerr-Effekt. Prakt. Metallographie **8** (1973) 462

2.2.5.2 Examples of how ferromagnetic domain structures can be revealed

Figs. 401 and 402: Cobalt-samarium alloy containing the intermetallic compounds $SmCo_5$ and Sm_2Co_7

401: Section polished and partly coated with ZnSe, zero order, partly polarised light, polar Kerr effect.

This figure illustrates the accentuation of the polar magneto-optical Kerr effect by deposition of an interference layer. In the left hand picture the polished metal surface was covered during evaporation and the magnetic domains are only vaguely indicated. In the right hand picture, on the other hand, the magnetic domains show up in strong contrast.

402: Polished section coated with ZnSe, zero order, partly polarised light, polar Kerr effect.

The two intermetallic compounds $SmCo_5$ and Sm_2Co_7 can easily be distinguished by the different formation of the ferromagnetic domains. The compound Sm_2Co_7 in the centre of the picture has a quite different structural pattern from the surrounding $SmCo_5$ phase.

Figs. 403–408: Ferromagnetic domain structures on a silicon steel

403: Silicon steel with 3.5% Si, annealed.

Polished section coated with TiO_2, zero order, reflection minimum of the α-iron lies at 540 nm, polarised light, meridional Kerr effect.

The photograph shows intricately formed compensating zones on a cube plane.

404: Schematic representation of the same area of the microstructure as in Fig. 403. The directions of magnetisation of the individual magnetic domains are indicated by arrows.

405: Same material, same coating, photographed under the same conditions as Fig. 403.

The photograph shows triangular boundary zones at a grain boundary.

406: Schematic representation of the structures in Fig. 405. Here again the magnetic orientations of the individual domains are indicated by arrows.

407: Same material, same coating and observation conditions as Fig. 403.

The photograph shows so-called "Christmas tree structures" which appear when the surface is slightly inclined to the (100) plane.

408: Schematic representation of the domains in Fig. 407. The direction of magnetisation is shown by arrows.

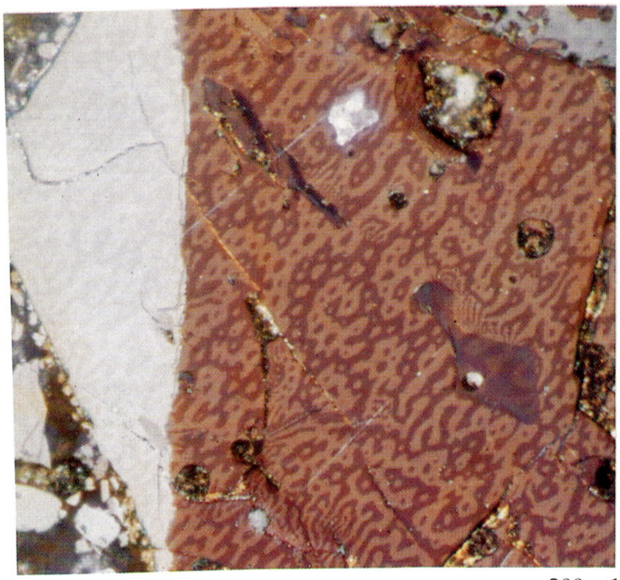

401 200 : 1

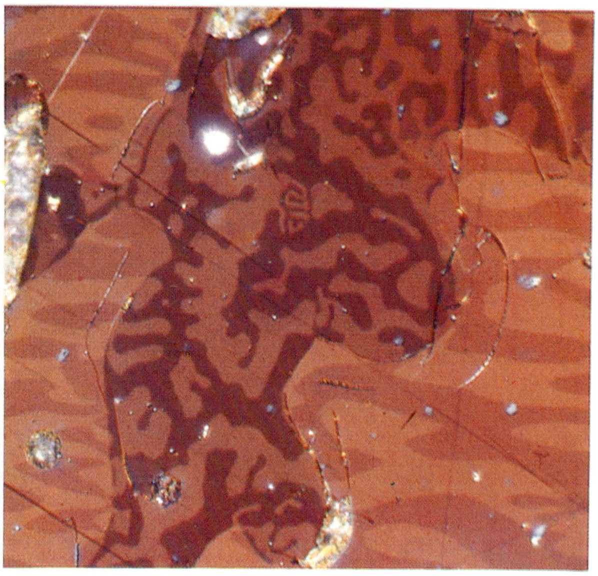

402 500 : 1

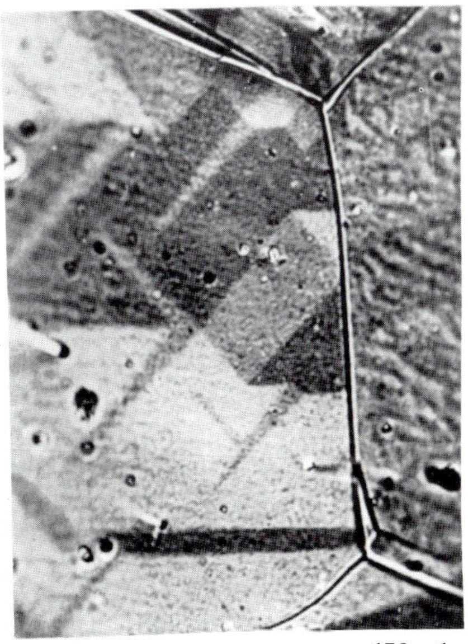

403 150 : 1

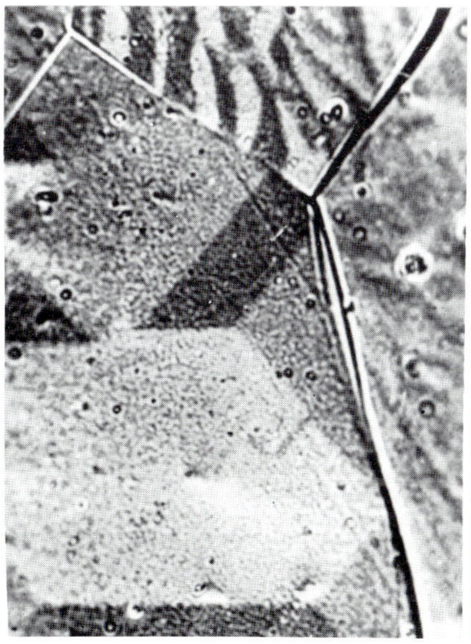

405 150 : 1

407 150 : 1

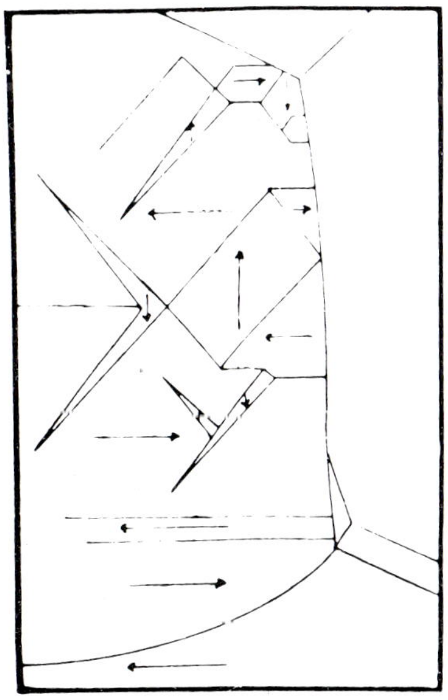

404

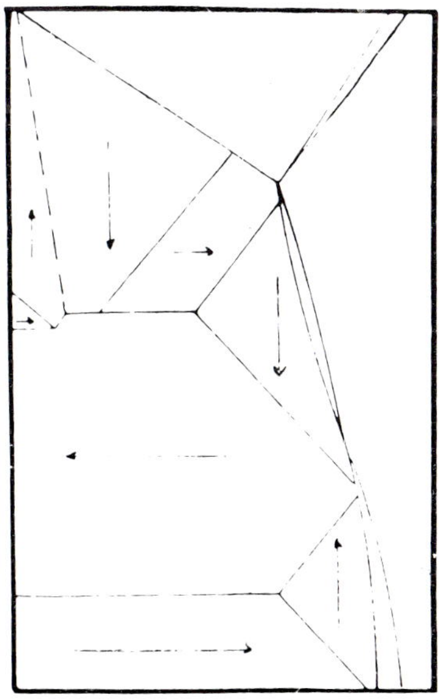

406

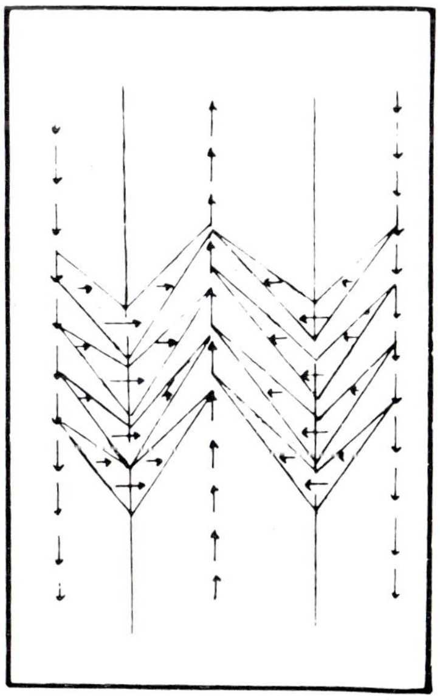

408

Figs. 409–411: **Ferromagnetic domain structures on polycrystalline cobalt materials**

409: Pure cobalt, section coated with evaporated TiO_2, zero order, reflection minimum of the cobalt lies at 540 nm, viewed in polarised light using the polar Kerr effect.

The photograph shows so-called star patterns and stripe structures. Stray fields are clearly visible along the grain boundary that runs across the centre of the picture. The star-shaped regions can always be expected on cobalt when the c-axis of the hexagonal crystal is perpendicular to the surface of the section. Striped regions occur when the c-axis lies parallel to the surface.

410: Same material and same coating and viewing conditions as in Fig. 409.

The grain in the middle of the picture has an ordered stripe structure which shows that the direction of spontaneous magnetisation, the c-axis, is parallel to the surface. Once again star patterns can be seen in neighbouring grains.

411: Cobalt-samarium alloy produced by melting, fine-grained.

Polished section, coated with ZnSe, zero order, reflection minimum of the matrix lies at 540 nm, viewed in polarised light using the polar Kerr effect.

The photograph shows clearly domain structures of different brightnesses consisting of intermetallic samarium-cobalt compounds in the very finely-grained matrix.

Figs. 412 and 413: **Ferromagnetic domain structures on iron carbide Fe_3C**

Coarse iron carbides in a special melt, polished section coated with TiO_2, zero order, reflection minimum of the matrix lies at 540 nm, viewed in polarised light using the polar Kerr effect.

The two photographs show complicated structures and no correlation can be made between magnetic and crystallographic order. The "plait pattern" visible particularly in Fig. 412 is characteristic of the magnetic domains on iron carbide Fe_3C.

Figs. 414–420: **Influence of an external magnetic field on the alignment of the ferromagnetic domain structures**

Polycrystalline cobalt, polished section coated with ZnS, zero order, viewed in polarised light using the meridional Kerr effect.

By applying an external magnetic field a high degree of alignment of the ferromagnetic domains can be achieved compared with the initial spontaneously magnetized state. In the photographic reproduction oppositely magnetized regions behave as positive to negative. Figs. 414 to 420 shows the magnetisation curve for a cobalt crystallite with its direction of spontaneous magnetisation parallel to the c-axis. Starting from the spontaneously magnetized condition (applied field = 0, Fig. 414) the specimen was magnetized by an external field of different strengths and opposing direction. It can be seen that the band patterns which indicate an antiparallel configuration of the magnetic domains in the grain, become increasingly more uniform with increasing field strength (Figs. 415 to 417 or 418 to 420). At very high externally applied fields, all the domains are aligned in one direction and this is expressed in the uniform grey tone (Figs. 417 or 420). The figures on the individual pictures give the coil current in mA.

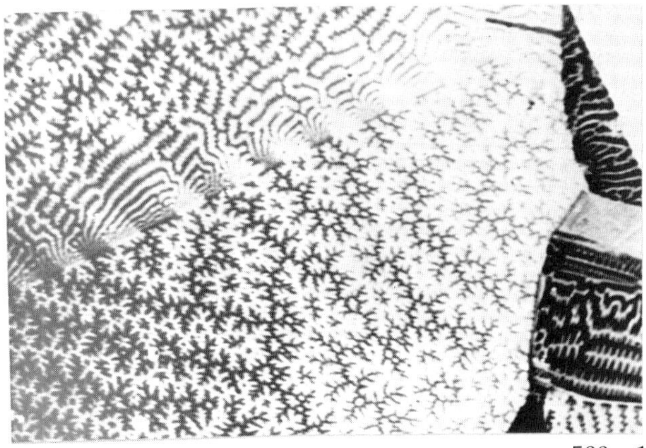

409 500 : 1

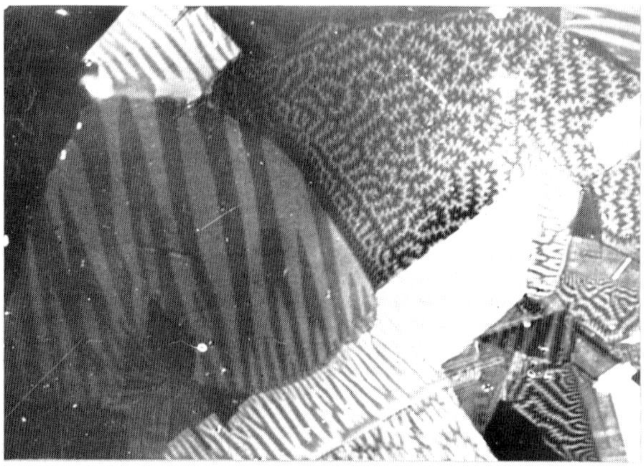

410 500 : 1

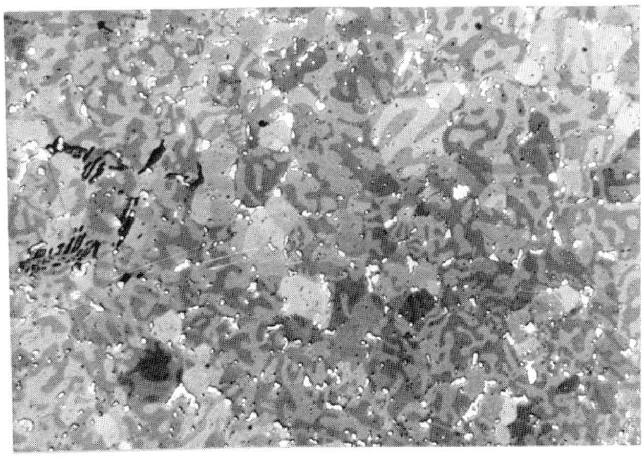

411 500 : 1

412 1000 : 1

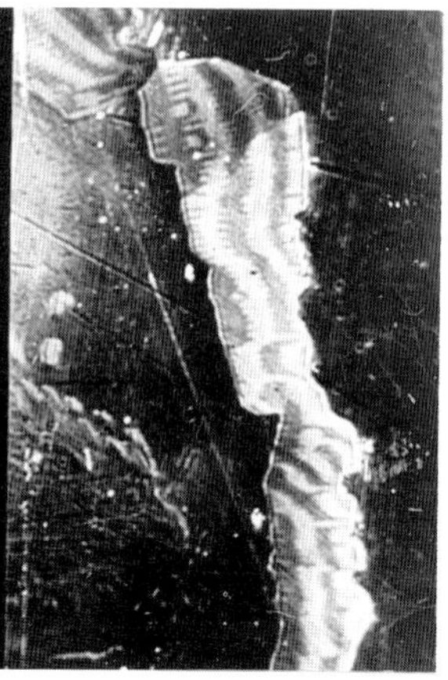

413 1000 : 1

414 200 : 1

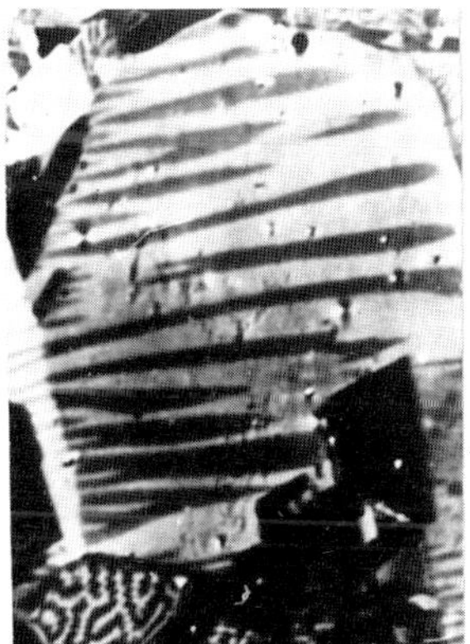

+1 ↑

415 200 : 1

+2 ↑

416 200 : 1

+3 ↑

417 200 : 1

↓ −1

418 200 : 1

↓ −2

419 200 : 1

↓ −3

420 200 : 1

Figs. 421–423: **Influence of an externally applied load on the formation of ferromagnetic domain structures in steel with 3.5% Si**

As well as internal stresses and external magnetic fields, external loads can also change the direction of magnetisation in crystals, particularly when the crystal anisotrophy energy is low. This is true for example of iron and its alloys. Even at low tensile stresses in the elastic region a definite alignment of the Bloch walls can be observed in the direction of the applied load (Figs. 422 and 423). In the case of materials with a high crystal anisotropy energy, e.g. cobalt, the arrangement of the domains is not influenced even by plastic deformation. In Figs. 421 to 423 the originally vertical direction of spontaneous magnetisation (Fig. 421) is rotated by an external load into the direction of the applied tensile stress which runs horizontally in the plane of the picture.

421: Material coated with ZnS, zero order, viewed in polarised light using the meridional Kerr effect, specimen loaded to $0.1^0/_{00}$ elastic strain.

422: Same material, same coating and viewing conditions as in Fig. 421, specimen loaded to $0.4^0/_{00}$ elastic strain.

423: Same material, same coating and viewing conditions as in Fig. 421, specimen loaded to $0.6^0/_{00}$ plastic strain.

Figs. 424–426: **Differentiation of tetragonal and cubic martensite**

424: Steel with 1.0% C and 0.6% Mn, 1 h at 1150^{0}C/ice water,coated with TiO_2, zero order, viewed in polarised light using the polar Kerr effect.

In the quenched state, finely banded structures can be seen in an irregular configuration characteristic of a crystallite subject to high internal stresses.

425: Same material, same area of the microstructure, same coating as in Fig. 424, annealed 1 h at 100^{0}C in air.

Band-like domains are produced in a regular configuration with the direction of magnetisation, perpendicular to the longitudinal axes of the martensite needles.

426: Same material, same area of the microstructure, same coating conditions as in Fig. 424, annealed 1 h at 180^{0}C in air. The regular arrangement of the domains perpendicular to the longitudinal axes of the martensite needles is complete. The tetragonal martensite has transformed into cubic martensite.

Figs. 427–429: **The influence of internal stresses on the magnetic domain structures in a iron-nickel alloy with 32.5% Ni**

427: Material as cast, quenched to –120^{0}C, coated with TiO_2, zero order, viewed in polarised light using the meridional Kerr effect.

Figs. 428 and 429: See page 220

421 50 : 1

422 50 : 1

423 50 : 1

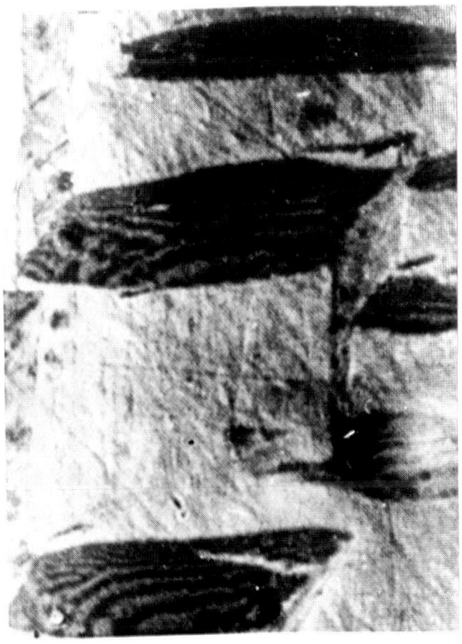

424 250 . 1

425 250 : 1

426 250 : 1

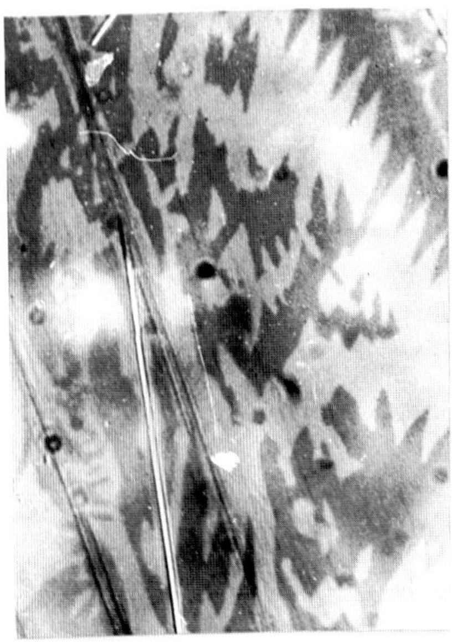

427 600 : 1

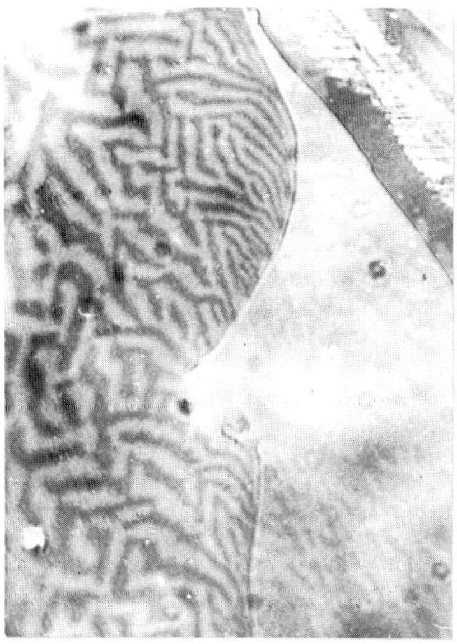

428 600 : 1

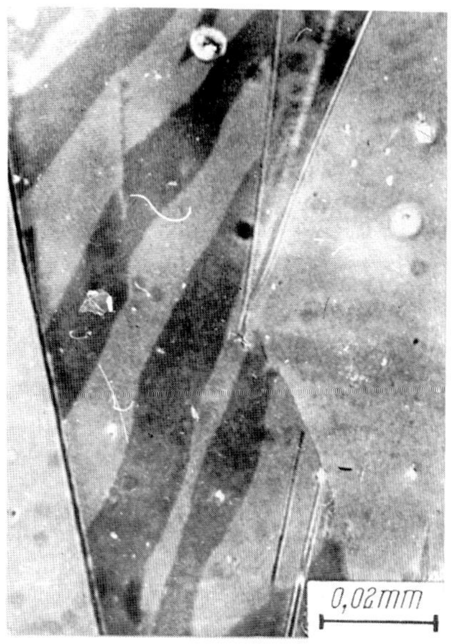

429 600 : 1

428: Same material, subsequently annealed at $+100^0$ C, same coating conditions as in Fig. 427.

429: Same material, subsequently annealed at 200^0 C, same coating conditions as in Fig. 427.

In the case of the iron-nickel alloy the rearrangement of domains under stress can be observed as the formation of more or less regular band structures reminiscent of the rearrangement of tetragonal martensite to cubic martensite. At higher annealing temperatures and with the associated removal of internal stresses the peculiar patterns in Figs. 427 and 428 are replaced by very broad band structures.

3.1 Name Index

Aldinger, S. 123, 141, 147, 173
Allen, T. H. 132
Altenhoff, H. 179
Arndt, R. 116
Aven, M. 237
Bachmann, H. 178
Back, A. 64, 132, 141
Bartz, G. 64, 132, 141, 237
Baumgartl, S. 23, 68, 86, 147
Beaglehole, D. 132
Beraha, E. 47
Beyer, J. 65
Bischoffberger, A. 178
Bond, W. L. 237
Brois, H. J. 65
Bühler, H.-E. 19, 23, 40, 65, 68, 69, 85, 86, 132, 147, 173, 179, 181, 185, 192, 202, 213, 237
Bungardt, W. 141
Cardona, M. 237
Caye, R. 23
Clarke, K. M. 123
Claussen, N. 173
Colbus, J. 158
Dautzenberg, N. 86
Demirsoy, S. 23
Dietrich, R. 64, 132, 141
Domazer, H.-G. 178
Donachie, M. I. 111
Draugelates, U. 40, 158
Eichler, K. 132
Elssner, G. 64, 123, 141, 147, 173, 197
Engler, G. 65, 69, 85, 86
Ettwig, H.-H. 19, 23, 46, 65, 68, 69, 181, 185, 213
Exner, H. E. 47, 64
Feldmann, D. 179, 213
Fischer, A. 40, 69, 116
Frühling, H.-G. 40
Gahm, I. 47
Geiger, T. 202
Gell, M. 111
Gräf, I. 40, 158
Guard, R. W. 158
Hanemann, H. 132
Harbeke, G. 237
Hartmann, K. H. 158
Heareus, O. S. 237
Heide, E. 40
Henkhaus, R. 173
Henricks, R. J. 111
Herget, C. 178, 179
Hessel, J. 173
Hillnhagen, E. 116
Hofer, A. 178
Hofmann, S. 64
Holm, R. 64
Horn, E. 179, 213
Hougardy, H. P. 40
Jackel, G. 19, 23, 46, 68, 86, 145, 185, 202, 237
Jackson, J. J. 111
Jeglitsch, F. 64, 132
Jestädt, M. 173
Judtmann, H. 86
Jung, O. 116
Kerkveld, G. E. 65
Kessler, H. 85
Kiesheyer, H. 202
Knosp, H. 46, 47
Kohlhaas, E. 19, 23, 40, 69, 85, 116
Kopp, W.-U. 173
Kossel, D. 237
Lang, W. 158
Lange, H. 158
Lemaire, R. 178
Liewald, G. 141
Lowery, H. 123
Lyman, T. 132
Marple, D. F. 237
Medenbach, K. 23
Menzel, D. 64
Metzler, H. 197
Meyer, L. 40, 69, 86, 147, 185
Mitsche, R. 23, 46, 47, 64, 86, 237
Morck, G. 86, 147
Müller, E. 173
Müller, H. 85
Münch, W. 40
Nikoleizig, A. 86, 147
Nilson, H. 68
Nitsche, D. 197
Novak, H. 64, 132
Nutting, J. 111
Ondracek, G. 64, 141
Pabst, R. 64, 132, 141, 173
de Paoli, A. 158
Paufler, P. 132
Pearson, W. B. 132
Pepperhoff, W. 19, 23, 46, 47, 65, 68, 69, 85, 116, 179, 185, 192, 213
Peter, W. 85, 116
Petzow, G. 47, 64, 147, 158, 173, 197
Pitsch, W. 85
Pusch, R. 19
Raether, H. 46
Ray, A. E. 178
Richter, M. 40
Riedel, S. 64
Roberts, E. F. J. 123
Robusch, G. 19, 23, 46, 65, 69, 85, 86, 147
Rösch, S. 40
Röser, H. 173
Scheidl, H. 23, 86, 237

Schlüter, P. 123, 141, 147
Schorr, M. 179
Schrader, A. 132
Schubert, F. 111
Schüller, H.-J. 68, 85
Schultze, W. 40
Schulz, J. 179, 213
Schulze, G. E. R. 132
Schwaab, P. 68
Schwarz, M. 173
Scouler, W. J. 132
Segall, B. 237
Semlitsch, M. 19, 47, 86, 141, 202
Sepold, G. 147
Shpigler, B. 47
Siegel, S. 132
Silber, F. A. 86
Smare, D. L. 123
Smith, E. A. 158
Spieler, K. 64, 141
Steffens, H.-D. 147, 158
Stewart, R. L. 64
Strnat, K.-J. 178
Thiemann, E. 19, 23, 68, 202
Tool, A.-Q. 123
Velicescu, M. 179
Warlimont, H. 132, 237
Wesemeyer, H. 46
Weber, S. 132, 237
Wielage, B. 158
Wihl, M. 132
Wilkinson, H. 123
Wulff, J. 132
Zogg, H. 132, 237

3.2 Subject Index

Absorption, absorption coefficient 15, 16, 60, 63
Achromatic point 26, 27, 28
Alumina 41 to 44
Aluminium 128
Aluminium alloys 124 to 143
Aluminium-iron-carbon system 148, 183 to 187
Aluminized sheet (hot) 146, 148
Amplitude 10
Amplitude condition 11, 16 ff
Angle of incidence 11, 210
Anisotropy, optical 192 to 195
Antiferromagnetic order 212
Antimony sulphide 60, 61
Austenite 68, 70 to 86

Bainite 84
Barium ferrite 178 to 180
Bloch walls see magnetic materials, magnetic domains
Body colours 25, 28
Brass 122
Brazing materials 158 to 172
Brightness 24
Bronze 126

Carbides 84 to 120
Carbonitrides 100 to 110
Carbosulphides 100
Cast alloys, see chromium steel, chill cast and carbonitrides
Cathode 53 to 55
Cathode materials 61 to 64
Cathodic sputtering 53 to 55, 61 to 64
Cementite 42, 43
Ceramics 173 to 175
Cerium alloys 144
Cerium fluoride 59, 60, 61
Cermets 146, 147, 154 to 156, 173 to 175
Chromium steel, chill cast 81 to 82, 104 to 106
Coating 52 to 58
Cobalt 114 to 120, 142, 177 to 182, 214 to 217
Cobalt bonding phase 114, 116 to 120
Cold work 45, 48, 49
Cold worked layer 41, 45, 48
Colour 24 to 38
Colour chart 26, 27, 29
Colour contrast 24 to 38
Colour difference 33
Colour displacement 33, 38
Colour perception 24
Colour samples 28, 30, 31
Complementary colours 28
Composites 146 to 172
Conductivity, electrical 184
Constants, optical 20 to 23, 129, 229 ff
Contrast 12, 13, 14, 17 to 19, 24 to 40
Copper 122 to 127, 170 to 172
Corrosion 200 to 209
Corrosion products 201
Cracks 43, 46, 50, 51
Cryolite, see layer materials

Darkness level 24 to 38
Deformation depth 41
Diamond polish 41 to 44
Diffusion 183 to 189
Dispersion 19, 60
Drying stains 46, 50

Edge sharpness, see preparation
Enamel 198
Equivalent wavelength 28
Etching 72, 74, 76, 84, 87, 88, 116, 124, 126, 138, 142, 148, 160 to 171
Europium sulphide 212
Evaporation 52, 231
Evaporation coating 52, 53
Evaporation equipment 53, 56, 57
Evaporation source 52, 231
Evaporation temperature 231

Failure 115
Ferrite 70 to 89
Ferrites 180, 182, 183
Ferromagnetism 177, 210 to 220
Ferrotitanide 107, 108
Filters, see light, monochromatic, and interference filters
Fluorides, see layer materials

Galvanizing 146, 188
Gas discharge 53 to 55
Gas ion reaction coating, see sputtering, reactive
Glass 21, 22
Gold 62, 63, 184, 190
Graphite 88
Grinding materials 41

Half-value width 21, 22
Hard metals 114 to 120
High temperature corrosion 200 to 209
Homogeneity, see diffusion and segregation
Hot gas corrosion 204 to 209
Hue 24, 28

Immersion oil 21, 43
Impurity layers 41, 43, 45, 48
Inclusions, non-metallic 90 to 101
Indium 124
Indium phosphide 59 to 61, 79, 80
Infiltration 197
Interface, optical 11
Interference 11
Interference filter 12
Interference layer 10 to 23
Interference minimum, see reflection minimum
Interference, order of 14, 21, 22
Intermediate etch 41
Ion etching 43
Iron 68 to 71
Iron-aluminium carbide 185 to 187
Iron carbide, see cementite
Iron cathode 54, 55, 62, 63
Iron oxides 90 to 94, 200 to 204
Iron zinc carbide 188

Kerr effect 210 to 220

Layer materials 59 to 64, 231, 232, 236
Layer materials, absorbing 59 to 64, 231, 232, 236
Layer materials, non-absorbing 59 to 64, 231, 232
Layer refractive index, see layer materials
Layer thickness 15, see also interference, order of
Layer thickness measurement 56 to 58
Lead fluoride 59, 60, 61
Light 10
Light, monochromatic 12 to 19, 21 to 23, 38, 39
Light, polarized 192 to 195, 210
Light, velocity of 10
Light wave 10

Magnesium 136
Magnetic domains 210 to 220
Magnetic materials 177 to 182
Magnetization 210 to 220
Magnetization direction 210 to 220
Martensite 68 to 83
Martensite, cubic 218
Martensite, tetragonal 193, 195, 218
Materials, transparent 20 to 22, 147, 197
Microtome 44
Moisture 43, 46, 50
Molten metal corrosion 200
Molybdenum 156

Needle valve 54
Néel temperature 212
Nickel 111 to 113, 141 to 143
Nickel alloys 111 to 113, 204, 209
Niobium 156
Nitrides 100 to 105, 108

Oblique section 115
Order, see interference, order of
Order, degree of crystallographic 183, 184
Order, degree of magnetic 210 to 212
Oscillating quartz 56
Oxalic acid, see etching
Oxidation 200 to 209
Oxides 198, 200 to 204
Oxygen, see reaction gas

Palladium 170 to 172
Phase (of wave) 11
Phase change 11
Phase condition 11, 14 to 16
Phosphide 88
Photometer 58
Plasma sprayed layers 146, 147, 150 to 153
Plastics 173 to 175
Platinum 184, 190
Polarization (of light) 192 to 195, 210 to 212
Polarization plane 210
Polishing 41 to 46
Polishing, alumina 41 to 46, see esp. 42
Polishing, diamond 41 to 46, see esp. 42
Polishing, electrolytic 41 to 46, see esp. 42
Polishing materials 41 to 46
Pore geometry 197 to 199
Pores 43, 46, 115, 197 to 199
Porosity 46, 48, 49
Powder metallurgy 107, 109, 114
Preparation 41 to 47
Preparation defects 45 to 51
Purple line 26

Reaction gas 54
Reflection enhancement 15, 16, 197, 198
Reflection maximum 15
Reflection minimum 12, 15
Reflection reduction 11, 16
Reflectivity 12, 15, 42, 44, 236
Reflectivity ratio 17, 18, 76 to 79
Refractive index 14 to 17
Residual austenite 72 to 85
Resin 174, 193

Sand casting, see aluminium alloys

Saturation (of colour) 25
Scaling, scale layers, see oxidation
Scratches 41, 45, 48, 49
Segregation 183, 184, 190 to 192
Semiconductor materials, see layer materials 59 to 61
Sigma phase 108 to 110
Silicides, see brazing materials
Silver 155, 156, 170 to 172
Smeared layers 45, 48, 49
Solubility 183 to 189
Specimen preparation 41 to 51
Spectral colour 24 to 28
Spectral colour line 26
Sputtering 53 to 55
Sputtering apparatus 53 to 55
Sputtering, reactive 53 to 55, 61 to 64
Sputtering rate 54
Stains, see drying stains 43
Steels, austenitic 106, 108
Steels, case-hardening 74
Steels, ferritic 84
Steels, high-alloy 84
Steels, high-speed tool 102 to 104
Steels, high-temperature 111 to 113
Steels, low-alloy 84
Steels, medium-alloy 84
Steels, micro-alloy 84
Steels, surface-treated 146–148
Steels, unalloyed 84
Steel-zinc system 188
Stress (internal) 218
Sulphides 90 to 101
Superconductors 212

Tantalum 155 to 156, 170
Temper etch 116 to 118
Tensile stress 218, 219
Titanium, see ferrotitanide, hard metals
Titanium oxide, see layer materials
Transmissivity (of filter) 25
Tungsten, tungsten carbide, see hard metals

Vacuum 52
Vacuum chamber 52

Wavelength 10 ff
Wavelength, illumination, see wavelength, observation
Wavelength, observation 12 to 14, 38, 39, 78 to 82, 98
Wood's metal 142

Y-phase, see magnetic materials, ferrites

Zinc-coated steel 146, 188, 189
Zinc selenide, see layer materials 61, 231, 232
Zinc sulphide, see layer materials 61, 231, 232
Zinc telluride, see layer materials 61, 231, 232

4. Source list

The micrographs and figures used in this book were kindly supplied by the following companies and institutes:

Source	Figures
Allianz-Zentrum für Technik GmbH, Ismaning:	73
Daimler-Benz AG, Mannheim:	195
Dechema-Institut, Frankfurt:	401, 402
Th. Goldschmidt AG, Essen:	248–254, 335–340
W. C. Heraeus GmbH, Hanau:	322–324, 356–361
IBM – Deutschland GmbH, Sindelfingen:	327–329
Institut für angewandte Materialforschung (IFAM), Bremen:	24–35, 269–276, 285–290, 297–318
Krupp-Forschungsinstitut GmbH, Essen:	181–194, 331–334, 371, 372, 409
Mannesmann-Forschungsinstitut GmbH, Duisburg:	68, 76–85, 88–91, 169, 117, 169, 362–367, 403, 405, 407, 409, 410, 414–429
Max-Planck-Institut für Eisenforschung GmbH, Düsseldorf:	112–116, 125–131, 133–142, 145, 170–174, 178–180, 291–296
Max-Planck-Institut für Metallforschung, Stuttgart:	199–212, 241–247, 262–268, 277–284, 319–321, 325, 326, 373–375
Schweizerische Aluminium AG, Neuhausen/Rheinfall:	215–239
Gebr. Sulzer AG, Winterthur:	387
Thyssen AG, Duisburg:	41–56, 69–71, 92, 93, 95–100, 104, 105, 107, 108, 111, 112, 143, 144, 147–150, 157–162, 166–168, 343, 344, 347–355, 368–372, 376, 377, 380–386
Thyssen-Edelstahlwerke AG, Krefeld:	151–156, 163–165, 175–177, 387–398

5. OPTICAL CONSTANTS OF PHASES AND LAYER MATERIALS

Table 7
Properties of materials for the production of evaporated layers

No.	Material	Refractive index n_s (λ = 550 nm)	Absorption coefficient k_s (λ = 550 nm)	Evaporation temperature range (^{0} C)	Support sheet material	Remarks
1	Na_3AlF_6	1,35	0	900–1200	Mo	very suitable
2	MgF_2	1,38	0	1200–1600	Mo, Ta	evaporates from the melt, very suitable
3	ThF_4	1,52	0	1300–1600	Mo, Ta	suitable but radioactive, care needed with handling
4	LaF_3	1,6	0	1300–1600	Mo, Ta	evaporates from the melt, suitable
5	CeF_3	1,6	0	1300–1600	Mo, Ta	evaporates from the melt, very suitable
6	PbF_2	1,75	0	800–1100	W, Al_2O_3	evaporates from the melt, very suitable
7	ZnS	2,4	0	800–1200	Mo	reproducible refractive index, very suitable
8	ZnSe	2,65	> 0	500–800	Mo, W	very suitable, slightly hygroscopic
9	TiO_2	2,7	0,04	> 1500	W	suitable but evaporation not possible directly, very high evaporation temp., white heat
10	ZnTe	3,25	0,4	800–1200	Mo, Ta	very suitable, slightly hygroscopic
11	Sb_2S_3	2,3–2,4	0,07	300–400	Mo, Ta	strong absorp. at < 600 nm, otherwise suitable
12	CdS	2,4	0,14	800–1200	W, Mo, Ta	does not condense well on polished surface, not so suitable
13	CdSe	~ 2,6	> 0,05	800–1000	Mo, Ta	decomposition can be avoided by careful evaporation, in this case suitable
14	CdTe	3,3	0,2	800–1000	Mo, Ta	partial decomposition possible, unsuitable
15	InP	3,4–3,5	> 0,2	—	W	evaporates from the melt, partial decomposition, unsuitable
16	GaAs	3,4	1,6	—	—	decomposes on evaporation, unsuitable
17	GaP	3,45	0	—	—	decomposes on evaporation, unfortunately unsuitable

Note: Care must be taken when working with compounds of cadmium because of their toxic nature.

Table 8
Dependence of the optical constants of evaporated layers on wavelength

λ (nm)	ZnO[1)5)]		CdS[1)]		ZnS[5)]		Sb_2S_3[4)]		ZnSe[2)3)]		TiO_2[1)5)]		ZnTe[1)]		CdTe[1)]		GaAs[1)]		GaP[1)5)]		InP	
	n_s	k_s	n_s	k_s	n_s	k_s	n_s	k_s	n_s	k_s	n_s	k_s	n_s	k_s	n_s	k_s	n_s	k_s	n_s	k_s	h_s	k_s
400			2,01	0,54								0,05	3,48	1,1	3,9	1	2,45	2,36		0		
450	2,11	0	2,04	0,48	2,48						2,99	0,06	3,37	0,75	3,63	0,6	2,92	2,26		0	3,7	
500	2,06	0	2,28	0,33	2,42				2,74		2,7	0,05	3,31	0,54	3,43	0,32	3,29	1,94		0	3,5	
550	2,02	0	2,13	0,14	2,4	0	2,5	0,07	2,65		2,73	0,04	3,24	0,4	3,33	0,2	3,4	1,63	3,44	0		
600	2,0	0	2,01	0,12	2,36				2,61		2,75	0,03	3,2	0,28	3,3	0,12	3,45	1,42	3,35	0		
650	1,99	0	2,0	0,1			2,5	0,01			2,72	0,01	3,06	0,21	3,23	0,08	3,48	1,2	3,29	0		
700	1,90	0	1,96	0,1	2,33						2,69	0,01	2,98	0,19	3,18	0,05	3,5	1,05	3,24	0	3,4	

Table 9

Optical constants and recommendations for the contrasting of phases occurring predominantly in steels 6).

No.	(1) Phase	(2a) Chemical composition	(2b) Heat treatment of specimen material	(3) Coating layer deposited for the determination of the optical constants	(4) Order of interference	(5a) Reflectivity un-coated (R)	(5b) Reflectivity coated at minimum (R_{Min})	(6a) Refractive index of phase [n_P][1]	(6b) Absorption coefficient of phase [kp][1]	(6c) Wavelength of reflection minimum (λ_{Min}) nm	(7) Refractive index (n_s) necessary to satisfy the amplitude condition[1]	(8) Recommended material for single layer evaporation
1	α–Fe	1 % C, 0,6 % Mn	1100 °C/ ice bath	ZnS	1	0,578	0,055	2,09	3,20	500	3,38	ZnSe, ZnTe
2						0,581	0,058	2,0	3,20	550	3,5	
3						0,584	0,059	2,1	3,25	575	3,45	
4	α–Fe	Fe + 32,5 % Ni	1300 °C/–190 °C	ZnS	1	0,582	0,072	1,68	2,98	535	3,83	ZnSe, ZnTe
5	α–Fe	X 20 CrNiSi 25 4	1050 °C 1 h/furnace	ZnS	1	0,62	0,072	2,35	3,60	500	3,45	ZnSe, ZnTe
6						0,612	0,065	2,64	3,73	550	3,33	
7						0,623	0,075	2,54	3,79	570	3,45	
8	α–Fe	Fe + 14,8 % Mn	as cast	ZnS	1	0,584	0,075	1,64	2,96	460	3,91	ZnSe, ZnTe
9						0,586	0,075	1,67	3,00	540	3,88	
10						0,590	0,080	1,61	2,98	610	4,02	
11	α–Fe	Fe + > 0,1 to 1,0% Al	900 °C 1 h/furnace	ZnSe		0,58	0,022	2,24	3,29	520	3,30	ZnSe, ZnTe
12		Fe + 33% Al				0,445	0,002	1,77	2,25	580	2,88	
13	γ–Fe	1 % C, 0,6 % Mn	1100 °C/ ice bath	ZnS	1	0,609	0,078	2,00	3,38	500	3,66	ZnSe, ZnTe
14						0,619	0,082	2,11	3,53	550	3,65	
15						0,622	0,088	1,99	3,48	575	3,77	
16	γ–Fe	Fe + 32,5 Ni	1300 °C/–190 °C	ZnS	1	0,625	0,103	1,66	3,26	535	4,21	ZnSe, ZnTe
17	γ–Fe	20 CrNiSi 25 4	1050 °C 1 h/furnace	ZnS	1	0,635	0,083	2,62	3,95	500	3,50	ZnSe, ZnTe
18						0,625	0,098	1,78	3,35	550	4,02	
19						0,655	0,10	2,65	4,17	575	3,63	
20	ε–Fe	Fe + 14,8 % Mn	as cast	ZnS	1			1,25	3,20	550	6,5	ZnSe, ZnTe
21								1,3	3,50	600	6,3	
22	Fe_3AlC_x ($x \sim 0{,}8$)		900 °C 2 h/furnace	ZnSe	1	0,456	0,0035	1,84	2,36	545[2]	2,91	ZnSe
23						0,46	0,0025	1,85	2,35	555	2,90	
24	Fe_2Al_5		900 °C 2 h/furnace	ZnSe	1	0,493	0,004	2,03	2,61	500[2]	2,95	ZnSe
25						0,490	0,0045	1,96	2,57	505	2,97	
26	$FeAl_2$		950 °C 2 h/furnace	ZnSe	1	0,464	0,0035	1,80	2,36	505[2]	2,96	ZnSe
27						0,467	0,005	1,74	2,35	515	3,03	
27 a	α–Fe	1 % C, 0,6 % Mn	1100 °C/ ice bath	ThF_4	1	0,581	0,296	2,29	3,32	550	3,3	ZnSe, ZnTe
28	TiN			ZnS[4]	1	0,30		1,23	1,50	400	3,31	ZnSe, ZnTe
29						0,45		1,20	1,80	450	4,17	
30						0,52		1,15	2,23	500	5,86	
31						0,58		1,15	2,50	550	6,54	
32						0,62		1,15	2,75	600	7,18	
33						0,66		1,15	2,90	650	7,56	
34	$TiN_{0,8}C_{0,2}$			ZnS[4]	1	0,29		1,70	1,35	400	2,70	ZnSe, ZnS
35						0,315		1,50	1,58	450	2,54	ZnS, ZnSe
36						0,39		1,45	1,85	500	3,00	ZnSe, ZnTe
37						0,46		1,30	2,15	550	4,80	ZnTe
38						0,535		1,25	2,38	600	4,88	ZnTe
39						0,58		1,20	2,52	650	5,74	ZnTe
40	$TiN_{0,6}C_{0,4}$			ZnS[4]	1	0,35		1,19	1,58	400	3,78	ZnTe
41						0,35		1,50	1,70	450	2,69	ZnSe, ZnTe
42						0,36		1,70	1,80	500	2,51	ZnS, ZnSe
43						0,395		1,65	1,97	550	2,76	ZnSe,
44						0,44		1,50	2,1	600	3,21	ZnSe, ZnTe
45						0,49		1,35	2,25	650	3,97	ZnTe
46	$TiN_{0,4}C_{0,6}$			ZnS[4]	1	0,40		1,65	2,0	400	2,79	ZnSe, ZnS
47						0,39		1,75	1,95	450	2,61	ZnS
48						0,38		1,8	1,93	500	2,54	ZnS
49						0,37		1,85	1,90	550	2,46	ZnSe
50						0,385		2,05	2,00	600	2,42	ZnSe, ZnTe
51						0,40		1,95	2,10	650	2,56	ZnSe, ZnTe
52	$TiN_{0,2}C_{0,8}$			ZnS[4]	1	0 43		2,0	2,3	400	2,70	ZnSe
53						0,425		2,45	2,35	450	2,50	ZnS, ZnSe
54						0,42		2,65	2,3	500	2,41	ZnS
55						0,42		2,60	2,3	550	2,43	ZnS
56						0,415		2,55	2,3	600	2,44	ZnS
57						0,41		2,45	2,25	650	2,43	ZnS
58	Au			ZnS	1			1,15	1,75	450	4,46	ZnTe
59								0,55	1,4	500	not satisfied	
60								0,35	1,5	585	not satisfied	
61								0,25	2,1	550	not satisfied	
62								0,2	3,0	600	not satisfied	
63								0,15	3,6	650	not satisfied	
64	α–Fe_2O_3		commercial purity iron, oxidised in air, 1200 C, 10 h, furnace	ZnS	1	0,29	0,054	3,08	0,84	535[3]	1,85	PbF_2, LaF_3
65				ZnSe		0,29	0,099	3,10	0,82	510/515[3]	1,85	
66				ZnTe		0,29	0,182	3,09	0,83	535[3]	1,85	
67				ThF_4		0,29	0,035	3,08	0,83	525[3]	1,85	
68				PbF_2		0,29	0,007	3,05	0,95	500[3]	1,87	
69	Fe_3O_4		commercial purity iron, oxidised in air, 1200 C, 10 h, furnace	ZnS	1	0,19	0,080	2,30	0,68	575/580[3]	1,63	LaF_3, PbF_2
70				ZnSe		0,19	0,098	2,27	0,71	555/560[3]	1,63	
71				ZnTe		0,19	0,155	2,41	0,51	590[3]	1,61	
72				ThF_4		0,19	0,005	2,21 ?	0,77 ?	540[3]	1,64	
73				PbF_2		0,19	0,02	?	?	520[3]	?	
74	calcium diferrite		basic sinter, degree of basicity ~ 2.5	ZnS	1	0,149	0,116	2,2	0,32	505[3]	1,51	ThF_4
75						0,147	0,121	2,2	0,275	515[3]	1,50	
76						0,1425	0,135	2,2	0,135	545[3]	1,49	

[1]) Values rounded up or down to two decimal places. – [2]) λ min Fe standard ~ 510 to 520 nm. – [3]) λ min Fe standard ~ 540 to 550 nm. – [4]) For phases 28 to 57 the calculation of the optical constants in the TiN-TiC system is based on a constant refractive index of 2.36 for ZnS. The values need a slight correction because of slight dispersion by zinc sulphide in the visible range and at short and long wavelengths (< 500 nm and > 600 nm).

Table 10a
Optical constants of Al and phases in Al-alloys 7).

Phase	Wavelength λ (nm)	Reflectivity R (in air or vacuum)	n	k
Pure Al in ultra high vac.	638	91%	1,6	7,9
Pure Al	div.	72–91,5%	0,4–1,5	2,9–7
Al-Cu solid solution	400–700	$(R_{Al\text{-}Cu}\text{-}R_{Al}) \sim -1\%/at\%Cu$		
Al-Mn solid solution	400–700	$(R_{Al\text{-}Mn}\text{-}R_{Al}) \sim -1\%/at\%Mn$		
Al-Mg solid solution	500–580	$(R_{Al\text{-}Mg}\text{-}R_{Al}) \sim 0{,}2\%/at\%Mg$		
Polished surface in air of Al-Mn and Al-Si solid solutions	420–700	85–92%	0,5–1	~ 5–6
Al_3Mg_2	430–580	85%		
$MgZn_2$	500	85%	0,9	4,5
	620	78%	1,9	5,2
Al_6Mn	470		1–1,5	2,7–3,3
α-AlFeSi	550		1,5–2,5	2,8–3,5
β-AlFeSi	550		1,5–2,5	2,8–3,5
Si	400		4,5	0,15
	520		4,3	0,04
	700		3,8	0,01
Mg_2Si	400		~ 4	~ 5
	520		~ 7	~ 1
	700		~ 4,2	~ 0,2

Table 10b
Appearance of the phases in Al-alloys coated with ZnTe, zero order, (layer thickness = 24 nm), 7).

Phase	Structure*	Homogeneity range wt. %	Colour for a 24 nm thick ZnTe layer (cf. section 2.1.6, Figs. 213–239)	Birefringence
Al-solid solution	cF 4			no
Al_4Ca	tI 10		violet blue	very strong
Al_7Cr	mC 104	22–23 Cr	reddish lilac	medium
$Al_{11}Cr_2$	mC 104	24–26,5 Cr	reddish lilac-violet	weak
G_1-Al_{12}(Cr, Mn)	cI 26	2–4 Cr, 10–12 Mn	purplish red	no
α-$Al_{13}Cr_4Si_4$	cF 84		oxide red	no
Θ-Al_2Cu	tI 12	46,4–47,5 Cu	light blue-skyblue	very strong
$(Al, Cu)_6(Fe, Cu)$ (α-$Al_{23}CuFe_4$)	oC 28	Cu:Fe = 0,34–0,41	olive green	weak-medium
Al_7Cu_2Fe	tP 40	13–29 Cu, 12–18 Fe	reddish violet	weak
T-Al_6CuMg_4 ($(Al, Cu)_{49}Mg_{32}$)	cI 162	~ 15–25 Cu, 30–35 Mg	azure	no
S-Al_2CuMg	oC 16	~ 46 Cu, ~ 17 Mg	turquoise	weak
T-$Al_{20}Cu_2Mn_3$	oC 152	17,5–22 Cu, 25–31,5 Mn	reddish violet	weak
Θ-Al_3Fe ($Al_{13}Fe_4$)	mC 101	36–39 Fe	bluish-grey	weak
$Al_8FeMg_3Si_6$	hP 18		reddish-lilac	very strong
Al_6(Fe, Mn)	oC 28	Fe:Mn = 0–0,5	violet	strong
α-$Al_{12}(Fe, Mn)_3Si$	cI 138	30–33 (Fe + Mn), Mn > 5, 7–11 Si	reddish lilac	no
T-Al_9FeNi	mP 22	5–15 Fe, 15–27 Ni	copper	no
α-$Al_{12}Fe_3Si$	hP	30–33 Fe, 7,4–11 Si	reddish lilac	no
β-$Al_9Fe_2Si_2$	m	25,5–27 Fe, 12,1–15 Si	reddish lilac	medium
β-Al_3Mg_2(Al_8Mg_5)	cP 108	35,1–39 Mg	light blue-dove grey	no
T-$Al_{18}Mg_3Mn_2$	cF 184	~ 12–13 Mg, 14–15 Mn	gentian	no
$(Al, Zn)_{49}Mg_{32}$	cI 162	20–35 Mg, 22–65 Zn	reddish lilac	no
Al_6Mn	oC 28		brown ochre	strong
α-$Al_{12}Mn_3Si$	cP 138	25–29 Mn, 8–13 Si	brown	no
β-Al_9Mn_3Si	hP 26	33–37 Mn, 5–8 Si	brown	weak
Al_3Ni	oP 16		reddish brown	medium
Al_3Ti	tI 8	~ 1 at. %	reddish lilac	weak
$Al_{10}V$	cF 184	~ 0,5 at. %	violet	no
Al_3Zr	tI 16		bluish lilac	strong
$CaSi_2$	hR 6		reddish violet	not observable
Mg_2Si	cF 12		pearl white-light blue	no
$MgZn_2$	hP 12	15,3–16,1 Mg	light blue	weak
Si	cF 8		pale green	no

* First letter = crystallographic system: c = cubic, h = hexagonal, t = tetragonal, o = orthorhombic, m = monoclinic; second letter = lattice type: P = primitive, I = body-centred, F = face-centred, C = base (c) centred, R = rhombohedral; number = number of atoms per unit cell.

Note: The colours given in this table are translated from the original work 7). The values for shade, brightness and saturation after DIN 6164 are explained in Section 2.1.6 and Figs. 213 to 239.

Table 11
Optical constants of metallic oxide layers, produced by sputtering in oxygen, 8).

Oxide	450 nm		550 nm		650 nm	
	n_s	k_s	n_s	k_s	n_s	k_s
Hg	2,77	0,21	2,66	0,04	2,66	0,01
Pb	2,68	0,25	2,55	0,06	2,43	0,01
Fe	2,55	0,30	2,42	0,10	2,32	0,02
Pt	2,50	0,24	2,45	0,12	2,40	0,06
Au	2,28	0,34	2,25	0,23	2,22	0,12
Cu	2,03	0,44	2,09	0,37	2,11	0,31
Ni	1,88	0,22	1,87	0,19	1,87	0,18
Error	± 3%	± 20%	± 2%	± 15%	± 1%	± 10%

Table 12
Reflectivities of metallic and non-metallic phases (for $\lambda = 589.3$ nm), 9).

Phase	Reflectivity in %	Remarks
silver (Ag)	94	
magnesium (Mg)	93	
copper (Cu)	83	
aluminium (Al)	82,7	
platinum (Pt)	73	
manganese (Mn)	64	
nickel (Ni)	62	
iron (Fe)	59	
nickel sulphide (NiS)	54	
iron sulphide (FeS)	37	
copper sulphide (CuS)	15/24	strong optical anisotropy
manganese sulphide (MnS)	21	
iron oxide (Fe_2O_3)	24/27,5	strong optical anisotropy
iron oxide (Fe_3O_4)	21	
iron oxide (FeO)	19	
chromium oxide (Cr_2O_3)	18	
chromide ($FeO \cdot Cr_2O_3$)	12,5	
fayalite ($FeO \cdot SiO_2$)	8,9	
herzynite ($FeO \cdot Al_2O_3$)	8,6	
corundum (Al_2O_3)	7,6	
tephroite ($2MnO \cdot SiO_2$)	8,2	
rhodonite ($MnO \cdot SiO_2$)	7,1	
sillimanite ($Al_2O_3 \cdot SiO_2$)	6,4	
quartz (SiO_2)	4,58	

Literature to 5.

1) Cardona, M. and Harbeke, G.: Optical properties and band structure of wurtzite-type crystals and rutile. Phys. Rev. **137** (1965), No. 5A, March 1965

2) Aven, M., Marple, D. F. and Segall, B.: Some electrical and optical properties of ZnSe. J. Appl. Phys. **32** Suppl. (1961), No. 10, Oct. 1961

3) Marple, D.T.F.: Refractive index of ZnSe, ZnTe, CdTe. J. Appl. Phys. **33** (1964), No. 3 Part 1, March 1964

4) Heareus, O. S.: Optical properties of thin solid films. Butterworth, London 1955

5) Bond, W. L.: Measurement of reactive indices of several crystals. J. Appl. Phys. **6** (1965), No. 5, May 1965

6) Bühler, H.-E. and Jackel, G.: Tafeln zur quantitativen Metallographie mit aufgedampften Interferenzschichten. Arch. Eisenhüttenwes. **41** (1970) 859

7) Zogg, H., Weber, S. and Warlimont, H.: Optische Gefügeentwicklung von Al-Legierungen durch aufgedampfte ZnTe-Interferenzschichten. Prakt. Metallographie **14** (1977), 553

8) Bartz, G. and Kossel, D.: unpublished work of Ernst Leitz GmbH, Wetzlar

9) Scheidl, H. and Mitsche, R.: Neue Beispiele aus der Mikroreflexionsmessung. Radex-Rundschau **3/4** (1967), 596